I0065852

Understanding Climate Change

Understanding Climate Change

Edited by
Clarence Richmond

Larsen & Keller
www.larsen-keller.com

Understanding Climate Change
Edited by Clarence Richmond
ISBN: 978-1-63549-068-8 (Hardback)

© 2017 Larsen & Keller

🖧 Larsen & Keller

Published by Larsen and Keller Education,
5 Penn Plaza,
19th Floor,
New York, NY 10001, USA

Cataloging-in-Publication Data

Understanding climate change / edited by Clarence Richmond.
    p. cm.
Includes bibliographical references and index.
ISBN 978-1-63549-068-8
1. Climatic changes. 2. Climatology. 3. Climate change mitigation. I. Richmond, Clarence.
QC903 .U53 2017
363.738 74--dc23

This book contains information obtained from authentic and highly regarded sources. All chapters are published with permission under the Creative Commons Attribution Share Alike License or equivalent. A wide variety of references are listed. Permissions and sources are indicated; for detailed attributions, please refer to the permissions page. Reasonable efforts have been made to publish reliable data and information, but the authors, editors and publisher cannot assume any responsibility for the vailidity of all materials or the consequences of their use.

Trademark Notice: All trademarks used herein are the property of their respective owners. The use of any trademark in this text does not vest in the author or publisher any trademark ownership rights in such trademarks, nor does the use of such trademarks imply any affiliation with or endorsement of this book by such owners.

The publisher's policy is to use permanent paper from mills that operate a sustainable forestry policy. Furthermore, the publisher ensures that the text paper and cover boards used have met acceptable environmental accreditation standards.

Printed and bound in the United States of America.

For more information regarding Larsen and Keller Education and its products, please visit the publisher's website www.larsen-keller.com

Table of Contents

Preface

As a sub-field of atmospheric sciences, climate change refers to the changes which happen to the weather patterns of an area over a period of time. It happens due to the changes in solar radiation, volcanic activities, biotic processes and plate tectonics, etc. In this book, constant effort has been made to make the understanding of the different concepts of climate change as easy and informative as possible for the readers. It is a compilation of chapters that discuss the most vital concepts of this field. The topics introduced in this text are bound to provide great insights about the basic and complex areas of this subject. Different approaches, evaluations and methodologies have been included in it. Those with an interest in climate change would find this textbook helpful.

A short introduction to every chapter is written below to provide an overview of the content of the book:

Chapter 1 - The change in the weather patterns over a period of years is referred to as climate change. Climate change is mainly caused by either human activities or by biotic processes; plate tectonics and volcanic eruptions also play a key role in climate change. This section will provide an integrated understanding of climate change; **Chapter 2 -** Methods and techniques are an important component of any field of study. Some of these techniques are climate model, general circulation model, numerical weather prediction and tropical cyclone forecast model. Climate models are used to simulate the important aspects of climate like atmosphere, oceans and land whereas general circulation model is a mathematical model. It helps in the demonstration of the circulation of the ocean. The following chapter elucidates the methods and techniques that are related to climate change; **Chapter 3 -** The recent developments in the climate of our Earth have caused numerous concerns and challenges. Business action on climate change includes global warming and a range of activities related to it. Likewise, other challenges of climate change are land surface effects on climate, deforestation and climate change etc. This text is a compilation of the concerns and challenges faced in today's time related to climate change; **Chapter 4 -** Climate change mitigation is a number of precautions taken to limit the extent or rate of long- term climate change. It usually involves reduction of greenhouse gas effects. Examples for climate change mitigation include discontinuation of fossil fuels by using low carbon energy sources and by growing forests to remove greater amounts of carbon dioxide. The topics discussed in the section are of great importance to broaden the knowledge on climate change mitigation; **Chapter 5 -** This chapter concentrates on the various observations on climate change. Some of these are scientific opinion on climate change, attribution of recent climate change, climate change denial and Milankovitch cycles. Scientific opinion on climate is the judgment among scientists in regard to global warming. The scientific opinions are explained in synthesis reports. All the various observations of climate change have been carefully analyzed in this text; **Chapter 6 -** Climate change science can be traced back to the 19th century. Scientists argued that human emissions of greenhouse gasses are altering the climate. Research has expanded our knowledge on pollution and the reasons for climate change. The following chapter explains to the reader the importance of history of climate change science and its significance in contemporary times.

I extend my sincere thanks to the publisher for considering me worthy of this task. Finally, I thank my family for being a source of support and help.

Editor

Introduction to Climate Change

The change in the weather patterns over a period of years is referred to as climate change. Climate change is mainly caused by either human activities or by biotic processes; plate tectonics and volcanic eruptions also play a key role in climate change. This section will provide an integrated understanding of climate change.

Climate Change

Climate change is a change in the statistical distribution of weather patterns when that change lasts for an extended period of time (i.e., decades to millions of years). Climate change may refer to a change in average weather conditions, or in the time variation of weather around longer-term average conditions (i.e., more or fewer extreme weather events). Climate change is caused by factors such as biotic processes, variations in solar radiation received by Earth, plate tectonics, and volcanic eruptions. Certain human activities have also been identified as significant causes of recent climate change, often referred to as *global warming*.

Scientists actively work to understand past and future climate by using observations and theoretical models. A climate record—extending deep into the Earth's past—has been assembled, and continues to be built up, based on geological evidence from borehole temperature profiles, cores removed from deep accumulations of ice, floral and faunal records, glacial and periglacial processes, stable-isotope and other analyses of sediment layers, and records of past sea levels. More recent data are provided by the instrumental record. General circulation models, based on the physical sciences, are often used in theoretical approaches to match past climate data, make future projections, and link causes and effects in climate change.

Terminology

The most general definition of *climate change* is a change in the statistical properties (principally its mean and spread) of the climate system when considered over long periods of time, regardless of cause. Accordingly, fluctuations over periods shorter than a few decades, such as El Niño, do not represent climate change.

The term sometimes is used to refer specifically to climate change caused by human activity, as opposed to changes in climate that may have resulted as part of Earth's natural processes. In this sense, especially in the context of environmental policy, the term *climate change* has become synonymous with *anthropogenic global warming*. Within scientific journals, *global warming* refers to surface temperature increases while *climate change* includes global warming and everything else that increasing greenhouse gas levels affect.

Climatic Change Versus Climate Change

In 1966, the World Meteorological Organization (WMO) proposed the term climatic change to encompass all forms of climatic variability on time-scales longer than 10 years, whether the cause was natural or anthropogenic. Change was a given and climatic was used as an adjective to describe this kind of change (as opposed to political or economic change). When it was realized that human activities had a potential to drastically alter the climate, the term climate change replaced climatic change as the dominant term to reflect an anthropogenic cause. Climate change was incorporated in the title of the Intergovernmental Panel on Climate Change (IPCC) and the UN Framework Convention on Climate Change (UNFCCC). Climate change, used as a noun, became an issue rather than the technical description of changing weather.

Causes

On the broadest scale, the rate at which energy is received from the Sun and the rate at which it is lost to space determine the equilibrium temperature and climate of Earth. This energy is distributed around the globe by winds, ocean currents, and other mechanisms to affect the climates of different regions.

Factors that can shape climate are called climate forcings or "forcing mechanisms". These include processes such as variations in solar radiation, variations in the Earth's orbit, variations in the albedo or reflectivity of the continents and oceans, mountain-building and continental drift and changes in greenhouse gas concentrations. There are a variety of climate change feedbacks that can either amplify or diminish the initial forcing. Some parts of the climate system, such as the oceans and ice caps, respond more slowly in reaction to climate forcings, while others respond more quickly. There are also key threshold factors which when exceeded can produce rapid change.

Forcing mechanisms can be either "internal" or "external". Internal forcing mechanisms are natural processes within the climate system itself (e.g., the thermohaline circulation). External forcing mechanisms can be either natural (e.g., changes in solar output) or anthropogenic (e.g., increased emissions of greenhouse gases).

Whether the initial forcing mechanism is internal or external, the response of the climate system might be fast (e.g., a sudden cooling due to airborne volcanic ash reflecting sunlight), slow (e.g. thermal expansion of warming ocean water), or a combination (e.g., sudden loss of albedo in the arctic ocean as sea ice melts, followed by more gradual thermal expansion of the water). Therefore, the climate system can respond abruptly, but the full response to forcing mechanisms might not be fully developed for centuries or even longer.

Internal Forcing Mechanisms

Scientists generally define the five components of earth's climate system to include atmosphere, hydrosphere, cryosphere, lithosphere (restricted to the surface soils, rocks, and sediments), and biosphere. Natural changes in the climate system ("internal forcings") result in internal "climate variability". Examples include the type and distribution of species, and changes in ocean currents.

Ocean Variability

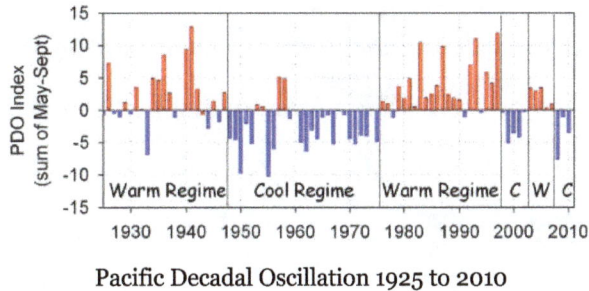

Pacific Decadal Oscillation 1925 to 2010

The ocean is a fundamental part of the climate system, some changes in it occurring at longer timescales than in the atmosphere, as it has hundreds of times more mass and thus very high thermal inertia, with effects such as the ocean depths still lagging today in temperature adjustment from effects of the Little Ice Age of past centuries).

Short-term fluctuations (years to a few decades) such as the El Niño-Southern Oscillation, the Pacific decadal oscillation, the North Atlantic oscillation, and the Arctic oscillation, represent climate variability rather than climate change. On longer time-scales, alterations to ocean processes such as thermohaline circulation play a key role in redistributing heat by carrying out a very slow and extremely deep movement of water and the long-term redistribution of heat in the world's oceans.

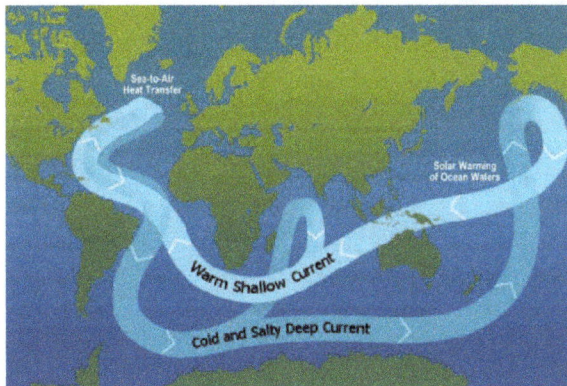

A schematic of modern thermohaline circulation. Tens of millions of years ago, continental-plate movement formed a land-free gap around Antarctica, allowing the formation of the ACC, which keeps warm waters away from Antarctica.

Life

Life affects climate through its role in the carbon and water cycles and through such mechanisms as albedo, evapotranspiration, cloud formation, and weathering. Examples of how life may have affected past climate include:

1. glaciation 2.3 billion years ago triggered by the evolution of oxygenic photosynthesis, which depleted the atmosphere of the greenhouse gas carbon dioxide and introduced free oxygen.

2. another glaciation 300 million years ago ushered in by long-term burial of decomposition-resistant detritus of vascular land-plants (creating a carbon sink and forming coal)

3. termination of the Paleocene-Eocene Thermal Maximum 55 million years ago by flourishing marine phytoplankton

4. reversal of global warming 49 million years ago by 800,000 years of arctic azolla blooms

5. global cooling over the past 40 million years driven by the expansion of grass-grazer ecosystems

External Forcing Mechanisms

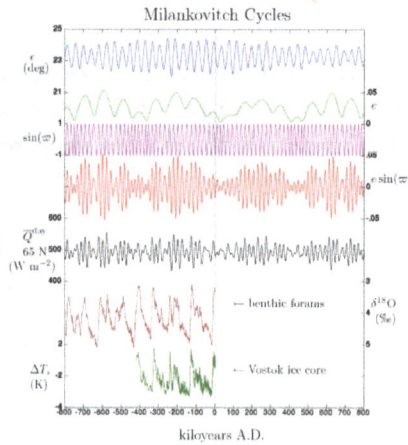

Milankovitch cycles from 800,000 years ago in the past to 800,000 years in the future.

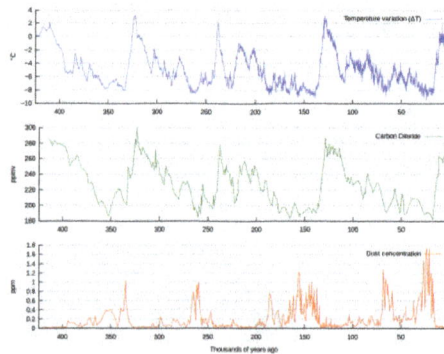

Variations in CO_2, temperature and dust from the Vostok ice core over the last 450,000 years

Orbital Variations

Slight variations in Earth's orbit lead to changes in the seasonal distribution of sunlight reaching the Earth's surface and how it is distributed across the globe. There is very little change to the area-averaged annually averaged sunshine; but there can be strong changes in the geographical and seasonal distribution. The three types of orbital variations are variations in Earth's eccentricity, changes in the tilt angle of Earth's axis of rotation, and precession of Earth's axis. Combined together, these produce Milankovitch cycles which have a large impact on climate and are notable for their correlation to glacial and interglacial periods, their correlation with the advance and retreat of the Sahara, and for their appearance in the stratigraphic record.

The IPCC notes that Milankovitch cycles drove the ice age cycles, CO_2 followed temperature change "with a lag of some hundreds of years," and that as a feedback amplified temperature change. The depths of the ocean have a lag time in changing temperature (thermal inertia on such scale). Upon seawater temperature change, the solubility of CO_2 in the oceans changed, as well as other factors impacting air-sea CO_2 exchange.

Solar Output

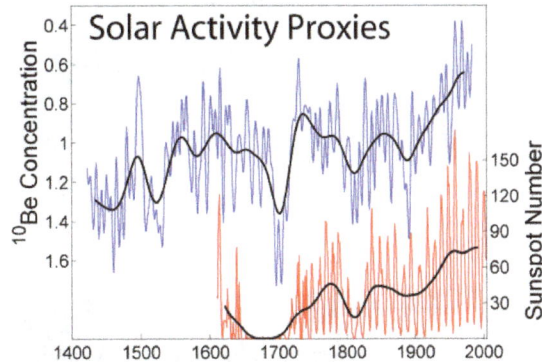

Variations in solar activity during the last several centuries based on observations of sunspots and beryllium isotopes. The period of extraordinarily few sunspots in the late 17th century was the Maunder minimum.

The Sun is the predominant source of energy input to the Earth. Other sources include geothermal energy from the Earth's core, and heat from the decay of radioactive compounds. Both long- and short-term variations in solar intensity are known to affect global climate.

Three to four billion years ago, the Sun emitted only 70% as much power as it does today. If the atmospheric composition had been the same as today, liquid water should not have existed on Earth. However, there is evidence for the presence of water on the early Earth, in the Hadean and Archean eons, leading to what is known as the faint young Sun paradox. Hypothesized solutions to this paradox include a vastly different atmosphere, with much higher concentrations of greenhouse gases than currently exist. Over the following approximately 4 billion years, the energy output of the Sun increased and atmospheric composition changed. The Great Oxygenation Event – oxygenation of the atmosphere around 2.4 billion years ago – was the most notable alteration. Over the next five billion years, the Sun's ultimate death as it becomes a red giant and then a white dwarf will have large effects on climate, with the red giant phase possibly ending any life on Earth that survives until that time.

Solar output also varies on shorter time scales, including the 11-year solar cycle and longer-term modulations. Solar intensity variations possibly as a result of the Wolf, Spörer and Maunder Minimum are considered to have been influential in triggering the Little Ice Age, and some of the warming observed from 1900 to 1950. The cyclical nature of the Sun's energy output is not yet fully understood; it differs from the very slow change that is happening within the Sun as it ages and evolves. Research indicates that solar variability has had effects including the Maunder minimum from 1645 to 1715 A.D., part of the Little Ice Age from 1550 to 1850 A.D. that was marked by relative cooling and greater glacier extent than the centuries before and afterward. Some studies point toward solar radiation increases from cyclical sunspot activity affecting global warming, and climate may be influenced by the sum of all effects (solar variation, anthropogenic radiative forcings, etc.).

Interestingly, a 2010 study *suggests*, "that the effects of solar variability on temperature throughout the atmosphere may be contrary to current expectations."

In an Aug 2011 Press Release, CERN announced the publication in the Nature journal the initial results from its CLOUD experiment. The results indicate that ionisation from cosmic rays significantly enhances aerosol formation in the presence of sulfuric acid and water, but in the lower atmosphere where ammonia is also required, this is insufficient to account for aerosol formation and additional trace vapours must be involved. The next step is to find more about these trace vapours, including whether they are of natural or human origin.

Volcanism

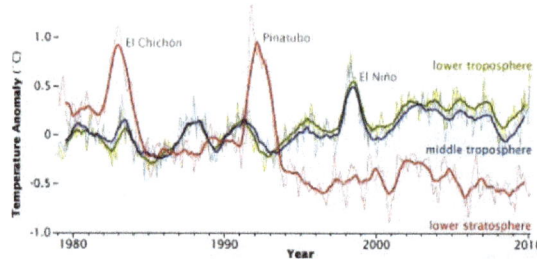

In atmospheric temperature from 1979 to 2010, determined by MSU NASA satellites, effects appear from aerosols released by major volcanic eruptions (El Chichón and Pinatubo). El Niño is a separate event, from ocean variability.

The eruptions considered to be large enough to affect the Earth's climate on a scale of more than 1 year are the ones that inject over 100,000 tons of SO_2 into the stratosphere. This is due to the optical properties of SO_2 and sulfate aerosols, which strongly absorb or scatter solar radiation, creating a global layer of sulfuric acid haze. On average, such eruptions occur several times per century, and cause cooling (by partially blocking the transmission of solar radiation to the Earth's surface) for a period of a few years.

The eruption of Mount Pinatubo in 1991, the second largest terrestrial eruption of the 20th century, affected the climate substantially, subsequently global temperatures decreased by about 0.5 °C (0.9 °F) for up to three years. Thus, the cooling over large parts of the Earth reduced surface temperatures in 1991-93, the equivalent to a reduction in net radiation of 4 watts per square meter. The Mount Tambora eruption in 1815 caused the Year Without a Summer. Much larger eruptions, known as large igneous provinces, occur only a few times every fifty - hundred million years - through flood basalt, and caused in Earth past global warming and mass extinctions.

Small eruptions, with injections of less than 0.1 Mt of sulfur dioxide into the stratosphere, impact the atmosphere only subtly, as temperature changes are comparable with natural variability. However, because smaller eruptions occur at a much higher frequency, they too have a significant impact on Earth's atmosphere.

Seismic monitoring maps current and future trends in volcanic activities, and tries to develop early warning systems. In climate modelling the aim is to study the physical mechanisms and feedbacks of volcanic forcing.

Volcanoes are also part of the extended carbon cycle. Over very long (geological) time periods, they release carbon dioxide from the Earth's crust and mantle, counteracting the uptake by sedimentary rocks and other geological carbon dioxide sinks. The US Geological Survey estimates are that volcanic emissions are at a much lower level than the effects of current human activities, which generate 100–300 times the amount of carbon dioxide emitted by volcanoes. A review of published studies indicates that annual volcanic emissions of carbon dioxide, including amounts released from mid-ocean ridges, volcanic arcs, and hot spot volcanoes, are only the equivalent of 3 to 5 days of human caused output. The annual amount put out by human activities may be greater than the amount released by supererruptions, the most recent of which was the Toba eruption in Indonesia 74,000 years ago.

Although volcanoes are technically part of the lithosphere, which itself is part of the climate system, the IPCC explicitly defines volcanism as an external forcing agent.

Plate Tectonics

Over the course of millions of years, the motion of tectonic plates reconfigures global land and ocean areas and generates topography. This can affect both global and local patterns of climate and atmosphere-ocean circulation.

The position of the continents determines the geometry of the oceans and therefore influences patterns of ocean circulation. The locations of the seas are important in controlling the transfer of heat and moisture across the globe, and therefore, in determining global climate. A recent example of tectonic control on ocean circulation is the formation of the Isthmus of Panama about 5 million years ago, which shut off direct mixing between the Atlantic and Pacific Oceans. This strongly affected the ocean dynamics of what is now the Gulf Stream and may have led to Northern Hemisphere ice cover. During the Carboniferous period, about 300 to 360 million years ago, plate tectonics may have triggered large-scale storage of carbon and increased glaciation. Geologic evidence points to a "megamonsoonal" circulation pattern during the time of the supercontinent Pangaea, and climate modeling suggests that the existence of the supercontinent was conducive to the establishment of monsoons.

The size of continents is also important. Because of the stabilizing effect of the oceans on temperature, yearly temperature variations are generally lower in coastal areas than they are inland. A larger supercontinent will therefore have more area in which climate is strongly seasonal than will several smaller continents or islands.

Human Influences

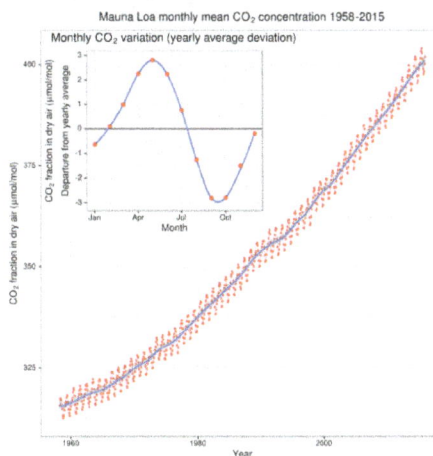

Increase in atmospheric CO_2 levels

In the context of climate variation, anthropogenic factors are human activities which affect the climate. The scientific consensus on climate change is "that climate is changing and that these changes are in large part caused by human activities," and it "is largely irreversible."

"Science has made enormous inroads in understanding climate change and its causes, and is beginning to help develop a strong understanding of current and potential impacts that will affect people today and in coming decades. This understanding is crucial because it allows decision makers to place climate change in the context of other large challenges facing the nation and the world. There are still some uncertainties, and there always will be in understanding a complex system like Earth's climate. Nevertheless, there is a strong, credible body of evidence, based on multiple lines of research, documenting that climate is changing and that these changes are in large part caused by human activities. While much remains to be learned, the core phenomenon, scientific questions, and hypotheses have been examined thoroughly and have stood firm in the face of serious scientific debate and careful evaluation of alternative explanations."

— United States National Research Council, Advancing the Science of Climate Change

Of most concern in these anthropogenic factors is the increase in CO_2 levels due to emissions from fossil fuel combustion, followed by aerosols (particulate matter in the atmosphere) and the CO_2 released by cement manufacture. Other factors, including land use, ozone depletion, animal agriculture and deforestation, are also of concern in the roles they play – both separately and in conjunction with other factors – in affecting climate, microclimate, and measures of climate variables.

Physical Evidence

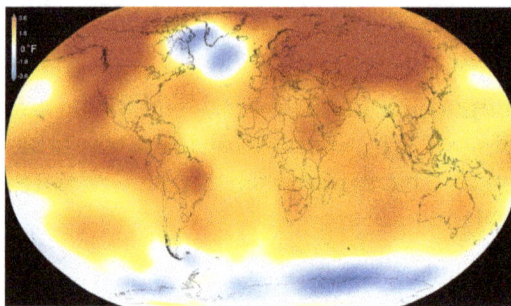

2015 – Warmest Global Year on Record (since 1880) – Colors indicate temperature anomalies (NASA/NOAA; 20 January 2016).

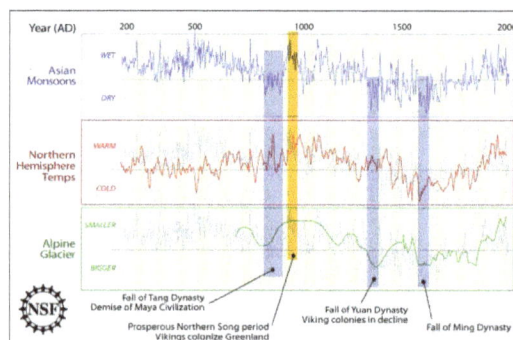

Comparisons between Asian Monsoons from 200 AD to 2000 AD (staying in the background on other plots), Northern Hemisphere temperature, Alpine glacier extent (vertically inverted as marked), and human history as noted by the U.S. NSF.

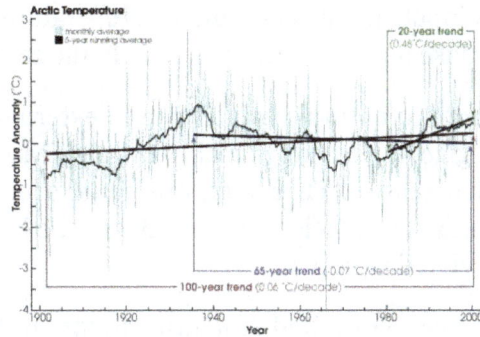

Arctic temperature anomalies over a 100-year period as estimated by NASA. Typical high monthly variance can be seen, while longer-term averages highlight trends.

Evidence for climatic change is taken from a variety of sources that can be used to reconstruct past climates. Reasonably complete global records of surface temperature are available beginning from the mid-late 19th century. For earlier periods, most of the evidence is indirect—climatic changes are inferred from changes in proxies, indicators that reflect climate, such as vegetation, ice cores, dendrochronology, sea level change, and glacial geology.

Temperature Measurements and Proxies

The instrumental temperature record from surface stations was supplemented by radiosonde balloons, extensive atmospheric monitoring by the mid-20th century, and, from the 1970s on, with global satellite data as well. The $^{18}O/^{16}O$ ratio in calcite and ice core samples used to deduce ocean temperature in the distant past is an example of a temperature proxy method, as are other climate metrics noted in subsequent categories.

Historical and Archaeological Evidence

Climate change in the recent past may be detected by corresponding changes in settlement and agricultural patterns. Archaeological evidence, oral history and historical documents can offer insights into past changes in the climate. Climate change effects have been linked to the collapse of various civilizations.

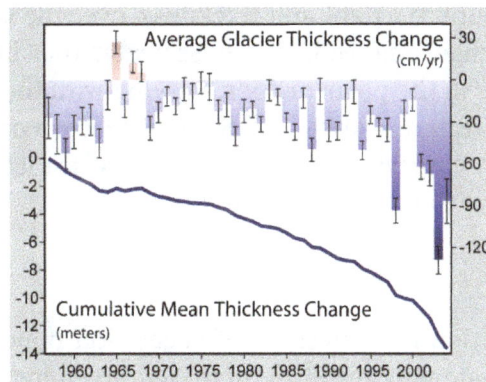

Decline in thickness of glaciers worldwide over the past half-century

Glaciers

Glaciers are considered among the most sensitive indicators of climate change. Their size is determined by a mass balance between snow input and melt output. As temperatures warm, glaciers retreat unless snow precipitation increases to make up for the additional melt; the converse is also true.

Glaciers grow and shrink due both to natural variability and external forcings. Variability in temperature, precipitation, and englacial and subglacial hydrology can strongly determine the evolution of a glacier in a particular season. Therefore, one must average over a decadal or longer time-scale and/or over many individual glaciers to smooth out the local short-term variability and obtain a glacier history that is related to climate.

A world glacier inventory has been compiled since the 1970s, initially based mainly on aerial photographs and maps but now relying more on satellites. This compilation tracks more than 100,000 glaciers covering a total area of approximately 240,000 km², and preliminary estimates indicate that the remaining ice cover is around 445,000 km². The World Glacier Monitoring Service collects data annually on glacier retreat and glacier mass balance. From this data, glaciers worldwide have been found to be shrinking significantly, with strong glacier retreats in the 1940s, stable or growing conditions during the 1920s and 1970s, and again retreating from the mid-1980s to present.

The most significant climate processes since the middle to late Pliocene (approximately 3 million years ago) are the glacial and interglacial cycles. The present interglacial period (the Holocene) has lasted about 11,700 years. Shaped by orbital variations, responses such as the rise and fall of continental ice sheets and significant sea-level changes helped create the climate. Other changes, including Heinrich events, Dansgaard–Oeschger events and the Younger Dryas, however, illustrate how glacial variations may also influence climate without the orbital forcing.

Glaciers leave behind moraines that contain a wealth of material—including organic matter, quartz, and potassium that may be dated—recording the periods in which a glacier advanced and retreated. Similarly, by tephrochronological techniques, the lack of glacier cover can be identified by the presence of soil or volcanic tephra horizons whose date of deposit may also be ascertained.

Arctic Sea Ice Loss

The decline in Arctic sea ice, both in extent and thickness, over the last several decades is further evidence for rapid climate change. Sea ice is frozen seawater that floats on the ocean surface. It covers millions of square miles in the polar regions, varying with the seasons. In the Arctic, some sea ice remains year after year, whereas almost all Southern Ocean or Antarctic sea ice melts away and reforms annually. Satellite observations show that Arctic sea ice is now declining at a rate of 13.3 percent per decade, relative to the 1981 to 2010 average.

Vegetation

A change in the type, distribution and coverage of vegetation may occur given a change in the climate. Some changes in climate may result in increased precipitation and warmth, resulting in improved plant growth and the subsequent sequestration of airborne CO_2. A gradual increase in warmth in a region will lead to earlier flowering and fruiting times, driving a change in the timing of life cycles of dependent organisms. Conversely, cold will cause plant bio-cycles to lag. Larger, faster or more radical changes, however, may result in vegetation stress, rapid plant loss and desertification in certain circumstances. An example of this occurred during the Carboniferous Rainforest Collapse (CRC), an extinction event 300 million years ago. At this time vast rainforests covered the equatorial region of Europe and America. Climate change devastated these tropical rainforests, abruptly fragmenting the habitat into isolated 'islands' and causing the extinction of many plant and animal species.

Pollen Analysis

Palynology is the study of contemporary and fossil palynomorphs, including pollen. Palynology is used to infer the geographical distribution of plant species, which vary under different climate conditions. Different groups of plants have pollen with distinctive shapes and surface textures, and since the outer surface of pollen is composed of a very resilient material, they resist decay. Changes in the type of pollen found in different layers of sediment in lakes, bogs, or river deltas indicate changes in plant communities. These changes are often a sign of a changing climate. As an example, palynological studies have been used to track changing vegetation patterns throughout the Quaternary glaciations and especially since the last glacial maximum.

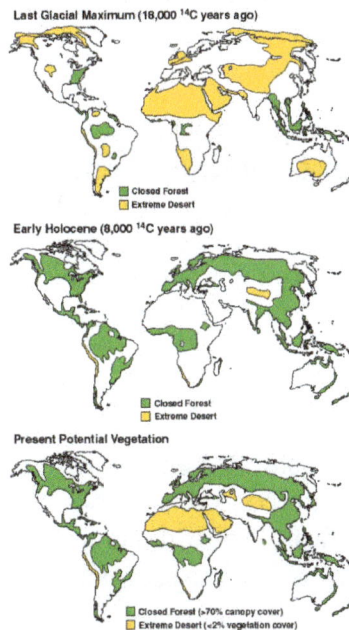

Top: Arid ice age climate
Middle: Atlantic Period, warm and wet
Bottom: Potential vegetation in climate now if not for human effects like agriculture.

Cloud Cover and Precipitation

Past precipitation can be estimated in the modern era with the global network of precipitation gauges. Surface coverage over oceans and remote areas is relatively sparse, but, reducing reliance on interpolation, satellite clouds and precipitation data has been available since the 1970s. Quantification of climatological variation of precipitation in prior centuries and epochs is less complete but approximated using proxies such as marine sediments, ice cores, cave stalagmites, and tree rings. In July 2016 scientists published evidence of increased cloud cover over polar regions, as predicted by climate models.

Climatological temperatures substantially affect cloud cover and precipitation. For instance, during the Last Glacial Maximum of 18,000 years ago, thermal-driven evaporation from the oceans onto continental landmasses was low, causing large areas of extreme desert, including polar deserts (cold but with low rates of cloud cover and precipitation). In contrast, the world's climate was cloudier and wetter than today near the start of the warm Atlantic Period of 8000 years ago.

Estimated global land precipitation increased by approximately 2% over the course of the 20th century, though the calculated trend varies if different time endpoints are chosen, complicated by ENSO and other oscillations, including greater global land cloud cover precipitation in the 1950s and 1970s than the later 1980s and 1990s despite the positive trend over the century overall. Similar slight overall increase in global river runoff and in average soil moisture has been perceived.

Dendroclimatology

Dendroclimatology is the analysis of tree ring growth patterns to determine past climate variations. Wide and thick rings indicate a fertile, well-watered growing period, whilst thin, narrow rings indicate a time of lower rainfall and less-than-ideal growing conditions.

Ice Cores

Analysis of ice in a core drilled from an ice sheet such as the Antarctic ice sheet, can be used to show a link between temperature and global sea level variations. The air trapped in bubbles in the ice can also reveal the CO_2 variations of the atmosphere from the distant past, well before modern environmental influences. The study of these ice cores has been a significant indicator of the changes in CO_2 over many millennia, and continues to provide valuable information about the differences between ancient and modern atmospheric conditions.

Animals

Remains of beetles are common in freshwater and land sediments. Different species of beetles tend to be found under different climatic conditions. Given the extensive lineage of beetles whose genetic makeup has not altered significantly over the millennia, knowledge of the present climatic range of the different species, and the age of the sediments in which remains are found, past climatic conditions may be inferred.

Similarly, the historical abundance of various fish species has been found to have a substantial relationship with observed climatic conditions. Changes in the primary productivity of autotrophs in the oceans can affect marine food webs.

Sea Level Change

Global sea level change for much of the last century has generally been estimated using tide gauge measurements collated over long periods of time to give a long-term average. More recently, altimeter measurements — in combination with accurately determined satellite orbits — have provided an improved measurement of global sea level change. To measure sea levels prior to instrumental measurements, scientists have dated coral reefs that grow near the surface of the ocean, coastal sediments, marine terraces, ooids in limestones, and nearshore archaeological remains. The predominant dating methods used are uranium series and radiocarbon, with cosmogenic radionuclides being sometimes used to date terraces that have experienced relative sea level fall. In the early Pliocene, global temperatures were 1–2 °C warmer than the present temperature, yet sea level was 15–25 meters higher than today.

Climate

Climate is the statistics of weather, usually over a 30-year interval. It is measured by assessing the patterns of variation in temperature, humidity, atmospheric pressure, wind, precipitation, atmospheric particle count and other meteorological variables in a given region over long periods of time. Climate differs from weather, in that weather only describes the short-term conditions of these variables in a given region.

A region's climate is generated by the climate system, which has five components: atmosphere, hydrosphere, cryosphere, lithosphere, and biosphere.

The climate of a location is affected by its latitude, terrain, and altitude, as well as nearby water bodies and their currents. Climates can be classified according to the average and the typical ranges of different variables, most commonly temperature and precipitation. The most commonly used classification scheme was Köppen climate classification originally developed by Wladimir Köppen. The Thornthwaite system, in use since 1948, incorporates evapotranspiration along with temperature and precipitation information and is used in studying biological diversity and the potential effects on it of climate changes. The Bergeron and Spatial Synoptic Classification systems focus on the origin of air masses that define the climate of a region.

Paleoclimatology is the study of ancient climates. Since direct observations of climate are not available before the 19th century, paleoclimates are inferred from *proxy variables* that include non-biotic evidence such as sediments found in lake beds and ice cores, and biotic evidence such as tree rings and coral. Climate models are mathematical models of past, present and future climates. Climate change may occur over long and short timescales from a variety of factors; recent warming is discussed in global warming.

Definition

Climate is commonly defined as the weather averaged over a long period. The standard averaging period is 30 years, but other periods may be used depending on the purpose. Climate also includes statistics other than the average, such as the magnitudes of day-to-day or year-to-year variations. The Intergovernmental Panel on Climate Change (IPCC) 2001 glossary definition is as follows:

Climate in a narrow sense is usually defined as the "average weather," or more rigorously, as the statistical description in terms of the mean and variability of relevant quantities over a period ranging from months to thousands or millions of years. The classical period is 30 years, as defined by the World Meteorological Organization (WMO). These quantities are most often surface variables such as temperature, precipitation, and wind. Climate in a wider sense is the state, including a statistical description, of the climate system.

The World Meteorological Organization (WMO) describes climate "normals" as "reference points used by climatologists to compare current climatological trends to that of the past or what is considered 'normal'. A Normal is defined as the arithmetic average of a climate element (e.g. temperature) over a 30-year period. A 30 year period is used, as it is long enough to filter out any interannual variation or anomalies, but also short enough to be able to show longer climatic trends." The WMO originated from the International Meteorological Organization which set up a technical commission for climatology in 1929. At its 1934 Wiesbaden meeting the technical commission designated the thirty-year period from 1901 to 1930 as the reference time frame for climatological standard normals. In 1982 the WMO agreed to update climate normals, and in these were subsequently completed on the basis of climate data from 1 January 1961 to 31 December 1990.

The difference between climate and weather is usefully summarized by the popular phrase "Climate is what you expect, weather is what you get." Over historical time spans there are a number of nearly constant variables that determine climate, including latitude, altitude, proportion of land to water, and proximity to oceans and mountains. These change only over periods of millions of years due to processes such as plate tectonics. Other climate determinants are more dynamic: the thermohaline circulation of the ocean leads to a 5 °C (9 °F) warming of the northern Atlantic Ocean compared to other ocean basins. Other ocean currents redistribute heat between land and water on a more regional scale. The density and type of vegetation coverage affects solar heat absorption, water retention, and rainfall on a regional level. Alterations in the quantity of atmospheric greenhouse gases determines the amount of solar energy retained by the planet, leading to global warming or global cooling. The variables which determine climate are numerous and the interactions complex, but there is general agreement that the broad outlines are understood, at least insofar as the determinants of historical climate change are concerned.

Climate Classification

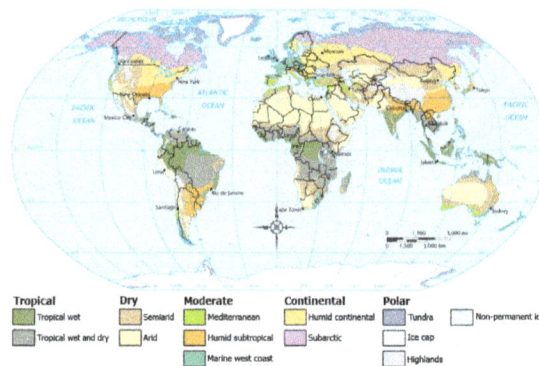

Worldwide climate classifications

There are several ways to classify climates into similar regimes. Originally, climes were defined in Ancient Greece to describe the weather depending upon a location's latitude. Modern climate classification methods can be broadly divided into *genetic* methods, which focus on the causes of climate, and *empiric* methods, which focus on the effects of climate. Examples of genetic classification include methods based on the relative frequency of different air mass types or locations within synoptic weather disturbances. Examples of empiric classifications include climate zones defined by plant hardiness, evapotranspiration, or more generally the Köppen climate classification which was originally designed to identify the climates associated with certain biomes. A common short-coming of these classification schemes is that they produce distinct boundaries between the zones they define, rather than the gradual transition of climate properties more common in nature.

Bergeron and Spatial Synoptic

The simplest classification is that involving air masses. The Bergeron classification is the most widely accepted form of air mass classification. Air mass classification involves three letters. The first letter describes its moisture properties, with c used for continental air masses (dry) and m for maritime air masses (moist). The second letter describes the thermal characteristic of its source region: T for tropical, P for polar, A for Arctic or Antarctic, M for monsoon, E for equatorial, and S for superior air (dry air formed by significant downward motion in the atmosphere). The third letter is used to designate the stability of the atmosphere. If the air mass is colder than the ground below it, it is labeled k. If the air mass is warmer than the ground below it, it is labeled w. While air mass identification was originally used in weather forecasting during the 1950s, climatologists began to establish synoptic climatologies based on this idea in 1973.

Based upon the Bergeron classification scheme is the Spatial Synoptic Classification system (SSC). There are six categories within the SSC scheme: Dry Polar (similar to continental polar), Dry Moderate (similar to maritime superior), Dry Tropical (similar to continental tropical), Moist Polar (similar to maritime polar), Moist Moderate (a hybrid between maritime polar and maritime tropical), and Moist Tropical (similar to maritime tropical, maritime monsoon, or maritime equatorial).

Köppen

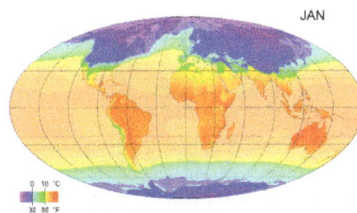

Monthly average surface temperatures from 1961–1990. This is an example of how climate varies with location and season

Monthly global images from NASA Earth Observatory (interactive SVG)

The Köppen classification depends on average monthly values of temperature and precipitation. The most commonly used form of the Köppen classification has five primary types labeled A through E. These primary types are A, tropical; B, dry; C, mild mid-latitude; D, cold mid-latitude; and E, polar. The five primary classifications can be further divided into secondary classifications such as rain forest, monsoon, tropical savanna, humid subtropical, humid continental, oceanic climate, Mediterranean climate, steppe, subarctic climate, tundra, polar ice cap, and desert.

Rain forests are characterized by high rainfall, with definitions setting minimum normal annual rainfall between 1,750 millimetres (69 in) and 2,000 millimetres (79 in). Mean monthly temperatures exceed 18 °C (64 °F) during all months of the year.

A monsoon is a seasonal prevailing wind which lasts for several months, ushering in a region's rainy season. Regions within North America, South America, Sub-Saharan Africa, Australia and East Asia are monsoon regimes.

The world's cloudy and sunny spots. NASA Earth Observatory map using data collected between July 2002 and April 2015.

A tropical savanna is a grassland biome located in semiarid to semi-humid climate regions of sub-tropical and tropical latitudes, with average temperatures remain at or above 18 °C (64 °F) year round and rainfall between 750 millimetres (30 in) and 1,270 millimetres (50 in) a year. They are widespread on Africa, and are found in India, the northern parts of South America, Malaysia, and Australia.

Cloud cover by month for 2014. NASA Earth Observatory

The humid subtropical climate zone where winter rainfall (and sometimes snowfall) is associated with large storms that the westerlies steer from west to east. Most summer rainfall occurs during thunderstorms and from occasional tropical cyclones. Humid subtropical climates lie on the east side continents, roughly between latitudes 20° and 40° degrees away from the equator.

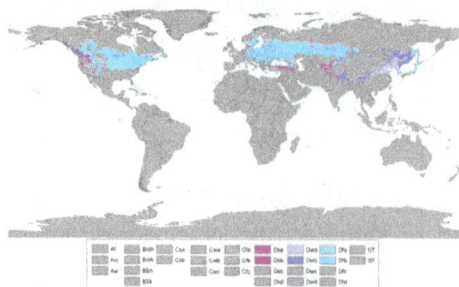

Humid continental climate, worldwide

A humid continental climate is marked by variable weather patterns and a large seasonal temperature variance. Places with more than three months of average daily temperatures above 10 °C (50 °F) and a coldest month temperature below –3 °C (27 °F) and which do not meet the criteria for an arid or semiarid climate, are classified as continental.

An oceanic climate is typically found along the west coasts at the middle latitudes of all the world's continents, and in southeastern Australia, and is accompanied by plentiful precipitation year round.

The Mediterranean climate regime resembles the climate of the lands in the Mediterranean Basin, parts of western North America, parts of Western and South Australia, in southwestern South Africa and in parts of central Chile. The climate is characterized by hot, dry summers and cool, wet winters.

A steppe is a dry grassland with an annual temperature range in the summer of up to 40 °C (104 °F) and during the winter down to –40 °C (–40 °F).

A subarctic climate has little precipitation, and monthly temperatures which are above 10 °C (50 °F) for one to three months of the year, with permafrost in large parts of the area due to the cold winters. Winters within subarctic climates usually include up to six months of temperatures averaging below 0 °C (32 °F).

Map of arctic tundra

Tundra occurs in the far Northern Hemisphere, north of the taiga belt, including vast areas of northern Russia and Canada.

A polar ice cap, or polar ice sheet, is a high-latitude region of a planet or moon that is covered in ice. Ice caps form because high-latitude regions receive less energy as solar radiation from the sun than equatorial regions, resulting in lower surface temperatures.

A desert is a landscape form or region that receives very little precipitation. Deserts usually have a large diurnal and seasonal temperature range, with high or low, depending on location daytime temperatures (in summer up to 45 °C or 113 °F), and low nighttime temperatures (in winter down to 0 °C or 32 °F) due to extremely low humidity. Many deserts are formed by rain shadows, as mountains block the path of moisture and precipitation to the desert.

Thornthwaite

Devised by the American climatologist and geographer C. W. Thornthwaite, this climate classification method monitors the soil water budget using evapotranspiration. It monitors the portion of total precipitation used to nourish vegetation over a certain area. It uses indices such as a humidity index and an aridity index to determine an area's moisture regime based upon its average temperature, average rainfall, and average vegetation type. The lower the value of the index in any given area, the drier the area is.

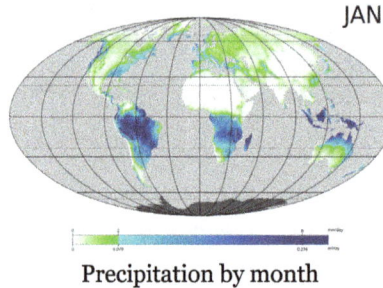

JAN

Precipitation by month

The moisture classification includes climatic classes with descriptors such as hyperhumid, humid, subhumid, subarid, semi-arid (values of −20 to −40), and arid (values below −40). Humid regions experience more precipitation than evaporation each year, while arid regions experience greater evaporation than precipitation on an annual basis. A total of 33 percent of the Earth's landmass is considered either arid or semi-arid, including southwest North America, southwest South America, most of northern and a small part of southern Africa, southwest and portions of eastern Asia, as well as much of Australia. Studies suggest that precipitation effectiveness (PE) within the Thornthwaite moisture index is overestimated in the summer and underestimated in the winter. This index can be effectively used to determine the number of herbivore and mammal species numbers within a given area. The index is also used in studies of climate change.

Thermal classifications within the Thornthwaite scheme include microthermal, mesothermal, and megathermal regimes. A microthermal climate is one of low annual mean temperatures, generally between 0 °C (32 °F) and 14 °C (57 °F) which experiences short summers and has a potential evaporation between 14 centimetres (5.5 in) and 43 centimetres (17 in). A mesothermal climate lacks persistent heat or persistent cold, with potential evaporation between 57 centimetres (22 in) and 114 centimetres (45 in). A megathermal climate is one with persistent high temperatures and abundant rainfall, with potential annual evaporation in excess of 114 centimetres (45 in).

Record

Modern

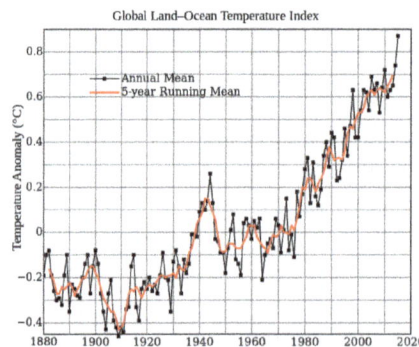

Global mean surface temperature change since 1880. Source: NASA GISS

Details of the modern climate record are known through the taking of measurements from such weather instruments as thermometers, barometers, and anemometers during the past few centuries. The instruments used to study weather over the modern time scale, their known error, their immediate environment, and their exposure have changed over the years, which must be considered when studying the climate of centuries past.

Paleoclimatology

Paleoclimatology is the study of past climate over a great period of the Earth's history. It uses evidence from ice sheets, tree rings, sediments, coral, and rocks to determine the past state of the climate. It demonstrates periods of stability and periods of change and can indicate whether changes follow patterns such as regular cycles.

Climate Change

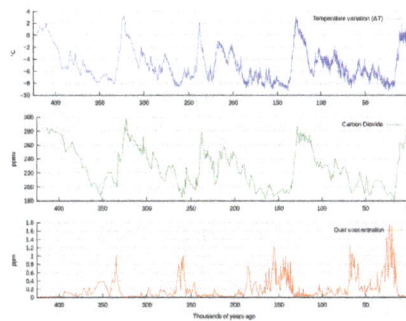

Variations in CO_2, temperature and dust from the Vostok ice core over the past 450,000 years

Climate change is the variation in global or regional climates over time. It reflects changes in the variability or average state of the atmosphere over time scales ranging from decades to millions of years. These changes can be caused by processes internal to the Earth, external forces (e.g. variations in sunlight intensity) or, more recently, human activities.

In recent usage, especially in the context of environmental policy, the term "climate change" often refers only to changes in modern climate, including the rise in average surface temperature known as global warming. In some cases, the term is also used with a presumption of human causation, as in the United Nations Framework Convention on Climate Change (UNFCCC). The UNFCCC uses "climate variability" for non-human caused variations.

Earth has undergone periodic climate shifts in the past, including four major ice ages. These consisting of glacial periods where conditions are colder than normal, separated by interglacial periods. The accumulation of snow and ice during a glacial period increases the surface albedo, reflecting more of the Sun's energy into space and maintaining a lower atmospheric temperature. Increases in greenhouse gases, such as by volcanic activity, can increase the global temperature and produce an interglacial period. Suggested causes of ice age periods include the positions of the continents, variations in the Earth's orbit, changes in the solar output, and volcanism.

Climate Models

Climate models use quantitative methods to simulate the interactions of the atmosphere, oceans, land surface and ice. They are used for a variety of purposes; from the study of the dynamics of the weather and climate system, to projections of future climate. All climate models balance, or very nearly balance, incoming energy as short wave (including visible) electromagnetic radiation to the earth with outgoing energy as long wave (infrared) electromagnetic radiation from the earth. Any imbalance results in a change in the average temperature of the earth.

The most talked-about applications of these models in recent years have been their use to infer the consequences of increasing greenhouse gases in the atmosphere, primarily carbon dioxide. These models predict an upward trend in the global mean surface temperature, with the most rapid increase in temperature being projected for the higher latitudes of the Northern Hemisphere.

Models can range from relatively simple to quite complex:

1. Simple radiant heat transfer model that treats the earth as a single point and averages outgoing energy

2. this can be expanded vertically (radiative-convective models), or horizontally

3. finally, (coupled) atmosphere–ocean–sea ice global climate models discretise and solve the full equations for mass and energy transfer and radiant exchange.

Climate forecasting is a way by some scientists are using to predict climate change. In 1997 the prediction division of the International Research Institute for Climate and Society at Columbia University began generating seasonal climate forecasts on a real-time basis. To produce these forecasts an extensive suite of forecasting tools was developed, including a multimodel ensemble approach that required thorough validation of each model's accuracy level in simulating interannual climate variability.

References

*   Fredlund, D.G.; Rahardjo, H. (1993). Soil Mechanics for Unsaturated Soils (PDF). Wiley-Interscience. ISBN 978-0-471-85008-3. OCLC 26543184. Retrieved 2008-05-21.

*   Brown, Dwayne; Cabbage, Michael; McCarthy, Leslie; Norton, Karen (20 January 2016). "NASA, NOAA Analyses Reveal Record-Shattering Global Warm Temperatures in 2015". NASA. Retrieved 21 January 2016.

*   Planton, Serge (France; editor) (2013). "Annex III. Glossary: IPCC - Intergovernmental Panel on Climate Change" (PDF). IPCC Fifth Assessment Report. p. 1450. Retrieved 25 July 2016.

*   Shepherd, Dr. J. Marshall; Shindell, Drew; O'Carroll, Cynthia M. (1 February 2005). "What's the Difference Between Weather and Climate?". NASA. Retrieved 13 November 2015.

*   "Commission For Climatology: Over Eighty Years of Service" (pdf). World Meteorological Organization. 2011. pp. 6, 8, 10, 21, 26. Retrieved 1 September 2015.

*   Gillis, Justin (28 November 2015). "Short Answers to Hard Questions About Climate Change". New York Times. Retrieved 29 November 2015.

2

Climate Change Prediction: Methods and Techniques

Methods and techniques are an important component of any field of study. Some of these techniques are climate model, general circulation model, numerical weather prediction and tropical cyclone forecast model. Climate models are used to simulate the important aspects of climate like atmosphere, oceans and land whereas general circulation model is a mathematical model. It helps in the demonstration of the circulation of the ocean. The following chapter elucidates the methods and techniques that are related to climate change.

Climate Model

Climate models use quantitative methods to simulate the interactions of the important drivers of climate, including atmosphere, oceans, land surface and ice. They are used for a variety of purposes from study of the dynamics of the climate system to projections of future climate.

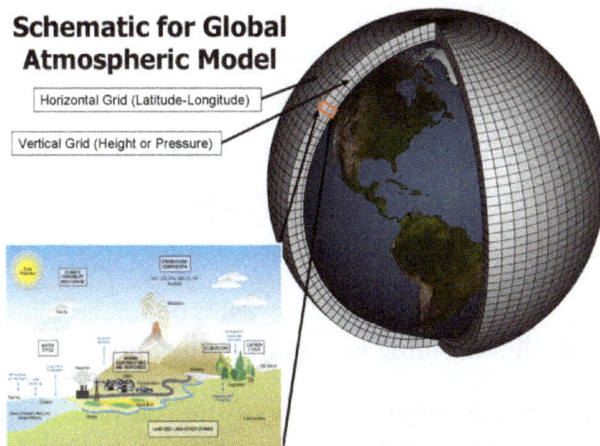

Schematic for Global Atmospheric Model

Horizontal Grid (Latitude-Longitude)

Vertical Grid (Height or Pressure)

Climate models are systems of differential equations based on the basic laws of physics, fluid motion, and chemistry. To "run" a model, scientists divide the planet into a 3-dimensional grid, apply the basic equations, and evaluate the results. Atmospheric models calculate winds, heat transfer, radiation, relative humidity, and surface hydrology within each grid and evaluate interactions with neighboring points.

All climate models take account of incoming energy from the sun as short wave electromagnetic radiation, chiefly visible and short-wave (near) infrared, as well as outgoing long wave (far) infrared electromagnetic. Any imbalance results in a change in temperature.

Models vary in complexity:

A simple radiant heat transfer model treats the earth as a single point and averages outgoing energy

- This can be expanded vertically (radiative-convective models) and/or horizontally

- Finally, (coupled) atmosphere–ocean–sea ice global climate models solve the full equations for mass and energy transfer and radiant exchange.

- Box models can treat flows across and within ocean basins.

- Other types of modelling can be interlinked, such as land use, allowing researchers to predict the interaction between climate and ecosystems.

Box Models

Box models are simplified versions of complex systems, reducing them to boxes (or reservoirs) linked by fluxes. The boxes are assumed to be mixed homogeneously. Within a given box, the concentration of any chemical species is therefore uniform. However, the abundance of a species within a given box may vary as a function of time due to the input to (or loss from) the box or due to the production, consumption or decay of this species within the box.

Simple box models, i.e. box model with a small number of boxes whose properties (e.g. their volume) do not change with time, are often useful to derive analytical formulas describing the dynamics and steady-state abundance of a species. More complex box models are usually solved using numerical techniques.

Box models are used extensively to model environmental systems or ecosystems and in studies of ocean circulation and the carbon cycle.

Zero-dimensional Models

A very simple model of the radiative equilibrium of the Earth is

$$(1-a)S\pi r^2 = 4\pi r^2 \epsilon \sigma T^4$$

where

- the left hand side represents the incoming energy from the Sun

- the right hand side represents the outgoing energy from the Earth, calculated from the Stefan-Boltzmann law assuming a model-fictive temperature, T, sometimes called the 'equilibrium temperature of the Earth', that is to be found,

and

- S is the solar constant – the incoming solar radiation per unit area—about 1367 W·m^{-2}

- a is the Earth's average albedo, measured to be 0.3.

- r is Earth's radius—approximately 6.371×10^6m

- π is the mathematical constant (3.141...)

- σ is the Stefan-Boltzmann constant—approximately 5.67×10^{-8} J·K^{-4}·m^{-2}·s^{-1}

• ϵ is the effective emissivity of earth, about 0.612

The constant πr^2 can be factored out, giving

$$(1-a)S = 4\epsilon\sigma T^4$$

Solving for the temperature,

$$T = \sqrt[4]{\frac{(1-a)S}{4\epsilon\sigma}}$$

This yields an apparent effective average earth temperature of 288 K (15 °C; 59 °F). This is because the above equation represents the effective *radiative* temperature of the Earth (including the clouds and atmosphere). The use of effective emissivity and albedo account for the greenhouse effect.

This very simple model is quite instructive, and the only model that could fit on a page. For example, it easily determines the effect on average earth temperature of changes in solar constant or change of albedo or effective earth emissivity.

The average emissivity of the earth is readily estimated from available data. The emissivities of terrestrial surfaces are all in the range of 0.96 to 0.99 (except for some small desert areas which may be as low as 0.7). Clouds, however, which cover about half of the earth's surface, have an average emissivity of about 0.5 (which must be reduced by the fourth power of the ratio of cloud absolute temperature to average earth absolute temperature) and an average cloud temperature of about 258 K (−15 °C; 5 °F). Taking all this properly into account results in an effective earth emissivity of about 0.64 (earth average temperature 285 K (12 °C; 53 °F)).

This simple model readily determines the effect of changes in solar output or change of earth albedo or effective earth emissivity on average earth temperature. It says nothing, however about what might cause these things to change. Zero-dimensional models do not address the temperature distribution on the earth or the factors that move energy about the earth.

Radiative-convective Models

The zero-dimensional model above, using the solar constant and given average earth temperature, determines the effective earth emissivity of long wave radiation emitted to space. This can be refined in the vertical to a one-dimensional radiative-convective model, which considers two processes of energy transport:

1. upwelling and downwelling radiative transfer through atmospheric layers that both absorb and emit infrared radiation

2. upward transport of heat by convection (especially important in the lower troposphere).

The radiative-convective models have advantages over the simple model: they can determine the effects of varying greenhouse gas concentrations on effective emissivity and therefore the surface temperature. But added parameters are needed to determine local emissivity and albedo and address the factors that move energy about the earth.

Effect of ice-albedo feedback on global sensitivity in a one-dimensional radiative-convective climate model.

Higher-dimension Models

The zero-dimensional model may be expanded to consider the energy transported horizontally in the atmosphere. This kind of model may well be zonally averaged. This model has the advantage of allowing a rational dependence of local albedo and emissivity on temperature – the poles can be allowed to be icy and the equator warm – but the lack of true dynamics means that horizontal transports have to be specified.

EMICs (Earth-system Models of Intermediate Complexity)

Depending on the nature of questions asked and the pertinent time scales, there are, on the one extreme, conceptual, more inductive models, and, on the other extreme, general circulation models operating at the highest spatial and temporal resolution currently feasible. Models of intermediate complexity bridge the gap. One example is the Climber-3 model. Its atmosphere is a 2.5-dimensional statistical-dynamical model with $7.5° \times 22.5°$ resolution and time step of half a day; the ocean is MOM-3 (Modular Ocean Model) with a $3.75° \times 3.75°$ grid and 24 vertical levels.

Gcms (Global Climate Models or General Circulation Models)

General Circulation Models (GCMs) discretise the equations for fluid motion and energy transfer and integrate these over time. Unlike simpler models, GCMs divide the atmosphere and/or oceans into grids of discrete "cells", which represent computational units. Unlike simpler models which make mixing assumptions, processes internal to a cell—such as convection—that occur on scales too small to be resolved directly are parameterised at the cell level, while other functions govern the interface between cells.

Atmospheric GCMs (AGCMs) model the atmosphere and impose sea surface temperatures as boundary conditions. Coupled atmosphere-ocean GCMs (AOGCMs, e.g. HadCM3, EdGCM, GFDL CM2.X, ARPEGE-Climat) combine the two models. The first general circulation climate model that combined both oceanic and atmospheric processes was developed in the late 1960s at the NOAA Geophysical Fluid Dynamics Laboratory AOGCMs represent the pinnacle of complexity in climate models and internalise as many processes as possible. However, they are still under development and uncertainties remain. They may be coupled to models of other processes, such as the carbon cycle, so as to better model feedback effects. Such integrated multi-system models are sometimes referred to as either "earth system models" or "global climate models."

Research and Development

There are three major types of institution where climate models are developed, implemented and used:

1.   National meteorological services. Most national weather services have a climatology section.

2. Universities. Relevant departments include atmospheric sciences, meteorology, climatology, and geography.

3. National and international research laboratories. Examples include the National Center for Atmospheric Research (NCAR, in Boulder, Colorado, USA), the Geophysical Fluid Dynamics Laboratory (GFDL, in Princeton, New Jersey, USA), the Hadley Centre for Climate Prediction and Research (in Exeter, UK), the Max Planck Institute for Meteorology in Hamburg, Germany, or the Laboratoire des Sciences du Climat et de l'Environnement (LSCE), France, to name but a few.

The World Climate Research Programme (WCRP), hosted by the World Meteorological Organization (WMO), coordinates research activities on climate modelling worldwide.

A 2012 U.S. National Research Council report discussed how the large and diverse U.S. climate modeling enterprise could evolve to become more unified. Efficiencies could be gained by developing a common software infrastructure shared by all U.S. climate researchers, and holding an annual climate modeling forum, the report found.

General Circulation Model

A general circulation model (GCM) is a type of climate model. It employs a mathematical model of the general circulation of a planetary atmosphere or ocean. It uses the Navier–Stokes equations on a rotating sphere with thermodynamic terms for various energy sources (radiation, latent heat). These equations are the basis for computer programs used to simulate the Earth's atmosphere or oceans. Atmospheric and oceanic GCMs (AGCM and OGCM) are key components along with sea ice and land-surface components.

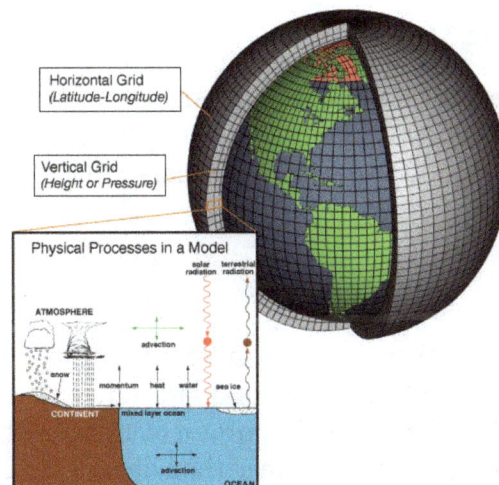

Climate models are systems of differential equations based on the basic laws of physics, fluid motion, and chemistry. To "run" a model, scientists divide the planet into a 3-dimensional grid, apply the basic equations, and evaluate the results. Atmospheric models calculate winds, heat transfer, radiation, relative humidity, and surface hydrology within each grid and evaluate interactions with neighboring points.

GCMs and global climate models are used for weather forecasting, understanding the climate and forecasting climate change.

Versions designed for decade to century time scale climate applications were originally created by Syukuro Manabe and Kirk Bryan at the Geophysical Fluid Dynamics Laboratory in Princeton, New Jersey. These models are based on the integration of a variety of fluid dynamical, chemical and sometimes biological equations.

Terminology

The acronym *GCM* originally stood for *General Circulation Model*. Recently, a second meaning came into use, namely *Global Climate Model*. While these do not refer to the same thing, General Circulation Models are typically the tools used for modelling climate, and hence the two terms are sometimes used interchangeably. However, the term "global climate model" is ambiguous and may refer to an integrated framework that incorporates multiple components including a general circulation model, or may refer to the general class of climate models that use a variety of means to represent the climate mathematically.

History

In 1956, Norman Phillips developed a mathematical model that could realistically depict monthly and seasonal patterns in the troposphere. It became the first successful climate model. Following Phillips's work, several groups began working to create GCMs. The first to combine both oceanic and atmospheric processes was developed in the late 1960s at the NOAA Geophysical Fluid Dynamics Laboratory. By the early 1980s, the United States' National Center for Atmospheric Research had developed the Community Atmosphere Model; this model has been continuously refined. In 1996, efforts began to model soil and vegetation types. Later the Hadley Centre for Climate Prediction and Research's HadCM3 model coupled ocean-atmosphere elements. The role of gravity waves was added in the mid-1980s. Gravity waves are required to simulate regional and global scale circulations accurately.

Atmospheric and Oceanic Models

Atmospheric (AGCMs) and oceanic GCMs (OGCMs) can be coupled to form an atmosphere-ocean coupled general circulation model (CGCM or AOGCM). With the addition of submodels such as a sea ice model or a model for evapotranspiration over land, AOGCMs become the basis for a full climate model.

Trends

A recent trend in GCMs is to apply them as components of Earth system models, e.g. by coupling ice sheet models for the dynamics of the Greenland and Antarctic ice sheets, and one or more chemical transport models (CTMs) for species important to climate. Thus a carbon CTM may allow a GCM to better predict anthropogenic changes in carbon dioxide concentrations. In addition, this approach allows accounting for inter-system feedback: e.g. chemistry-climate models allow the possible effects of climate change on ozone hole to be studied.

Climate prediction uncertainties depend on uncertainties in chemical, physical and social models. Significant uncertainties and unknowns remain, especially regarding the future course of human population, industry and technology.

Structure

Three-dimensional (more properly four-dimensional) GCMs apply discrete equations for fluid motion and integrate these forward in time. They contain parameterisations for processes such as convection that occur on scales too small to be resolved directly.

A simple general circulation model (SGCM) consists of a dynamic core that relates properties such as temperature to others such as pressure and velocity. Examples are programs that solve the primitive equations, given energy input and energy dissipation in the form of scale-dependent friction, so that atmospheric waves with the highest wavenumbers are most attenuated. Such models may be used to study atmospheric processes, but are not suitable for climate projections.

Atmospheric GCMs (AGCMs) model the atmosphere (and typically contain a land-surface model as well) using imposed sea surface temperatures (SSTs). They may include atmospheric chemistry.

AGCMs consist of a dynamical core which integrates the equations of fluid motion, typically for:

1. surface pressure

2. horizontal components of velocity in layers

3. temperature and water vapor in layers

4. radiation, split into solar/short wave and terrestrial/infra-red/long wave

5. parameters for:

    1. convection

    2. land surface processes

    3. albedo

    4. hydrology

    5. cloud cover

A GCM contains prognostic equations that are a function of time (typically winds, temperature, moisture, and surface pressure) together with diagnostic equations that are evaluated from them for a specific time period. As an example, pressure at any height can be diagnosed by applying the hydrostatic equation to the predicted surface pressure and the predicted values of temperature between the surface and the height of interest. Pressure is used to compute the pressure gradient force in the time-dependent equation for the winds.

OGCMs model the ocean (with fluxes from the atmosphere imposed) and may contain a sea ice model. For example, the standard resolution of HadOM3 is 1.25 degrees in latitude and longitude, with 20 vertical levels, leading to approximately 1,500,000 variables.

AOGCMs (e.g. HadCM3, GFDL CM2.X) combine the two submodels. They remove the need to specify fluxes across the interface of the ocean surface. These models are the basis for model predictions of future climate, such as are discussed by the IPCC. AOGCMs internalise as many processes as possible. They have been used to provide predictions at a regional scale. While the simpler models are generally susceptible to analysis and their results are easier to understand, AOGCMs may be nearly as hard to analyse as the climate itself.

Grid

The fluid equations for AGCMs are made discrete using either the finite difference method or the spectral method. For finite differences, a grid is imposed on the atmosphere. The simplest grid uses constant angular grid spacing (i.e., a latitude / longitude grid). However, non-rectangular grids (e.g., icosahedral) and grids of variable resolution are more often used. The LMDz model can be arranged to give high resolution over any given section of the planet. HadGEM1 (and other ocean models) use an ocean grid with higher resolution in the tropics to help resolve processes believed to be important for the El Niño Southern Oscillation (ENSO). Spectral models generally use a gaussian grid, because of the mathematics of transformation between spectral and grid-point space. Typical AGCM resolutions are between 1 and 5 degrees in latitude or longitude: HadCM3, for example, uses 3.75 in longitude and 2.5 degrees in latitude, giving a grid of 96 by 73 points (96 x 72 for some variables); and has 19 vertical levels. This results in approximately 500,000 "basic" variables, since each grid point has four variables (u, v, T, Q), though a full count would give more (clouds; soil levels). HadGEM1 uses a grid of 1.875 degrees in longitude and 1.25 in latitude in the atmosphere; HiGEM, a high-resolution variant, uses 1.25 x 0.83 degrees respectively. These resolutions are lower than is typically used for weather forecasting. Ocean resolutions tend to be higher, for example HadCM3 has 6 ocean grid points per atmospheric grid point in the horizontal.

For a standard finite difference model, uniform gridlines converge towards the poles. This would lead to computational instabilities and so the model variables must be filtered along lines of latitude close to the poles. Ocean models suffer from this problem too, unless a rotated grid is used in which the North Pole is shifted onto a nearby landmass. Spectral models do not suffer from this problem. Some experiments use geodesic grids and icosahedral grids, which (being more uniform) do not have pole-problems. Another approach to solving the grid spacing problem is to deform a Cartesian cube such that it covers the surface of a sphere.

Flux Buffering

Some early versions of AOGCMs required an *ad hoc* process of "flux correction" to achieve a stable climate. This resulted from separately prepared ocean and atmospheric models that each used an implicit flux from the other component different than that component could produce. Such a model failed to match observations. However, if the fluxes were 'corrected', the factors that led to these unrealistic fluxes might be unrecognised, which could affect model sensitivity. As a result, the vast majority of models used in the current round of IPCC reports do not use them. The model improvements that now make flux corrections unnecessary include improved ocean physics, improved resolution in both atmosphere and ocean, and more physically consistent coupling between atmosphere and ocean submodels. Improved models now maintain stable, multi-century simulations of surface climate that are considered to be of sufficient quality to allow their use for climate projections.

Convection

Moist convection releases latent heat and is important to the Earth's energy budget. Convection occurs on too small a scale to be resolved by climate models, and hence it must be handled via parameters. This has been done since the 1950s. Akio Arakawa did much of the early work, and variants of his scheme are still used, although a variety of different schemes are now in use. Clouds are also typically handled with a parameter, for a similar lack of scale. Limited understanding of clouds has limited the success of this strategy, but not due to some inherent shortcoming of the method.

Software

Most models include software to diagnose a wide range of variables for comparison with observations or study of atmospheric processes. An example is the 1.5-metre temperature, which is the standard height for near-surface observations of air temperature. This temperature is not directly predicted from the model but is deduced from surface and lowest-model-layer temperatures. Other software is used for creating plots and animations.

Projections

Projected annual mean surface air temperature from 1970-2100, based on SRES emissions scenario A1B, using the NOAA GFDL CM2.1 climate model (credit: NOAA Geophysical Fluid Dynamics Laboratory).

Coupled AOGCMs use transient climate simulations to project/predict climate changes under various scenarios. These can be idealised scenarios (most commonly, CO_2 emissions increasing at 1%/yr) or based on recent history (usually the "IS92a" or more recently the SRES scenarios). Which scenarios are most realistic remains uncertain.

The 2001 IPCC Third Assessment Report F igure 9.3 shows the global mean response of 19 different coupled models to an idealised experiment in which emissions increased at 1% per year. Figure 9.5 shows the response of a smaller number of models to more recent trends. For the 7 climate models shown there, the temperature change to 2100 varies from 2 to 4.5 °C with a median of about 3 °C.

Future scenarios do not include unknown events – for example, volcanic eruptions or changes in solar forcing. These effects are believed to be small in comparison to greenhouse gas (GHG) forcing in the long term, but large volcanic eruptions, for example, can exert a substantial temporary cooling effect.

Human GHG emissions are a model input, although it is possible to include an economic/technological submodel to provide these as well. Atmospheric GHG levels are usually supplied as an input, though it is possible to include a carbon cycle model that reflects vegetation and oceanic processes to calculate such levels.

Emissions Scenarios

Projected change in annual mean surface air temperature from the late 20th century to the middle

21st century, based on SRES emissions scenario A1B (credit: NOAA Geophysical Fluid Dynamics Laboratory).

For the six SRES marker scenarios, IPCC (2007:7–8) gave a "best estimate" of global mean temperature increase (2090–2099 relative to the period 1980–1999) of 1.8 °C to 4.0 °C. Over the same time period, the "likely" range (greater than 66% probability, based on expert judgement) for these scenarios was for a global mean temperature increase of 1.1 to 6.4 °C.

In 2008 a study made climate projections using several emission scenarios. In a scenario where global emissions start to decrease by 2010 and then declined at a sustained rate of 3% per year, the likely global average temperature increase was predicted to be 1.7 °C above pre-industrial levels by 2050, rising to around 2 °C by 2100. In a projection designed to simulate a future where no efforts are made to reduce global emissions, the likely rise in global average temperature was predicted to be 5.5 °C by 2100. A rise as high as 7 °C was thought possible, although less likely.

Another no-reduction scenario resulted in a median warming over land (2090–99 relative to the period 1980–99) of 5.1 °C. Under the same emissions scenario but with a different model, the predicted median warming was 4.1 °C.

Model Accuracy

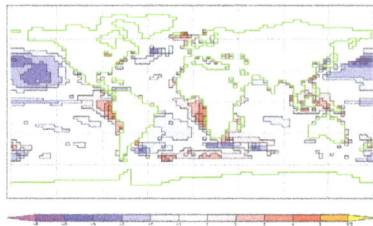

SST errors in HadCM3

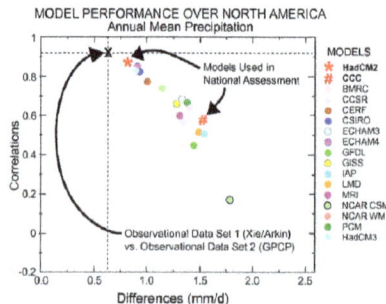

North American precipitation from various models.

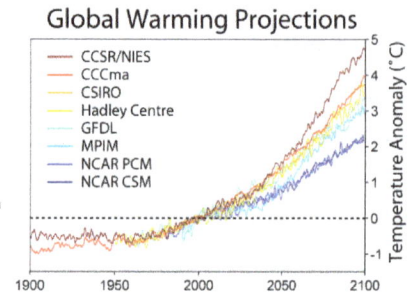

Temperature predictions from some climate models assuming the SRES A2 emissions scenario.

AOGCMs internalise as many processes as are sufficiently understood. However, they are still under development and significant uncertainties remain. They may be coupled to models of other processes, such as the carbon cycle, so as to better model feedbacks. Most recent simulations show "plausible" agreement with the measured temperature anomalies over the past 150 years, when driven by observed changes in greenhouse gases and aerosols. Agreement improves by including both natural and anthropogenic forcings.

Imperfect models may nevertheless produce useful results. GCMs are capable of reproducing the general features of the observed global temperature over the past century.

A debate over how to reconcile climate model predictions that upper air (tropospheric) warming should be greater than observed surface warming, some of which appeared to show otherwise, was resolved in favour of the models, following data revisions.

Cloud effects are a significant area of uncertainty in climate models. Clouds have competing effects on climate. They cool the surface by reflecting sunlight into space; they warm it by increasing the amount of infrared radiation transmitted from the atmosphere to the surface. In the 2001 IPCC report possible changes in cloud cover were highlighted as a major uncertainty in predicting climate.

Climate researchers around the world use climate models to understand the climate system. Thousands of papers have been published about model-based studies. Part of this research is to improve the models.

In 2000, a comparison between measurements and dozens of GCM simulations of ENSO-driven tropical precipitation, water vapor, temperature, and outgoing longwave radiation found similarity between measurements and simulation of most factors. However the simulated change in precipitation was about one-fourth less than what was observed. Errors in simulated precipitation imply errors in other processes, such as errors in the evaporation rate that provides moisture to create precipitation. The other possibility is that the satellite-based measurements are in error. Either indicates progress is required in order to monitor and predict such changes.

A more complete discussion of climate models is provided in the IPCC's Third Assessment Report.

1. The model mean exhibits good agreement with observations.

2. The individual models often exhibit worse agreement with observations.

3. Many of the non-flux adjusted models suffered from unrealistic climate drift up to about 1 °C/century in global mean surface temperature.

4. The errors in model-mean surface air temperature rarely exceed 1 °C over the oceans and 5 °C over the continents; precipitation and sea level pressure errors are relatively greater but the magnitudes and patterns of these quantities are recognisably similar to observations.

5. Surface air temperature is particularly well simulated, with nearly all models closely matching the observed magnitude of variance and exhibiting a correlation > 0.95 with the observations.

6. Simulated variance of sea level pressure and precipitation is within ±25% of observed.

7. All models have shortcomings in their simulations of the present day climate of the stratosphere, which might limit the accuracy of predictions of future climate change.

    1. There is a tendency for the models to show a global mean cold bias at all levels.

    2. There is a large scatter in the tropical temperatures.

    3. The polar night jets in most models are inclined poleward with height, in noticeable contrast to an equatorward inclination of the observed jet.

    3. There is a differing degree of separation in the models between the winter sub-tropical jet and the polar night jet.

8. For nearly all models the r.m.s. error in zonal- and annual-mean surface air temperature is small compared with its natural variability.

1.   There are problems in simulating natural seasonal variability. (2000)

      1.   In flux-adjusted models, seasonal variations are simulated to within 2 K of observed values over the oceans. The corresponding average over non-flux-adjusted models shows errors up to about 6 K in extensive ocean areas.

      2.   Near-surface land temperature errors are substantial in the average over flux-adjusted models, which systematically underestimates (by about 5 K) temperature in areas of elevated terrain. The corresponding average over non-flux-adjusted models forms a similar error pattern (with somewhat increased amplitude) over land.

      3.   In Southern Ocean mid-latitudes, the non-flux-adjusted models overestimate the magnitude of January-minus-July temperature differences by ~5 K due to an overestimate of summer (January) near-surface temperature. This error is common to five of the eight non-flux-adjusted models.

      4.   Over Northern Hemisphere mid-latitude land areas, zonal mean differences between July and January temperatures simulated by the non-flux-adjusted models show a greater spread (positive and negative) about observed values than results from the flux-adjusted models.

      5.   The ability of coupled GCMs to simulate a reasonable seasonal cycle is a necessary condition for confidence in their prediction of long-term climatic changes (such as global warming), but it is not a sufficient condition unless the seasonal cycle and long-term changes involve similar climatic processes.

9.   Coupled climate models do not simulate with reasonable accuracy clouds and some related hydrological processes (in particular those involving upper tropospheric humidity). Problems in the simulation of clouds and upper tropospheric humidity, remain worrisome because the associated processes account for most of the uncertainty in climate model simulations of anthropogenic change.

The precise magnitude of future changes in climate is still uncertain; for the end of the 21st century (2071 to 2100), for SRES scenario A2, the change of global average SAT change from AOGCMs compared with 1961 to 1990 is +3.0 °C (5.4 °F) and the range is +1.3 to +4.5 °C (+2.3 to 8.1 °F).

The IPCC's Fifth Assessment Report asserted "...very high confidence that models reproduce the general features of the global-scale annual mean surface temperature increase over the historical period." However, the report also observed that the rate of warming over the period 1998-2012 was lower than that predicted by 111 out of 114 Coupled Model Intercomparison Project climate models.

Relation to Weather Forecasting

The global climate models used for climate projections are similar in structure to (and often share computer code with) numerical models for weather prediction, but are nonetheless logically distinct.

Most weather forecasting is done on the basis of interpreting numerical model results. Since forecasts are short—typically a few days or a week—such models do not usually contain an ocean model but rely on imposed SSTs. They also require accurate initial conditions to begin the forecast—typically these are taken from the output of a previous forecast, blended with observations. Predictions must require only a few hours; but because they only cover a one-week the models can be run at higher resolution than in climate mode. Currently the ECMWF runs at 40 km (25 mi) resolution as opposed to the 100-to-200 km (62-to-124 mi) scale used by typical climate model runs. Often local models are run using global model results for boundary conditions, to achieve higher local resolution: for example, the Met Office runs a mesoscale model with an 11 km (6.8 mi) resolution covering the UK, and various agencies in the US employ models such as the NGM and NAM models. Like most global numerical weather prediction models such as the GFS, global climate models are often spectral models instead of grid models. Spectral models are often used for global models because some computations in modeling can be performed faster, thus reducing run times.

Computations

Climate models use quantitative methods to simulate the interactions of the atmosphere, oceans, land surface and ice.

All climate models take account of incoming energy as short wave electromagnetic radiation, chiefly visible and short-wave (near) infrared, as well as outgoing energy as long wave (far) infrared electromagnetic radiation from the earth. Any imbalance results in a change in temperature.

The most talked-about models of recent years relate temperature to emissions of greenhouse gases. These models project an upward trend in the surface temperature record, as well as a more rapid increase in temperature at higher altitudes.

Three (or more properly, four since time is also considered) dimensional GCM's discretise the equations for fluid motion and energy transfer and integrate these over time. They also contain parametrisations for processes such as convection that occur on scales too small to be resolved directly.

Atmospheric GCMs (AGCMs) model the atmosphere and impose sea surface temperatures as boundary conditions. Coupled atmosphere-ocean GCMs (AOGCMs, e.g. HadCM3, EdGCM, GFDL CM2.X, ARPEGE-Climat) combine the two models.

Models range in complexity:

1. A simple radiant heat transfer model treats the earth as a single point and averages outgoing energy

2. This can be expanded vertically (radiative-convective models), or horizontally

3. Finally, (coupled) atmosphere–ocean–sea ice global climate models discretise and solve the full equations for mass and energy transfer and radiant exchange.

4. Box models treat flows across and within ocean basins.

Other submodels can be interlinked, such as land use, allowing researchers to predict the interaction between climate and ecosystems.

Other Climate Models

Earth-system models of intermediate complexity (EMICs)

The Climber-3 model uses a 2.5-dimensional statistical-dynamical model with 7.5° × 22.5° resolution and time step of 1/2 a day. An oceanic submodel is MOM-3 (Modular Ocean Model) with a 3.75° × 3.75° grid and 24 vertical levels.

Radiative-convective Models (RCM)

One-dimensional, radiative-convective models were used to verify basic climate assumptions in the '80s and '90s.

Numerical Weather Prediction

Numerical weather prediction (NWP) uses mathematical models of the atmosphere and oceans to predict the weather based on current weather conditions. Though first attempted in the 1920s, it was not until the advent of computer simulation in the 1950s that numerical weather predictions produced realistic results. A number of global and regional forecast models are run in different countries worldwide, using current weather observations relayed from radiosondes, weather satellites and other observing systems as inputs.

Mathematical models based on the same physical principles can be used to generate either short-term weather forecasts or longer-term climate predictions; the latter are widely applied for understanding and projecting climate change. The improvements made to regional models have allowed for significant improvements in tropical cyclone track and air quality forecasts; however, atmospheric models perform poorly at handling processes that occur in a relatively constricted area, such as wildfires.

Manipulating the vast datasets and performing the complex calculations necessary to modern numerical weather prediction requires some of the most powerful supercomputers in the world. Even with the increasing power of supercomputers, the forecast skill of numerical weather models extends to only about six days. Factors affecting the accuracy of numerical predictions include the density and quality of observations used as input to the forecasts, along with deficiencies in the numerical models themselves. Post-processing techniques such as model output statistics (MOS) have been developed to improve the handling of errors in numerical predictions.

A more fundamental problem lies in the chaotic nature of the partial differential equations that govern the atmosphere. It is impossible to solve these equations exactly, and small errors grow with time (doubling about every five days). Present understanding is that this chaotic behavior limits accurate forecasts to about 14 days even with perfectly accurate input data and a flawless model. In addition, the partial differential equations used in the model need to be supplemented with parameterizations for solar radiation, moist processes (clouds and precipitation), heat exchange, soil, vegetation, surface water, and the effects of terrain. In an effort to quantify the large amount of inherent uncertainty remaining in numerical predictions, ensemble forecasts have been used since the 1990s to help gauge the confidence in the forecast, and to obtain useful results farther into the future than otherwise possible. This approach analyzes multiple forecasts created with an individual forecast model or multiple models.

History

The history of numerical weather prediction began in the 1920s through the efforts of Lewis Fry Richardson, who used procedures originally developed by Vilhelm Bjerknes to produce by hand a six-hour forecast for the state of the atmosphere over two points in central Europe, taking at least six weeks to do so. It was not until the advent of the computer and computer simulations that computation time was reduced to less than the forecast period itself. The ENIAC was used to create the first weather forecasts via computer in 1950, based on a highly simplified approximation to the atmospheric governing equations. In 1954, Carl-Gustav Rossby's group at the Swedish Meteorological and Hydrological Institute used the same model to produce the first operational forecast (i.e., a routine prediction for practical use). Operational numerical weather prediction in the United States began in 1955 under the Joint Numerical Weather Prediction Unit (JNWPU), a joint project by the U.S. Air Force, Navy and Weather Bureau. In 1956, Norman Phillips developed a mathematical model which could realistically depict monthly and seasonal patterns in the troposphere; this became the first successful climate model. Following Phillips' work, several groups began working to create general circulation models. The first general circulation climate model that combined both oceanic and atmospheric processes was developed in the late 1960s at the NOAA Geophysical Fluid Dynamics Laboratory.

The ENIAC main control panel at the Moore School of Electrical Engineering

As computers have become more powerful, the size of the initial data sets has increased and newer atmospheric models have been developed to take advantage of the added available computing power. These newer models include more physical processes in the simplifications of the equations of motion in numerical simulations of the atmosphere. In 1966, West Germany and the United States began producing operational forecasts based on primitive-equation models, followed by the United Kingdom in 1972 and Australia in 1977. The development of limited area (regional) models facilitated advances in forecasting the tracks of tropical cyclones as well as air quality in the 1970s and 1980s. By the early 1980s models began to include the interactions of soil and vegetation with the atmosphere, which led to more realistic forecasts.

The output of forecast models based on atmospheric dynamics is unable to resolve some details of the weather near the Earth's surface. As such, a statistical relationship between the output of a numerical weather model and the ensuing conditions at the ground was developed in the 1970s and

1980s, known as model output statistics (MOS). Starting in the 1990s, model ensemble forecasts have been used to help define the forecast uncertainty and to extend the window in which numerical weather forecasting is viable farther into the future than otherwise possible.

Initialization

Weather reconnaissance aircraft, such as this WP-3D Orion, provide data that is then used in numerical weather forecasts.

The atmosphere is a fluid. As such, the idea of numerical weather prediction is to sample the state of the fluid at a given time and use the equations of fluid dynamics and thermodynamics to estimate the state of the fluid at some time in the future. The process of entering observation data into the model to generate initial conditions is called *initialization*. On land, terrain maps available at resolutions down to 1 kilometer (0.6 mi) globally are used to help model atmospheric circulations within regions of rugged topography, in order to better depict features such as downslope winds, mountain waves and related cloudiness that affects incoming solar radiation. The main inputs from country-based weather services are observations from devices (called radiosondes) in weather balloons that measure various atmospheric parameters and transmits them to a fixed receiver, as well as from weather satellites. The World Meteorological Organization acts to standardize the instrumentation, observing practices and timing of these observations worldwide. Stations either report hourly in METAR reports, or every six hours in SYNOP reports. These observations are irregularly spaced, so they are processed by data assimilation and objective analysis methods, which perform quality control and obtain values at locations usable by the model's mathematical algorithms. The data are then used in the model as the starting point for a forecast.

A variety of methods are used to gather observational data for use in numerical models. Sites launch radiosondes in weather balloons which rise through the troposphere and well into the stratosphere. Information from weather satellites is used where traditional data sources are not available. Commerce provides pilot reports along aircraft routes and ship reports along shipping routes. Research projects use reconnaissance aircraft to fly in and around weather systems of interest, such as tropical cyclones. Reconnaissance aircraft are also flown over the open oceans during the cold season into systems which cause significant uncertainty in forecast guidance, or are expected to be of high impact from three to seven days into the future over the downstream continent. Sea ice began to be initialized in forecast models in 1971. Efforts to involve sea surface temperature in model initialization began in 1972 due to its role in modulating weather in higher latitudes of the Pacific.

Computation

An atmospheric model is a computer program that produces meteorological information for future times at given locations and altitudes. Within any modern model is a set of equations, known as the primitive equations, used to predict the future state of the atmosphere. These equations—along with the ideal gas law—are used to evolve the density, pressure, and potential temperature scalar fields and the air velocity (wind) vector field of the atmosphere through time. Additional transport equations for pollutants and other aerosols are included in some primitive-equation high-resolution models as well. The equations used are nonlinear partial differential equations which are impossible to solve exactly through analytical methods, with the exception of a few idealized cases. Therefore, numerical methods obtain approximate solutions. Different models use different solution methods: some global models and almost all regional models use finite difference methods for all three spatial dimensions, while other global models and a few regional models use spectral methods for the horizontal dimensions and finite-difference methods in the vertical.

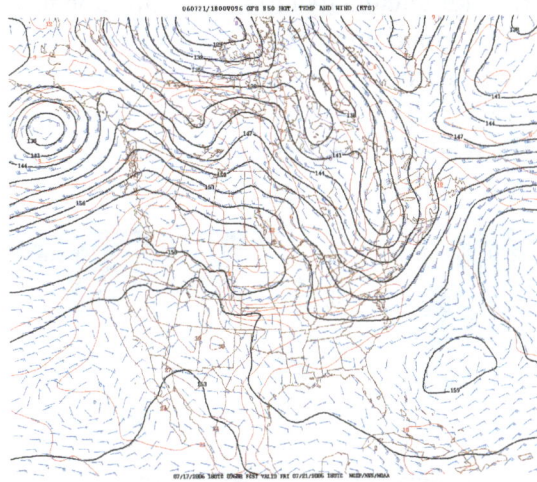

A prognostic chart of the 96-hour forecast of 850 mbar geopotential height and temperature from the Global Forecast System

These equations are initialized from the analysis data and rates of change are determined. These rates of change predict the state of the atmosphere a short time into the future; the time increment for this prediction is called a *time step*. This future atmospheric state is then used as the starting point for another application of the predictive equations to find new rates of change, and these new rates of change predict the atmosphere at a yet further time step into the future. This time stepping is repeated until the solution reaches the desired forecast time. The length of the time step chosen within the model is related to the distance between the points on the computational grid, and is chosen to maintain numerical stability. Time steps for global models are on the order of tens of minutes, while time steps for regional models are between one and four minutes. The global models are run at varying times into the future. The UKMET Unified Model is run six days into the future, while the European Centre for Medium-Range Weather Forecasts' Integrated Forecast System and Environment Canada's Global Environmental Multiscale Model both run out to ten days into the future, and the Global Forecast System model run by the Environmental Modeling Center is run sixteen days into the future. The visual output produced by a model solution is known as a prognostic chart, or *prog*.

Parameterization

Some meteorological processes are too small-scale or too complex to be explicitly included in numerical weather prediction models. *Parameterization* is a procedure for representing these processes by relating them to variables on the scales that the model resolves. For example, the gridboxes in weather and climate models have sides that are between 5 kilometers (3 mi) and 300 kilometers (200 mi) in length. A typical cumulus cloud has a scale of less than 1 kilometer (0.6 mi), and would require a grid even finer than this to be represented physically by the equations of fluid motion. Therefore, the processes that such clouds represent are parameterized, by processes of various sophistication. In the earliest models, if a column of air within a model gridbox was conditionally unstable (essentially, the bottom was warmer and moister than the top) and the water vapor content at any point within the column became saturated then it would be overturned (the warm, moist air would begin rising), and the air in that vertical column mixed. More sophisticated schemes recognize that only some portions of the box might convect and that entrainment and other processes occur. Weather models that have gridboxes with sides between 5 and 25 kilometers (3 and 16 mi) can explicitly represent convective clouds, although they need to parameterize cloud microphysics which occur at a smaller scale. The formation of large-scale (stratus-type) clouds is more physically based; they form when the relative humidity reaches some prescribed value. Sub-grid scale processes need to be taken into account. Rather than assuming that clouds form at 100% relative humidity, the cloud fraction can be related to a critical value of relative humidity less than 100%, reflecting the sub grid scale variation that occurs in the real world.

Field of cumulus clouds, which are parameterized since they are too small to be explicitly included within numerical weather prediction

The amount of solar radiation reaching the ground, as well as the formation of cloud droplets occur on the molecular scale, and so they must be parameterized before they can be included in the model. Atmospheric drag produced by mountains must also be parameterized, as the limitations in the resolution of elevation contours produce significant underestimates of the drag. This method of parameterization is also done for the surface flux of energy between the ocean and the atmosphere, in order to determine realistic sea surface temperatures and type of sea ice found near the ocean's surface. Sun angle as well as the impact of multiple cloud layers is taken into account. Soil type, vegetation type, and soil moisture all determine how much radiation goes into warming and how much moisture is drawn up into the adjacent atmosphere, and thus it is important to parameterize their contribution to these processes. Within air quality models, parameterizations take into account atmospheric emissions from multiple relatively tiny sources (e.g. roads, fields, factories) within specific grid boxes.

Domains

The horizontal domain of a model is either *global*, covering the entire Earth, or *regional*, covering only part of the Earth. Regional models (also known as *limited-area* models, or LAMs) allow for the use of finer grid spacing than global models because the available computational resources are focused on a specific area instead of being spread over the globe. This allows regional models to resolve explicitly smaller-scale meteorological phenomena that cannot be represented on the coarser grid of a global model. Regional models use a global model to specify conditions at the edge of their domain (boundary conditions) in order to allow systems from outside the regional model domain to move into its area. Uncertainty and errors within regional models are introduced by the global model used for the boundary conditions of the edge of the regional model, as well as errors attributable to the regional model itself.

Coordinate Systems

Horizontal Coordinates

Horizontal position may be expressed directly in geographic coordinates (latitude and longitude) for global models or in a map projection planar coordinates for regional models.

Vertical Coordinates

The vertical coordinate is handled in various ways. Lewis Fry Richardson's 1922 model used geometric height (z) as the vertical coordinate. Later models substituted the geometric z coordinate with a pressure coordinate system, in which the geopotential heights of constant-pressure surfaces become dependent variables, greatly simplifying the primitive equations. This correlation between coordinate systems can be made since pressure decreases with height through the Earth's atmosphere. The first model used for operational forecasts, the single-layer barotropic model, used a single pressure coordinate at the 500-millibar (about 5,500 m (18,000 ft)) level, and thus was essentially two-dimensional. High-resolution models—also called *mesoscale models*—such as the Weather Research and Forecasting model tend to use normalized pressure coordinates referred to as sigma coordinates. This coordinate system receives its name from the independent variable σ used to scale atmospheric pressures with respect to the pressure at the surface, and in some cases also with the pressure at the top of the domain.

Model Output Statistics

Because forecast models based upon the equations for atmospheric dynamics do not perfectly determine weather conditions, statistical methods have been developed to attempt to correct the forecasts. Statistical models were created based upon the three-dimensional fields produced by numerical weather models, surface observations and the climatological conditions for specific locations. These statistical models are collectively referred to as model output statistics (MOS), and were developed by the National Weather Service for their suite of weather forecasting models in the late 1960s.

Model output statistics differ from the *perfect prog* technique, which assumes that the output of numerical weather prediction guidance is perfect. MOS can correct for local effects that cannot be resolved by the model due to insufficient grid resolution, as well as model biases. Because MOS is run after its respective global or regional model, its production is known as post-processing. Forecast parameters within MOS include maximum and minimum temperatures, percentage chance of rain within a several hour period, precipitation amount expected, chance that the precipitation will be frozen in nature, chance for thunderstorms, cloudiness, and surface winds.

Ensembles

In 1963, Edward Lorenz discovered the chaotic nature of the fluid dynamics equations involved in weather forecasting. Extremely small errors in temperature, winds, or other initial inputs given to numerical models will amplify and double every five days, making it impossible for long-range forecasts—those made more than two weeks in advance—to predict the state of the atmosphere with any degree of forecast skill. Furthermore, existing observation networks have poor coverage in some regions (for example, over large bodies of water such as the Pacific Ocean), which introduces uncertainty into the true initial state of the atmosphere. While a set of equations, known as the Liouville equations, exists to determine the initial uncertainty in the model initialization, the equations are too complex to run in real-time, even with the use of supercomputers. These uncertainties limit forecast model accuracy to about five or six days into the future.

Edward Epstein recognized in 1969 that the atmosphere could not be completely described with a single forecast run due to inherent uncertainty, and proposed using an ensemble of stochastic Monte Carlo simulations to produce means and variances for the state of the atmosphere. Although this early example of an ensemble showed skill, in 1974 Cecil Leith showed that they produced adequate forecasts only when the ensemble probability distribution was a representative sample of the probability distribution in the atmosphere.

Since the 1990s, *ensemble forecasts* have been used operationally (as routine forecasts) to account for the stochastic nature of weather processes – that is, to resolve their inherent uncertainty. This method involves analyzing multiple forecasts created with an individual forecast model by using different physical parametrizations or varying initial conditions. Starting in 1992 with ensemble forecasts prepared by the European Centre for Medium-Range Weather Forecasts (ECMWF) and the National Centers for Environmental Prediction, model ensemble forecasts have been used to help define the forecast uncertainty and to extend the window in which numerical weather forecasting is viable farther into the future than otherwise possible. The ECMWF model, the Ensemble Prediction System, uses singular vectors to simulate the initial probability density, while the NCEP ensemble, the Global Ensemble Forecasting System, uses a technique known as vector breeding. The UK Met Office runs global and regional ensemble forecasts where perturbations to initial conditions are produced using a Kalman filter. There are 24 ensemble members in the Met Office Global and Regional Ensemble Prediction System (MOGREPS).

In a single model-based approach, the ensemble forecast is usually evaluated in terms of an average of the individual forecasts concerning one forecast variable, as well as the degree of agreement between various forecasts within the ensemble system, as represented by their overall spread. Ensemble spread is diagnosed through tools such as spaghetti diagrams, which show the dispersion of one quantity on prognostic charts for specific time steps in the future. Another tool where ensemble spread is used is a meteogram, which shows the dispersion in the forecast of one quantity for one specific location. It is common for the ensemble spread to be too small to include the weather that actually occurs, which can lead to forecasters misdiagnosing model uncertainty; this problem becomes particularly severe for forecasts of the weather about ten days in advance. When ensemble spread is small and the forecast solutions are consistent within multiple model runs, forecasters perceive more confidence in the ensemble mean, and the forecast in general. Despite this perception, a *spread-skill relationship* is often weak or not found, as spread-error correlations are normally less than 0.6, and only under special circumstances range between 0.6–0.7. The relationship between ensemble spread and forecast skill varies substantially depending on such factors as the forecast model and the region for which the forecast is made.

In the same way that many forecasts from a single model can be used to form an ensemble, multiple models may also be combined to produce an ensemble forecast. This approach is called *multi-model ensemble forecasting*, and it has been shown to improve forecasts when compared to a single model-based approach. Models within a multi-model ensemble can be adjusted for their various biases, which is a process known as *superensemble forecasting*. This type of forecast significantly reduces errors in model output.

Applications

Air Quality Modeling

Air quality forecasting attempts to predict when the concentrations of pollutants will attain levels that are hazardous to public health. The concentration of pollutants in the atmosphere is determined by their *transport*, or mean velocity of movement through the atmosphere, their diffusion, chemical transformation, and ground deposition. In addition to pollutant source and terrain information, these models require data about the state of the fluid flow in the atmosphere to determine its transport and diffusion. Meteorological conditions such as thermal inversions can prevent surface air from rising, trapping pollutants near the surface, which makes accurate forecasts of such events crucial for air quality modeling. Urban air quality models require a very fine computational mesh, requiring the use of high-resolution mesoscale weather models; in spite of this, the quality of numerical weather guidance is the main uncertainty in air quality forecasts.

Climate Modeling

A General Circulation Model (GCM) is a mathematical model that can be used in computer simulations of the global circulation of a planetary atmosphere or ocean. An atmospheric general circulation model (AGCM) is essentially the same as a global numerical weather prediction model, and some (such as the one used in the UK Unified Model) can be configured for both short-term weather forecasts and longer-term climate predictions. Along with sea ice and land-surface components, AGCMs and oceanic GCMs (OGCM) are key components of global climate models, and are widely applied for understanding the climate and projecting climate change. For aspects of climate change, a range of man-made chemical emission scenarios can be fed into the climate models to see how an enhanced greenhouse effect would modify the Earth's climate. Versions designed for climate applications with time scales of decades to centuries were originally created in 1969 by Syukuro Manabe and Kirk Bryan at the Geophysical Fluid Dynamics Laboratory in Princeton, New Jersey. When run for multiple decades, computational limitations mean that the models must use a coarse grid that leaves smaller-scale interactions unresolved.

Ocean Surface Modeling

The transfer of energy between the wind blowing over the surface of an ocean and the ocean's upper layer is an important element in wave dynamics. The spectral wave transport equation is used to describe the change in wave spectrum over changing topography. It simulates wave generation, wave movement (propagation within a fluid), wave shoaling, refraction, energy transfer between waves, and wave dissipation. Since surface winds are the primary forcing mechanism in the spectral wave transport equation, ocean wave models use information produced by numerical weather prediction models as inputs to determine how much energy is transferred from the atmosphere into the layer at the surface of the ocean. Along with dissipation of energy through whitecaps and resonance between waves, surface winds from numerical weather models allow for more accurate predictions of the state of the sea surface.

Tropical Cyclone Forecasting

Tropical cyclone forecasting also relies on data provided by numerical weather models. Three main classes of tropical cyclone guidance models exist: Statistical models are based on an analysis of storm behavior using climatology, and correlate a storm's position and date to produce a forecast that is not based on the physics of the atmosphere at the time. Dynamical models are numerical models that solve the governing equations of fluid flow in the atmosphere; they are based on the same principles as other limited-area numerical weather prediction models but may include special computational techniques such as refined spatial domains that move along with the cyclone. Models that use elements of both approaches are called statistical-dynamical models.

In 1978, the first hurricane-tracking model based on atmospheric dynamics—the movable fine-mesh (MFM) model—began operating. Within the field of tropical cyclone track forecasting, despite the ever-improving dynamical model guidance which occurred with increased computational power, it was not until the 1980s when numerical weather prediction showed skill, and until the 1990s when it consistently outperformed statistical or simple dynamical models. Predictions of the intensity of a tropical cyclone based on numerical weather prediction continue to be a challenge, since statistical methods continue to show higher skill over dynamical guidance.

Wildfire Modeling

On a molecular scale, there are two main competing reaction processes involved in the degradation of cellulose, or wood fuels, in wildfires. When there is a low amount of moisture in a cellulose fiber, volatilization of the fuel occurs; this process will generate intermediate gaseous products that will ultimately be the source of combustion. When moisture is present—or when enough heat is being carried away from the fiber, charring occurs. The chemical kinetics of both reactions indicate that there is a point at which the level of moisture is low enough—and/or heating rates high enough—for combustion processes become self-sufficient. Consequently, changes in wind speed, direction, moisture, temperature, or lapse rate at different levels of the atmosphere can have a significant impact on the behavior and growth of a wildfire. Since the wildfire acts as a heat source to the atmospheric flow, the wildfire can modify local advection patterns, introducing a feedback loop between the fire and the atmosphere.

A simplified two-dimensional model for the spread of wildfires that used convection to represent the effects of wind and terrain, as well as radiative heat transfer as the dominant method of heat transport led to reaction-diffusion systems of partial differential equations. More complex models join numerical weather models or computational fluid dynamics models with a wildfire component which allow the feedback effects between the fire and the atmosphere to be estimated. The additional complexity in the latter class of models translates to a corresponding increase in their computer power requirements. In fact, a full three-dimensional treatment of combustion via direct numerical simulation at scales relevant for atmospheric modeling is not currently practical because of the excessive computational cost such a simulation would require. Numerical weather models have limited forecast skill at spatial resolutions under 1 kilometer (0.6 mi), forcing complex wildfire models to parameterize the fire in order to calculate how the winds will be modified locally by the wildfire, and to use those modified winds to determine the rate at which the fire will spread locally. Although models such as Los Alamos' FIRETEC solve for the concentrations of fuel and oxygen, the computational grid cannot be fine enough to resolve the combustion reaction, so approximations must be made for the temperature distribution within each grid cell, as well as for the combustion reaction rates themselves.

Tropical Cyclone Forecast Model

A tropical cyclone forecast model is a computer program that uses meteorological data to forecast aspects of the future state of tropical cyclones. There are three types of models: statistical, dynamical, or combined statistical-dynamic. Dynamical models utilize powerful supercomputers with sophisticated mathematical modeling software and meteorological data to calculate future weather conditions. Statistical models forecast the evolution of a tropical cyclone in a simpler manner, by extrapolating from historical datasets, and thus can be run quickly on platforms such as personal computers. Statistical-dynamical models use aspects of both types of forecasting. Four primary types of forecasts exist for tropical cyclones: track, intensity, storm surge, and rainfall. Dynamical models were not developed until the 1970s and the 1980s, with earlier efforts focused on the storm surge problem.

Track models did not show forecast skill when compared to statistical models until the 1980s. Statistical-dynamical models were used from the 1970s into the 1990s. Early models use data from previous model runs while late models produce output after the official hurricane forecast has been sent. The use of consensus, ensemble, and superensemble forecasts lowers errors more than any individual forecast model. Both consensus and superensemble forecasts can use the guidance of global and regional models runs to improve the performance more than any of their respective components. Techniques used at the Joint Typhoon Warning Center indicate that superensemble forecasts are a very powerful tool for track forecasting.

Statistical Guidance

The first statistical guidance used by the National Hurricane Center was the Hurricane Analog Technique (HURRAN), which was available in 1969. It used the newly developed North Atlantic tropical cyclone database to find storms with similar tracks. It then shifted their tracks through the storm's current path, and used location, direction and speed of motion, and the date to find suitable analogs. The method did well with storms south of the 25th parallel which had not yet turned northward, but poorly with systems near or after recurvature. Since 1972, the Climatology and Persistence (CLIPER) statistical model has been used to help generate tropical cyclone track forecasts. In the era of skillful dynamical forecasts, CLIPER is now being used as the baseline to show model and forecaster skill. The Statistical Hurricane Intensity Forecast (SHIFOR) has been used since 1979 for tropical cyclone intensity forecasting. It uses climatology and persistence to predict future intensity, including the current Julian day, current cyclone intensity, the cyclone's intensity 12 hours ago, the storm's initial latitude and longitude, as well as its zonal (east-west) and meridional (north-south) components of motion.

A series of statistical-dynamical models, which used regression equations based upon CLIPER output and the latest output from primitive equation models run at the National Meteorological Center, then National Centers for Environmental Prediction, were developed between the 1970s and 1990s and were named NHC73, NHC83, NHC90, NHC91, and NHC98. Within the field of tropical cyclone track forecasting, despite the ever-improving dynamical model guidance which occurred with increased computational power, it was not until the decade of the 1980s when numerical weather prediction showed skill, and until the 1990s when it consistently outperformed statistical or simple dynamical models. In 1994, a version of SHIFOR was created for the northwest Pacific ocean for typhoon forecasting, known as the Statistical Typhoon Intensity Forecast (STIFOR), which used the 1971-1990 data for that region to develop intensity forecasts out to 72 hours into the future.

In regards to intensity forecasting, the Statistical Hurricane Intensity Prediction Scheme (SHIPS) utilizes relationships between environmental conditions from the Global Forecast System (GFS) such as vertical wind shear and sea surface temperatures, climatology, and persistence (storm behavior) via multiple regression techniques to come up with an intensity forecast for systems in the northern Atlantic and northeastern Pacific oceans. A similar model was developed for the northwest Pacific ocean and Southern Hemisphere known as the Statistical Intensity Prediction System (STIPS), which accounts for land interactions through the input environmental conditions from the Navy Operational Global Prediction System (NOGAPS) model. The version of SHIPS with an inland decay component is known as Decay SHIPS (DSHIPS). The Logistic Growth Equation Model (LGEM) uses the same input as SHIPS but within a simplified dynamical prediction system. Within tropical cyclone rainfall forecasting, the Rainfall Climatology and Persistence (r-CLIPER) model was developed using microwave rainfall data from polar orbiting satellites over the ocean and first-order rainfall measurements from the land, to come up with a realistic rainfall distribution for tropical cyclones based on the National Hurricane Center's track forecast. It has been operational since 2004. A statistical-parametric wind radii model has been developed for use at the National Hurricane Center and Joint Typhoon Warning Center which uses climatology and persistence to predict wind structure out to five days into the future.

Dynamical Guidance

During 1972, the first model to forecast storm surge along the continental shelf of the United States was developed, known as the Special Program to List the Amplitude of Surges from Hurricanes (SPLASH). In 1978, the first hurricane-tracking model based on atmospheric dynamics – the movable fine-mesh (MFM) model – began operating. The Quasi-Lagrangian Limited Area (QLM) model is a multi-level primitive equation model using a Cartesian grid and the Global Forecasting System (GFS) for boundary conditions. In the early 1980s, the assimilation of satellite-derived winds from water vapor, infrared, and visible satellite imagery was found to improve tropical cyclones track forecasting. The Geophysical Fluid Dynamics Laboratory (GFDL) hurricane model was used for research purposes between 1973 and the mid-1980s. Once it was determined that it could show skill in hurricane prediction, a multi-year transition transformed the research model into an operational model which could be used by the National Weather Service for both track and intensity forecasting in 1995. By 1985, the Sea Lake and Overland Surges from Hurricanes (SLOSH) Model had been developed for use in areas of the Gulf of Mexico and near the United States' East coast, which was more robust than the SPLASH model.

The Beta Advection Model (BAM) has been used operationally since 1987 using steering winds averaged through the 850 hPa to 200 hPa layer and the Beta effect which causes a storm to drift northwest due to differences in the coriolis effect across the tropical cyclone. The larger the cyclone, the larger the impact of the beta effect is likely to be. Starting in 1990, three versions of the BAM were run operationally: the BAM shallow (BAMS) average winds in an 850 hPa to 700 hPa layer, the BAM Medium (BAMM) which uses average winds in an 850 hPa to 400 hPa layer, and the BAM Deep (BAMD) which is the same as the pre-1990 BAM. For a weak hurricane without well-developed central thunderstorm activity, BAMS works well, because weak storms tend to be steered by low-level winds. As the storm grows stronger and associated thunderstorm activity near its center gets deeper, BAMM and BAMD become more accurate, as these types of storms are steered more by the winds in the upper-level. If the forecast from the three versions is similar, then the forecaster can conclude that there is minimal uncertainty, but if the versions vary by a great deal, then the forecaster has less confidence in the track predicted due to the greater uncertainty. Large differences between model predictions can also indicate wind shear in the atmosphere, which could affect the intensity forecast as well.

Tested in 1989 and 1990, The Vic Ooyama Barotropic (VICBAR) model used a cubic-B spline representation of variables for the objective analysis of observations and solutions to the shallow-water prediction equations on nested domains, with the boundary conditions defined as the global forecast model. It was implemented operationally as the Limited Area Sine Transform Barotropic (LBAR) model in 1992, using the GFS for boundary conditions. By 1990, Australia had developed its own storm surge model which was able to be run in a few minutes on a personal computer. The Japan Meteorological Agency (JMA) developed its own Typhoon Model (TYM) in 1994, and in 1998, the agency began using its own dynamic storm surge model.

The Hurricane Weather Research and Forecasting (HWRF) model is a specialized version of the Weather Research and Forecasting (WRF) model and is used to forecast the track and intensity of tropical cyclones. The model was developed by the National Oceanic and Atmospheric Administration (NOAA), the U.S. Naval Research Laboratory, the University of Rhode Island, and Florida State University. It became operational in 2007. Despite improvements in track forecasting, predictions of the intensity of a tropical cyclone based on numerical weather prediction continue to be a challenge, since statiscal methods continue to show higher skill over dynamical guidance. Other than the specialized guidance, global guidance such as the GFS, Unified Model (UKMET), NOGAPS, Japanese Global Spectral Model (GSM), European Centre for Medium-Range Weather Forecasts model, France's Action de Recherche Petite Echelle Grande Echelle (ARPEGE) and Aire

Limit´ee Adaptation Dynamique Initialisation (ALADIN) models, India's National Centre for Medium Range Weather Forecasting (NCMWRF) model, Korea's Global Data Assimilation and Prediction System (GDAPS) and Regional Data Assimilation and Prediction System (RDAPS) models, Hong Kong/China's Operational Regional Spectral Model (ORSM) model, and Canadian Global Environmental Multiscale Model (GEM) model are used for track and intensity purposes.

Timeliness

Some models do not produce output quickly enough to be used for the forecast cycle immediately after the model starts running (including HWRF, GFDL, and FSSE). Most of the above track models (except CLIPER) require data from global weather models, such as the GFS, which produce output about four hours after the synoptic times of 0000, 0600, 1200, and 1800 Universal Coordinated Time (UTC). For half of their forecasts, the NHC issues forecasts only three hours after that time, so some "early" models — NHC90, BAM, and LBAR — are run using a 12-hour-old forecast for the current time. "Late" models, such as the GFS and GFDL, finish after the advisory has already been issued. These models are interpolated to the current storm position for use in the following forecast cycle — for example, GFDI, the interpolated version of the GFDL model.

Consensus Methods

Using a consensus of forecast models reduces forecast error. Trackwise, the GUNA model is a consensus of the interpolated versions of the GFDL, UKMET with quality control applied to the cyclone tracker, United States Navy NOGAPS, and GFS models. The version of the GUNA corrected for model biases is known as the CGUN. The TCON consensus is the GUNA consensus plus the Hurricane WRF model. The version of the TCON corrected for model biases is known as the TCCN. A lagged average of the last two runs of the members within the TCON plus the ECMWF model is known as the TVCN consensus. The version of the TVCN corrected for model biases is the TVCC consensus.

In early 2013, The NAVGEM replaced the NOGAPS as the Navy's primary operational global forecast model. For the 2013 season, and until model verification can occur, it is not being utilized in the development of any consensus forecasts.

For intensity, a combination of the LGEM, interpolated GFDL, interpolated HWRF, and DSHIPS models is known as the ICON consensus. The lagged average of the last two runs of models within the ICON consensus is called the IVCN consensus. Across the northwest Pacific and Southern Hemisphere, a ten-member STIPS consensus is formed from the output of the NOGAPS, GFS, the Japanese GSM, the Coupled Ocean/Atmosphere Mesoscale Prediction System (COAMPS), the UKMET, the Japanese TYM, the GFDL with NOGAPS boundary conditions, the Air Force Weather Agency (AFWA) Model, the Australian Tropical Cyclone Local Area Prediction System, and the Weber Barotropic Model.

Ensemble Methods

No model is ever perfectly accurate because it is impossible to learn exactly everything about the atmosphere in a timely enough manner, and atmospheric measurements that are taken are not completely accurate. The use of the ensemble method of forecasting, whether it be a multi-model ensemble, or numerous ensemble members based on the global model, helps define the uncertainty and further limit errors.

The JMA has produced an 11-member ensemble forecast system for typhoons known as the Typhoon Ensemble Prediction System (TEPS) since February 2008, which is run out to 132 hours into the future. It uses a lower resolution version (with larger grid spacing) of its GSM, with ten perturbed members and one non-perturbed member. The system reduces errors by an average of 40 kilometres (25 mi) five days into the future when compared to its higher resolution GSM.

The Florida State Super Ensemble (FSSE) is produced from a suite of models which then uses statistical regression equations developed over a training phase to reduce their biases, which produces forecasts better than the member models or their mean solution. It uses 11 global models, including five developed at Florida State University, the Unified Model, the GFS, the NOGAPS, the United States Navy NOGAPS, the Australian Bureau of Meteorology Research Centre (BMRC) model, and Canadian Recherche en Prévision Numérique (RPN) model. It shows significant skill in track, intensity, and rainfall predictions of tropical cyclones.

The Systematic Approach Forecast Aid (SAFA) was developed by the Joint Typhoon Warning Center to create a selective consensus forecast which removed more erroneous forecasts at a 72hour time frame from consideration using the United States Navy NOGAPS model, the GFDL, the Japan Meteorological Agency's global and typhoon models, as well as the UKMET. All the models improved during SAFA's five-year history and removing erroneous forecasts proved difficult to do in operations.

Sunspot Theory

A 2010 report correlates low sunspot activity with high hurricane activity. Analyzing historical data, there was a 25% chance of at least one hurricane striking the continental United States during a peak sunspot year; a 64% chance during a low sunspot year. In June 2010, the hurricanes predictors in the US were not using this information.

Hurricane Forecast Model Accuracy

The accuracy of hurricane forecast models can vary significantly from storm to storm. For some storms, the factors affecting the hurricane track are relatively straightforward, and the models are not only accurate but they produce similar forecasts, while for other storms, the factors affecting the hurricane track and more complex, and different models produce very different forecasts.

References

- Lynch, Peter (2006). "The ENIAC Integrations". The Emergence of Numerical Weather Prediction. Cambridge University Press. pp. 206–208. ISBN 978-0-521-85729-1.

- Steyn, D. G. (1991). Air pollution modeling and its application VIII, Volume 8. Birkhäuser. pp. 241–242. ISBN 978-0-306-43828-8.

- Stensrud, David J. (2007). Parameterization schemes: keys to understanding numerical weather prediction models. Cambridge University Press. p. 56. ISBN 978-0-521-86540-1. Retrieved 2011-02-15.

- Houghton, John Theodore (1985). The Global Climate. Cambridge University Press archive. pp. 49–50. ISBN 978-0-521-31256-1. Retrieved 2011-01-08.

- Strikwerda, John C. (2004). Finite difference schemes and partial differential equations. SIAM. pp. 165–170. ISBN 978-0-89871-567-5. Retrieved 2010-12-31.

- Chan, Johnny C. L. & Jeffrey D. Kepert (2010). Global Perspectives on Tropical Cyclones: From Science to Mitigation. World Scientific. pp. 295–296. ISBN 978-981-4293-47-1. Retrieved 2011-02-24.

- Holton, James R. (2004). An introduction to dynamic meteorology, Volume 1. Academic Press. p. 480. ISBN 978-0-12-354015-7. Retrieved 2011-02-24.

- Brown, Molly E. (2008). Famine early warning systems and remote sensing data. Springer. p. 121. ISBN 978-3-540-75367-4. Retrieved 2011-02-24.

- Ahrens, C. Donald (2008). Essentials of meteorology: an invitation to the atmosphere. Cengage Learning. p. 244. ISBN 978-0-495-11558-8. Retrieved 2011-02-11.

- Stensrud, David J. (2007). Parameterization schemes: keys to understanding numerical weather prediction models. Cambridge University Press. p. 6. ISBN 978-0-521-86540-1. Retrieved 2011-02-15.

- Warner, Thomas Tomkins (2010). Numerical Weather and Climate Prediction. Cambridge University Press. p. 259. ISBN 978-0-521-51389-0. Retrieved 2011-02-11.

- Baum, Marsha L. (2007). When nature strikes: weather disasters and the law. Greenwood Publishing Group. p. 189. ISBN 978-0-275-22129-4. Retrieved 2011-02-11.

- Gultepe, Ismail (2007). Fog and boundary layer clouds: fog visibility and forecasting. Springer. p. 1144. ISBN 978-3-7643-8418-0. Retrieved 2011-02-11.

- Barry, Roger Graham; Chorley, Richard J. (2003). Atmosphere, weather, and climate. Psychology Press. p. 172. ISBN 978-0-415-27171-4. Retrieved 2011-02-11.

- Warner, Thomas Tomkins (2010). Numerical Weather and Climate Prediction. Cambridge University Press. pp. 266–275. ISBN 978-0-521-51389-0. Retrieved 2011-02-11.

3

Concerns and Challenges of Climate Change

The recent developments in the climate of our Earth have caused numerous concerns and challenges. Business action on climate change includes global warming and a range of activities related to it. Likewise, other challenges of climate change are land surface effects on climate, deforestation and climate change etc. This text is a compilation of the concerns and challenges faced in today's time related to climate change.

Global Warming

Global warming and climate change are terms for the observed century-scale rise in the average temperature of the Earth's climate system and its related effects. Multiple lines of scientific evidence show that the climate system is warming. Although the increase of near-surface atmospheric temperature is the measure of global warming often reported in the popular press, most of the additional energy stored in the climate system since 1970 has gone into the oceans. The rest has melted ice and warmed the continents and atmosphere. Many of the observed changes since the 1950s are unprecedented over tens to thousands of years.

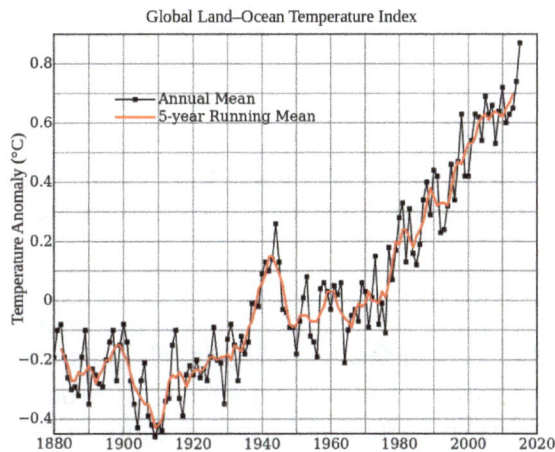

Global mean surface temperature change from 1880 to 2015, relative to the 1951–1980 mean. The black line is the annual mean and the red line is the 5-year running mean. Source: NASA GISS.

Scientific understanding of global warming is increasing. The Intergovernmental Panel on Climate Change (IPCC) reported in 2014 that scientists were more than 95% certain that global warming is mostly being caused by human (anthropogenic) activities, mainly increasing concentrations of greenhouse gases such as methane and carbon dioxide (CO_2). Human-made carbon dioxide continues to increase above levels not seen in hundreds of thousands of years. Methane and other, often much more potent, greenhouse gasses are also rising along with CO2. Currently, about half of the carbon dioxide released from the burning of fossil fuels remains in the atmosphere. The rest is absorbed by vegetation and the oceans. Climate model projections summarized in the report indicated that during the 21st century the global surface temperature is likely to rise a further 0.3 to 1.7 °C (0.5 to 3.1 °F) for their lowest emissions scenario and 2.6 to 4.8 °C (4.7 to 8.6 °F) for the highest emissions scenario. These findings have been recognized by the national science academies of the major industrialized nations and are not disputed by any scientific body of national or international standing.

Future climate change and associated impacts will differ from region to region around the globe. Anticipated effects include warming global temperature, rising sea levels, changing precipitation, and expansion of deserts in the subtropics. Warming is expected to be greater over land than over the oceans and greatest in the Arctic, with the continuing retreat of glaciers, permafrost and sea ice. Other likely changes include more frequent extreme weather events including heat waves, droughts, heavy rainfall with floods and heavy snowfall; ocean acidification; and species extinctions due to shifting temperature regimes. Effects significant to humans include the threat to food security from decreasing crop yields and the abandonment of populated areas due to rising sea levels. Because the climate system has a large "inertia" and greenhouse gasses will stay in the atmosphere for a long time, many of these effects will not only exist for decades or centuries, but will persist for tens of thousands of years.

Possible societal responses to global warming include mitigation by emissions reduction, adaptation to its effects, building systems resilient to its effects, and possible future climate engineering. Most countries are parties to the United Nations Framework Convention on Climate Change (UNFCCC), whose ultimate objective is to prevent dangerous anthropogenic climate change. Parties to the UNFCCC have agreed that deep cuts in emissions are required and that global warming should be limited to well below 2.0 °C (3.6 °F) relative to pre-industrial levels, with efforts made to limit warming to 1.5 °C (2.7 °F).

Public reactions to global warming and concern about its effects are also increasing. A global 2015 Pew Research Center report showed a median of 54% consider it "a very serious problem". There are significant regional differences, with Americans and Chinese (whose economies are responsible for the greatest annual CO_2 emissions) among the least concerned.

Observed Temperature Changes

The global average (land and ocean) surface temperature shows a warming of 0.85 [0.65 to 1.06] °C in the period 1880 to 2012, based on multiple independently produced datasets. Earth's average surface temperature rose by 70.74±0.18 °C over the period 1906–2005. The rate of warming almost doubled for the last half of that period 0.13±0.03 °C per decade, versus 0.07±0.02 °C per decade).

The average temperature of the lower troposphere has increased between 0.13 and 0.22 °C (0.23 and 0.40 °F) per decade since 1979, according to satellite temperature measurements. Climate proxies show the temperature to have been relatively stable over the one or two thousand years before 1850, with regionally varying fluctuations such as the Medieval Warm Period and the Little Ice Age.

The warming that is evident in the instrumental temperature record is consistent with a wide range of observations, as documented by many independent scientific groups. Examples include sea level rise, widespread melting of snow and land ice, increased heat content of the oceans, increased humidity, and the earlier timing of spring events, e.g., the flowering of plants. The probability that these changes could have occurred by chance is virtually zero.

Trends

Temperature changes vary over the globe. Since 1979, land temperatures have increased about twice as fast as ocean temperatures 0.25 °C per decade against 0.13 °C per decade). Ocean temperatures increase more slowly than land temperatures because of the larger effective heat capacity of the oceans and because the ocean loses more heat by evaporation. Since the beginning of industrialisation the temperature difference between the hemispheres has increased due to melting of sea ice and snow in the North. Average arctic temperatures have been increasing at almost twice the rate of the rest of the world in the past 100 years; however arctic temperatures are also highly variable. Although more greenhouse gases are emitted in the Northern than Southern Hemisphere this does not contribute to the difference in warming because the major greenhouse gases persist long enough to mix between hemispheres.

The thermal inertia of the oceans and slow responses of other indirect effects mean that climate can take centuries or longer to adjust to changes in forcing. One climate commitment study concluded that if greenhouse gases were stabilized at year 2000 levels, surface temperatures would still increase by about one-half degree Celsius, and another found that if they were stabilized at 2005 levels surface warming could exceed a whole degree Celsius. Some of this surface warming will be driven by past natural forcings which are still seeking equilibrium in the climate system. One study using a highly simplified climate model indicates these past natural forcings may account for as much as 64% of the committed 2050 surface warming and their influence will fade with time compared to the human contribution.

Global temperature is subject to short-term fluctuations that overlay long-term trends and can temporarily mask them. The relative stability in surface temperature from 2002 to 2009, which has been dubbed the global warming hiatus by the media and some scientists, is consistent with such an episode. 2015 updates to account for differing methods of measuring ocean surface temperature measurements show a positive trend over the recent decade.

Warmest Years

Fifteen of the top 16 warmest years have occurred since 2000. While record-breaking years can attract considerable public interest, individual years are less significant than the overall trend. Some climatologists have criticized the attention that the popular press gives to "warmest year" statistics; for example, Gavin Schmidt stated "the long-term trends or the expected sequence of records are far more important than whether any single year is a record or not." Ocean oscillations such as the El Niño Southern Oscillation (ENSO) can affect global average temperatures, causing temperatures of a given year to be abnormally warm or cold for reasons not directly related to the overall trend of climate change. For example, 1998 and 2015 temperatures were significantly enhanced by strong El Niño conditions.

Initial Causes of Temperature Changes (External Forcings)

The climate system can warm or cool in response to changes in *external forcings*. These are "external" to the climate system but not necessarily external to Earth. Examples of external forcings include changes in atmospheric composition (e.g., increased concentrations of greenhouse gases), solar luminosity, volcanic eruptions, and variations in Earth's orbit around the Sun.

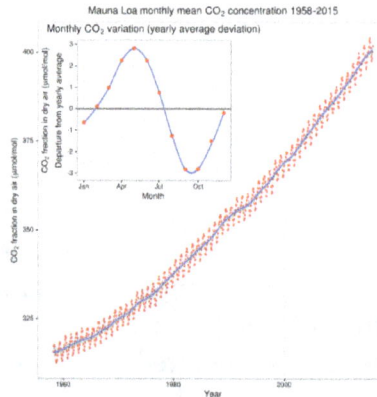

This graph, known as the Keeling Curve, documents the increase of atmospheric carbon dioxide concentrations from 1958–2015. Monthly CO_2 measurements display seasonal oscillations in an upward trend; each year's maximum occurs during the Northern Hemisphere's late spring, and declines during its growing season as plants remove some atmospheric CO_2.

Greenhouse Gases

The greenhouse effect is the process by which absorption and emission of infrared radiation by gases in a planet's atmosphere warm its lower atmosphere and surface. It was proposed by Joseph Fourier in 1824, discovered in 1860 by John Tyndall, was first investigated quantitatively by Svante Arrhenius in 1896, and was developed in the 1930s through 1960s by Guy Stewart Callendar.

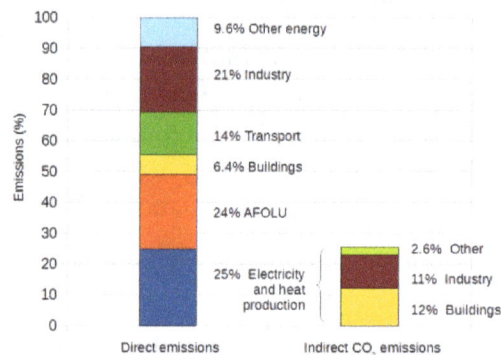

Annual world greenhouse gas emissions, in 2010, by sector.

Percentage share of global cumulative energy-related CO_2 emissions between 1751 and 2012 across different regions.

On Earth, naturally occurring amounts of greenhouse gases cause air temperature near the surface to be about 33 °C (59 °F) warmer than it would be in their absence. Without the Earth's atmosphere, the Earth's average temperature would be well below the freezing temperature of water. The major greenhouse gases are water vapour, which causes about 36–70% of the greenhouse effect; carbon dioxide (CO_2), which causes 9–26%; methane (CH_4), which causes 4–9%; and ozone (O_3), which causes 3–7%. Clouds also affect the radiation balance through cloud forcings similar to greenhouse gases.

Human activity since the Industrial Revolution has increased the amount of greenhouse gases in the atmosphere, leading to increased radiative forcing from CO_2, methane, tropospheric ozone, CFCs and nitrous oxide. According to work published in 2007, the concentrations of CO_2 and methane have increased by 36% and 148% respectively since 1750. These levels are much higher than at any time during the last 800,000 years, the period for which reliable data has been extracted from ice cores. Less direct geological evidence indicates that CO_2 values higher than this were last seen about 20 million years ago.

Fossil fuel burning has produced about three-quarters of the increase in CO_2 from human activity over the past 20 years. The rest of this increase is caused mostly by changes in land-use, particularly deforestation. Another significant non-fuel source of anthropogenic CO_2 emissions is the calcination of limestone for clinker production, a chemical process which releases CO_2. Estimates of global CO_2 emissions in 2011 from fossil fuel combustion, including cement production and gas flaring, was 34.8 billion tonnes (9.5 ± 0.5 PgC), an increase of 54% above emissions in 1990. Coal burning was responsible for 43% of the total emissions, oil 34%, gas 18%, cement 4.9% and gas flaring 0.7%

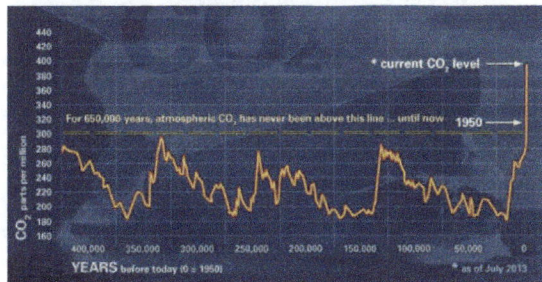

Atmospheric CO_2 concentration from 650,000 years ago to near present, using ice core proxy data and direct measurements.

In May 2013, it was reported that readings for CO_2 taken at the world's primary benchmark site in Mauna Loa surpassed 400 ppm. According to professor Brian Hoskins, this is likely the first time CO_2 levels have been this high for about 4.5 million years. Monthly global CO_2 concentrations exceeded 400 ppm in March 2015, probably for the first time in several million years. On 12 November 2015, NASA scientists reported that human-made carbon dioxide continues to increase above levels not seen in hundreds of thousands of years: currently, about half of the carbon dioxide released from the burning of fossil fuels is not absorbed by vegetation and the oceans and remains in the atmosphere.

Over the last three decades of the twentieth century, gross domestic product per capita and population growth were the main drivers of increases in greenhouse gas emissions. CO_2 emissions are continuing to rise due to the burning of fossil fuels and land-use change. Emissions can be attributed to different regions. Attributions of emissions due to land-use change are subject to considerable uncertainty.

Emissions scenarios, estimates of changes in future emission levels of greenhouse gases, have been projected that depend upon uncertain economic, sociological, technological, and natural developments. In most scenarios, emissions continue to rise over the century, while in a few, emissions are reduced. Fossil fuel reserves are abundant, and will not limit carbon emissions in the 21st century. Emission scenarios, combined with modelling of the carbon cycle, have been used to produce estimates of how atmospheric concentrations of greenhouse gases might change in the future. Using the six IPCC SRES "marker" scenarios, models suggest that by the year 2100, the atmospheric concentration of CO_2 could range between 541 and 970 ppm. This is 90–250% above the concentration in the year 1750.

The popular media and the public often confuse global warming with ozone depletion, i.e., the destruction of stratospheric ozone (e.g., the ozone layer) by chlorofluorocarbons. Although there are a few areas of linkage, the relationship between the two is not strong. Reduced stratospheric ozone has had a slight cooling influence on surface temperatures, while increased tropospheric ozone has had a somewhat larger warming effect.

Aerosols and Soot

Ship tracks can be seen as lines in these clouds over the Atlantic Ocean on the east coast of the United States. Atmospheric particles from these and other sources could have a large effect on climate through the aerosol indirect effect.

Global dimming, a gradual reduction in the amount of global direct irradiance at the Earth's surface, was observed from 1961 until at least 1990. Solid and liquid particles known as *aerosols*, produced by volcanoes and human-made pollutants, are thought to be the main cause of this dimming. They exert a cooling effect by increasing the reflection of incoming sunlight. The effects of the products of fossil fuel combustion – CO_2 and aerosols – have partially offset one another in recent decades, so that net warming has been due to the increase in non-CO_2 greenhouse gases such as methane. Radiative forcing due to aerosols is temporally limited due to the processes that remove aerosols from the atmosphere. Removal by clouds and precipitation gives tropospheric aerosols an atmospheric lifetime of only about a week, while stratospheric aerosols can remain for a few years. Carbon dioxide has a lifetime of a century or more, and as such, changes in aerosols will only delay climate changes due to carbon dioxide. Black carbon is second only to carbon dioxide for its contribution to global warming.

In addition to their direct effect by scattering and absorbing solar radiation, aerosols have indirect effects on the Earth's radiation budget. Sulfate aerosols act as cloud condensation nuclei and thus lead to clouds that have more and smaller cloud droplets. These clouds reflect solar radiation more efficiently than clouds with fewer and larger droplets, a phenomenon known as the Twomey effect. This effect also causes droplets to be of more uniform size, which reduces growth of raindrops and makes the cloud more reflective to incoming sunlight, known as the Albrecht effect. Indirect effects are most noticeable in marine stratiform clouds, and have very little radiative effect on convective clouds. Indirect effects of aerosols represent the largest uncertainty in radiative forcing.

Soot may either cool or warm Earth's climate system, depending on whether it is airborne or deposited. Atmospheric soot directly absorbs solar radiation, which heats the atmosphere and cools the surface. In isolated areas with high soot production, such as rural India, as much as 50% of surface warming due to greenhouse gases may be masked by atmospheric brown clouds. When deposited, especially on glaciers or on ice in arctic regions, the lower surface albedo can also directly heat the surface. The influences of atmospheric particles, including black carbon, are most pronounced in the tropics and sub-tropics, particularly in Asia, while the effects of greenhouse gases are dominant in the extratropics and southern hemisphere.

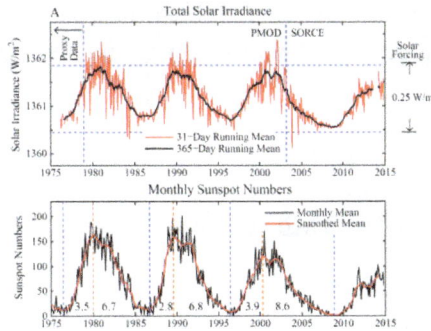

Changes in Total Solar Irradiance (TSI) and monthly sunspot numbers since the mid-1970s.

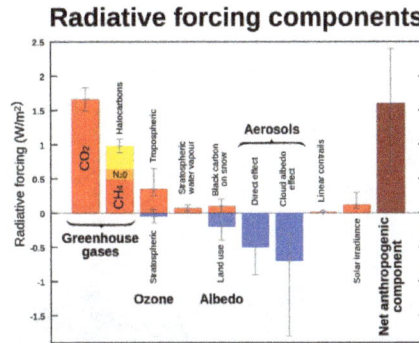

Contribution of natural factors and human activities to radiative forcing of climate change. Radiative forcing values are for the year 2005, relative to the pre-industrial era (1750). The contribution of solar irradiance to radiative forcing is 5% the value of the combined radiative forcing due to increases in the atmospheric concentrations of carbon dioxide, methane and nitrous oxide.

Solar Activity

Since 1978, solar irradiance has been measured by satellites. These measurements indicate that the Sun's radiative output has not increased since 1978, so the warming during the past 30 years cannot be attributed to an increase in solar energy reaching the Earth.

Climate models have been used to examine the role of the Sun in recent climate change. Models are unable to reproduce the rapid warming observed in recent decades when they only take into account variations in solar output and volcanic activity. Models are, however, able to simulate the observed 20th century changes in temperature when they include all of the most important external forcings, including human influences and natural forcings.

Another line of evidence against solar variations having caused recent climate change comes from looking at how temperatures at different levels in the Earth's atmosphere have changed. Models and observations show that greenhouse warming results in warming of the lower atmosphere (the troposphere) but cooling of the upper atmosphere (the stratosphere). Depletion of the ozone layer by chemical refrigerants has also resulted in a strong cooling effect in the stratosphere. If solar variations were responsible for observed warming, warming of both the troposphere and stratosphere would be expected.

Variations in Earth's Orbit

The tilt of the Earth's axis and the shape of its orbit around the Sun vary slowly over tens of thousands of years and are a natural source of climate change, by changing the seasonal and latitudinal distribution of solar insolation.

During the last few thousand years, this phenomenon contributed to a slow cooling trend at high latitudes of the Northern Hemisphere during summer, a trend that was reversed by greenhouse-gas-induced warming during the 20th century.

Variations in orbital cycles may initiate a new glacial period in the future, though the timing of this depends on greenhouse gas concentrations as well as the orbital forcing. A new glacial period is not expected within the next 50,000 years if atmospheric CO_2 concentration remains above 300 ppm.

Feedback

Sea ice, shown here in Nunavut, in northern Canada, reflects more sunshine, while open ocean absorbs more, accelerating melting.

The climate system includes a range of *feedbacks*, which alter the response of the system to changes in external forcings. Positive feedbacks increase the response of the climate system to an initial forcing, while negative feedbacks reduce it.

There are a range of feedbacks in the climate system, including water vapour, changes in ice-albedo (snow and ice cover affect how much the Earth's surface absorbs or reflects incoming sunlight), clouds, and changes in the Earth's carbon cycle (e.g., the release of carbon from soil). The main negative feedback is the energy the Earth's surface radiates into space as infrared radiation. According to the Stefan-Boltzmann law, if the absolute temperature (as measured in kelvin) doubles, radiated energy increases by a factor of 16 (2 to the 4th power).

Feedbacks are an important factor in determining the sensitivity of the climate system to increased atmospheric greenhouse gas concentrations. Other factors being equal, a higher *climate sensitivity* means that more warming will occur for a given increase in greenhouse gas forcing. Uncertainty over the effect of feedbacks is a major reason why different climate models project different magnitudes of warming for a given forcing scenario. More research is needed to understand the role of clouds and carbon cycle feedbacks in climate projections.

The IPCC projections previously mentioned span the "likely" range (greater than 66% probability, based on expert judgement) for the selected emissions scenarios. However, the IPCC's projections do not reflect the full range of uncertainty. The lower end of the "likely" range appears to be better constrained than the upper end.

Climate Models

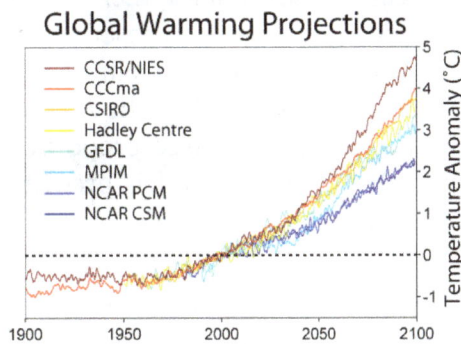

Calculations of global warming prepared in or before 2001 from a range of climate models under the SRES A2 emissions scenario, which assumes no action is taken to reduce emissions and regionally divided economic development.

Projected change in annual mean surface air temperature from the late 20th century to the middle 21st century, based on a medium emissions scenario (SRES A1B). This scenario assumes that no future policies are adopted to limit greenhouse gas emissions. Image credit: NOAA GFDL.

A climate model is a representation of the physical, chemical and biological processes that affect the climate system. Such models are based on scientific disciplines such as fluid dynamics and thermodynamics as well as physical processes such as radiative transfer. The models may be used to predict a range of variables such as local air movement, temperature, clouds, and other atmospheric properties; ocean temperature, salt content, and circulation; ice cover on land and sea; the transfer of heat and moisture from soil and vegetation to the atmosphere; and chemical and biological processes, among others.

Although researchers attempt to include as many processes as possible, simplifications of the actual climate system are inevitable because of the constraints of available computer power and limitations in knowledge of the climate system. Results from models can also vary due to different greenhouse gas inputs and the model's climate sensitivity. For example, the uncertainty in IPCC's 2007 projections is caused by (1) the use of multiple models with differing sensitivity to greenhouse gas concentrations, (2) the use of differing estimates of humanity's future greenhouse gas emissions, (3) any additional emissions from climate feedbacks that were not included in the models IPCC used to prepare its report, i.e., greenhouse gas releases from permafrost.

The models do not assume the climate will warm due to increasing levels of greenhouse gases. Instead the models predict how greenhouse gases will interact with radiative transfer and other physical processes. Warming or cooling is thus a result, not an assumption, of the models.

Clouds and their effects are especially difficult to predict. Improving the models' representation of clouds is therefore an important topic in current research. Another prominent research topic is expanding and improving representations of the carbon cycle.

Models are also used to help investigate the causes of recent climate change by comparing the observed changes to those that the models project from various natural and human causes. Although these models do not unambiguously attribute the warming that occurred from approximately 1910 to 1945 to either natural variation or human effects, they do indicate that the warming since 1970 is dominated by anthropogenic greenhouse gas emissions.

The physical realism of models is tested by examining their ability to simulate contemporary or past climates. Climate models produce a good match to observations of global temperature changes over the last century, but do not simulate all aspects of climate. Not all effects of global warming are accurately predicted by the climate models used by the IPCC. Observed Arctic shrinkage has been faster than that predicted. Precipitation increased proportionally to atmospheric humidity, and hence significantly faster than global climate models predict. Since 1990, sea level has also risen considerably faster than models predicted it would.

Observed and Expected Environmental Effects

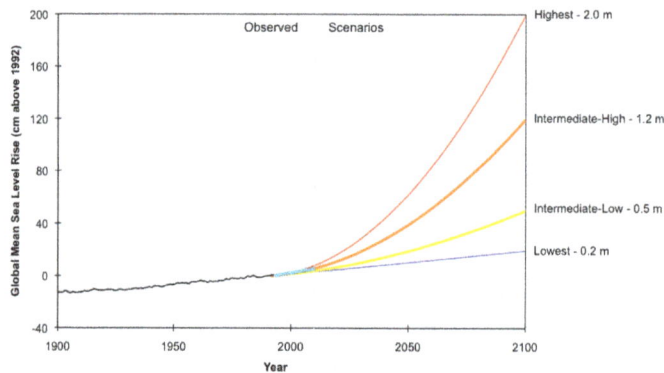

Projections of global mean sea level rise by Parris and others. Probabilities have not been assigned to these projections. Therefore, none of these projections should be interpreted as a "best estimate" of future sea level rise. Image credit: NOAA.

Anthropogenic forcing has likely contributed to some of the observed changes, including sea level rise, changes in climate extremes (such as the number of warm and cold days), declines in Arctic sea ice extent, glacier retreat, and greening of the Sahara.

During the 21st century, glaciers and snow cover are projected to continue their widespread retreat. Projections of declines in Arctic sea ice vary. Recent projections suggest that Arctic summers could be ice-free (defined as ice extent less than 1 million square km) as early as 2025-2030.

"Detection" is the process of demonstrating that climate has changed in some defined statistical sense, without providing a reason for that change. Detection does not imply attribution of the detected change to a particular cause. "Attribution" of causes of climate change is the process of establishing the most likely causes for the detected change with some defined level of confidence. Detection and attribution may also be applied to observed changes in physical, ecological and social systems.

Extreme Weather

Changes in regional climate are expected to include greater warming over land, with most warming at high northern latitudes, and least warming over the Southern Ocean and parts of the North Atlantic Ocean.

Future changes in precipitation are expected to follow existing trends, with reduced precipitation over subtropical land areas, and increased precipitation at subpolar latitudes and some equatorial regions. Projections suggest a probable increase in the frequency and severity of some extreme weather events, such as heat waves.

A 2015 study published in *Nature Climate Change*, states:

> " About 18% of the moderate daily precipitation extremes over land are attributable to the observed temperature increase since pre-industrial times, which in turn primarily results from human influence. For 2 °C of warming the fraction of precipitation extremes attributable to human influence rises to about 40%. Likewise, today about 75% of the moderate daily hot extremes over land are attributable to warming. It is the most rare and extreme events for which the largest fraction is anthropogenic, and that contribution increases nonlinearly with further warming. "

Data analysis of extreme events from 1960 till 2010 suggests that droughts and heat waves appear simultaneously with increased frequency. Extremely wet or dry events within the monsoon period have increased since 1980.

Sea Level Rise

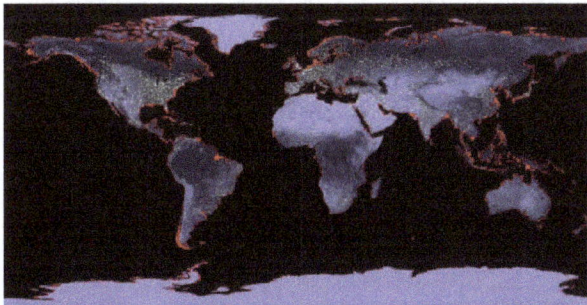

Map of the Earth with a six-metre sea level rise represented in red.

The sea level rise since 1993 has been estimated to have been on average 2.6 mm and 2.9 mm per year ± 0.4 mm. Additionally, sea level rise has accelerated from 1995 to 2015. Over the 21st century, the IPCC projects for a high emissions scenario, that global mean sea level could rise by 52–98 cm. The IPCC's projections are conservative, and may underestimate future sea level rise. Other estimates suggest that for the same period, global mean sea level could rise by 0.2 to 2.0 m (0.7–6.6 ft), relative to mean sea level in 1992.

Widespread coastal flooding would be expected if several degrees of warming is sustained for millennia. For example, sustained global warming of more than 2 °C (relative to pre-industrial levels) could lead to eventual sea level rise of around 1 to 4 m due to thermal expansion of sea water and the melting of glaciers and small ice caps. Melting of the Greenland ice sheet could contribute an additional 4 to 7.5 m over many thousands of years. It has been estimated that we are already committed to a sea-level rise of approximately 2.3 metres for each degree of temperature rise within the next 2,000 years.

Warming beyond the 2 °C target would potentially lead to rates of sea-level rise dominated by ice loss from Antarctica. Continued CO_2 emissions from fossil sources could cause additional tens of metres of sea level rise, over the next millennia and eventually ultimately eliminate the entire Antarctic ice sheet, causing about 58 metres of sea level rise.

Ecological Systems

In terrestrial ecosystems, the earlier timing of spring events, as well as poleward and upward shifts in plant and animal ranges, have been linked with high confidence to recent warming. Future climate change is expected to affect particular ecosystems, including tundra, mangroves, and coral reefs. It is expected that most ecosystems will be affected by higher atmospheric CO_2 levels, combined with higher global temperatures. Overall, it is expected that climate change will result in the extinction of many species and reduced diversity of ecosystems.

Increases in atmospheric CO_2 concentrations have led to an increase in ocean acidity. Dissolved CO_2 increases ocean acidity, measured by lower pH values. Between 1750 and 2000, surface-ocean pH has decreased by ≈0.1, from ≈8.2 to ≈8.1. Surface-ocean pH has probably not been below ≈8.1 during the past 2 million years. Projections suggest that surface-ocean pH could decrease by an additional 0.3–0.4 units by 2100. Future ocean acidification could threaten coral reefs, fisheries, protected species, and other natural resources of value to society.

Ocean deoxygenation is projected to increase hypoxia by 10%, and triple suboxic waters (oxygen concentrations 98% less than the mean surface concentrations), for each 1 °C of upper ocean warming.

Long-term Effects

On the timescale of centuries to millennia, the magnitude of global warming will be determined primarily by anthropogenic CO_2 emissions. This is due to carbon dioxide's very long lifetime in the atmosphere.

Stabilizing the global average temperature would require large reductions in CO_2 emissions, as well as reductions in emissions of other greenhouse gases such as methane and nitrous oxide. Emissions of CO_2 would need to be reduced by more than 80% relative to their peak level. Even if this were achieved, global average temperatures would remain close to their highest level for many centuries.

Long-term effects also include a response from the Earth's crust, due to ice melting and deglaciation, in a process called post-glacial rebound, when land masses are no longer depressed by the weight of ice. This could lead to landslides and increased seismic and volcanic activities. Tsunamis could be generated by submarine landslides caused by warmer ocean water thawing ocean-floor permafrost or releasing gas hydrates. Some world regions, such as the French Alps, already show signs of an increase in landslide frequency.

Large-scale and Abrupt Impacts

Climate change could result in global, large-scale changes in natural and social systems. Examples include the possibility for the Atlantic Meridional Overturning Circulation to slow- or shutdown, which in the instance of a shutdown would change weather in Europe and North America considerably, ocean acidification caused by increased atmospheric concentrations of carbon dioxide, and the long-term melting of ice sheets, which contributes to sea level rise.

Some large-scale changes could occur abruptly, i.e., over a short time period, and might also be irreversible. Examples of abrupt climate change are the rapid release of methane and carbon dioxide from permafrost, which would lead to amplified global warming, or the shutdown of thermohaline circulation. Scientific understanding of abrupt climate change is generally poor. The probability of abrupt change for some climate related feedbacks may be low. Factors that may increase the probability of abrupt climate change include higher magnitudes of global warming, warming that occurs more rapidly, and warming that is sustained over longer time periods.

Observed and Expected Effects on Social Systems

The effects of climate change on human systems, mostly due to warming or shifts in precipitation patterns, or both, have been detected worldwide. Production of wheat and maize globally has been impacted by climate change. While crop production has increased in some mid-latitude regions such as the UK and Northeast China, economic losses due to extreme weather events have increased globally. There has been a shift from cold- to heat-related mortality in some regions as a result of warming. Livelihoods of indigenous peoples of the Arctic have been altered by climate change, and there is emerging evidence of climate change impacts on livelihoods of indigenous peoples in other regions. Regional impacts of climate change are now observable at more locations than before, on all continents and across ocean regions.

The future social impacts of climate change will be uneven. Many risks are expected to increase with higher magnitudes of global warming. All regions are at risk of experiencing negative impacts. Low-latitude, less developed areas face the greatest risk. A study from 2015 concluded that economic growth (gross domestic product) of poorer countries is much more impaired with projected future climate warming, than previously thought.

A meta-analysis of 56 studies concluded in 2014 that each degree of temperature rise will increase violence by up to 20%, which includes fist fights, violent crimes, civil unrest or wars.

Examples of impacts include:

1. *Food*: Crop production will probably be negatively affected in low latitude countries, while effects at northern latitudes may be positive or negative. Global warming of around 4.6 °C relative to pre-industrial levels could pose a large risk to global and regional food security.

2. *Health*: Generally impacts will be more negative than positive. Impacts include: the effects of extreme weather, leading to injury and loss of life; and indirect effects, such as undernutrition brought on by crop failures.

Habitat Inundation

In small islands and mega deltas, inundation as a result of sea level rise is expected to threaten vital infrastructure and human settlements. This could lead to issues of homelessness in countries with low-lying areas such as Bangladesh, as well as statelessness for populations in countries such as the Maldives and Tuvalu.

Economy

Estimates based on the IPCC A1B emission scenario from additional CO_2 and CH_4 greenhouse gases released from permafrost, estimate associated impact damages by US$43 trillion.

Infrastructure

Continued permafrost degradation will likely result in unstable infrastructure in Arctic regions, or Alaska before 2100. Thus, impacting roads, pipelines and buildings, as well as water distribution, and cause slope failures.

Possible Responses to Global Warming

Mitigation

The graph on the right shows three "pathways" to meet the UNFCCC's 2 °C target, labelled "global technology", "decentralized solutions", and "consumption change". Each pathway shows how various measures (e.g., improved energy efficiency, increased use of renewable energy) could contribute to emissions reductions. Image credit: PBL Netherlands Environmental Assessment Agency.

Mitigation of climate change are actions to reduce greenhouse gas emissions, or enhance the capacity of carbon sinks to absorb GHGs from the atmosphere. There is a large potential for future reductions in emissions by a combination of activities, including: energy conservation and increased energy efficiency; the use of low-carbon energy technologies, such as renewable energy, nuclear energy, and carbon capture and storage; and enhancing carbon sinks through, for example, reforestation and preventing deforestation. A 2015 report by Citibank concluded that transitioning to a low carbon economy would yield positive return on investments.

Near- and long-term trends in the global energy system are inconsistent with limiting global warming at below 1.5 or 2 °C, relative to pre-industrial levels. Pledges made as part of the Cancún agreements are broadly consistent with having a likely chance (66 to 100% probability) of limiting global warming (in the 21st century) at below 3 °C, relative to pre-industrial levels.

In limiting warming at below 2 °C, more stringent emission reductions in the near-term would allow for less rapid reductions after 2030. Many integrated models are unable to meet the 2 °C target if pessimistic assumptions are made about the availability of mitigation technologies.

Adaptation

Other policy responses include adaptation to climate change. Adaptation to climate change may be planned, either in reaction to or anticipation of climate change, or spontaneous, i.e., without government intervention. Planned adaptation is already occurring on a limited basis. The barriers, limits, and costs of future adaptation are not fully understood.

A concept related to adaptation is *adaptive capacity*, which is the ability of a system (human, natural or managed) to adjust to climate change (including climate variability and extremes) to moderate potential damages, to take advantage of opportunities, or to cope with consequences. Unmitigated climate change (i.e., future climate change without efforts to limit greenhouse gas emissions) would, in the long term, be likely to exceed the capacity of natural, managed and human systems to adapt.

Environmental organizations and public figures have emphasized changes in the climate and the risks they entail, while promoting adaptation to changes in infrastructure needs and emissions reductions.

Climate Engineering

Climate engineering (sometimes called *geoengineering* or *climate intervention*) is the deliberate modification of the climate. It has been investigated as a possible response to global warming, e.g. by NASA and the Royal Society. Techniques under research fall generally into the categories solar radiation management and carbon dioxide removal, although various other schemes have been suggested. A study from 2014 investigated the most common climate engineering methods and concluded they are either ineffective or have potentially severe side effects and cannot be stopped without causing rapid climate change.

Discourse about Global Warming

Political Discussion

Article 2 of the UN Framework Convention refers explicitly to "stabilization of greenhouse gas concentrations." To stabilize the atmospheric concentration of CO_2, emissions worldwide would need to be dramatically reduced from their present level.

Most countries in the world are parties to the United Nations Framework Convention on Climate Change (UNFCCC). The ultimate objective of the Convention is to prevent dangerous human interference of the climate system. As stated in the Convention, this requires that GHG concentrations are stabilized in the atmosphere at a level where ecosystems can adapt naturally to climate change, food production is not threatened, and economic development can proceed in a sustainable fashion. The Framework Convention was agreed in 1992, but since then, global emissions have risen.

During negotiations, the G77 (a lobbying group in the United Nations representing 133 developing nations) pushed for a mandate requiring developed countries to "[take] the lead" in reducing their emissions. This was justified on the basis that: the developed world's emissions had contributed most to the cumulation of GHGs in the atmosphere; per-capita emissions (i.e., emissions per head of population) were still relatively low in developing countries; and the emissions of developing countries would grow to meet their development needs.

This mandate was sustained in the Kyoto Protocol to the Framework Convention, which entered into legal effect in 2005. In ratifying the Kyoto Protocol, most developed countries accepted legally binding commitments to limit their emissions. These first-round commitments expired in 2012.

United States President George W. Bush rejected the treaty on the basis that "it exempts 80% of the world, including major population centres such as China and India, from compliance, and would cause serious harm to the US economy."

At the 15th UNFCCC Conference of the Parties, held in 2009 at Copenhagen, several UNFCCC Parties produced the Copenhagen Accord. Parties associated with the Accord (140 countries, as of November 2010) aim to limit the future increase in global mean temperature to below 2 °C. The 16th Conference of the Parties (COP16) was held at Cancún in 2010. It produced an agreement, not a binding treaty, that the Parties should take urgent action to reduce greenhouse gas emissions to meet a goal of limiting global warming to 2 °C above pre-industrial temperatures. It also recognized the need to consider strengthening the goal to a global average rise of 1.5 °C.

Scientific Discussion

There is continuing discussion through published peer-reviewed scientific papers, which are assessed by scientists working in the relevant fields taking part in the Intergovernmental Panel on Climate Change. The scientific consensus as of 2013 stated in the IPCC Fifth Assessment Report is that it "is extremely likely that human influence has been the dominant cause of the observed warming since the mid-20th century". A 2008 report by the U.S. National Academy of Sciences stated that most scientists by then agreed that observed warming in recent decades was primarily caused by human activities increasing the amount of greenhouse gases in the atmosphere. In 2005 the Royal Society stated that while the overwhelming majority of scientists were in agreement on the main points, some individuals and organizations opposed to the consensus on urgent action needed to reduce greenhouse gas emissions have tried to undermine the science and work of the IPCC. National science academies have called on world leaders for policies to cut global emissions.

In the scientific literature, there is a strong consensus that global surface temperatures have increased in recent decades and that the trend is caused mainly by human-induced emissions of greenhouse gases. No scientific body of national or international standing disagrees with this view.

Discussion by the Public and in Popular Media

The global warming controversy refers to a variety of disputes, substantially more pronounced in the popular media than in the scientific literature, regarding the nature, causes, and consequences of global warming. The disputed issues include the causes of increased global average air temperature, especially since the mid-20th century, whether this warming trend is unprecedented or within normal climatic variations, whether humankind has contributed significantly to it, and whether the increase is completely or partially an artefact of poor measurements. Additional disputes concern estimates of climate sensitivity, predictions of additional warming, and what the consequences of global warming will be.

From 1990 to 1997, right-wing conservative think tanks in the United States mobilized to challenge the legitimacy of global warming as a social problem. They challenged the scientific evidence, argued that global warming will have benefits, and asserted that proposed solutions would do more harm than good. Some people dispute aspects of climate change science. Organizations such as the libertarian Competitive Enterprise Institute, conservative commentators, and some companies such as ExxonMobil have challenged IPCC climate change scenarios, funded scientists who disagree with the scientific consensus, and provided their own projections of the economic cost of stricter controls. On the other hand, some fossil fuel companies have scaled back their efforts in recent years, or even called for policies to reduce global warming. Global oil companies have begun to acknowledge climate change exists and is caused by human activities and the burning of fossil fuels.

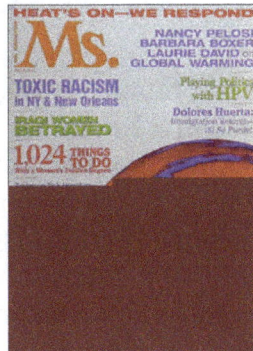

Global warming was the cover story in this 2007 issue of Ms. magazine

Surveys of Public Opinion

The world public, or at least people in economically advanced regions, became broadly aware of the global warming problem in the late 1980s. Polling groups began to track opinions on the subject, at first mainly in the United States. The longest consistent polling, by Gallup in the US, found relatively small deviations of 10% or so from 1998 to 2015 in opinion on the seriousness of global warming, but with increasing polarization between those concerned and those unconcerned.

The first major worldwide poll, conducted by Gallup in 2008-2009 in 127 countries, found that some 62% of people worldwide said they knew about global warming. In the advanced countries of North America, Europe and Japan, 90% or more knew about it (97% in the U.S., 99% in Japan); in less developed countries, especially in Africa, fewer than a quarter knew about it, although many had noticed local weather changes. Among those who knew about global warming, there was a wide variation between nations in belief that the warming was a result of human activities.

By 2010, with 111 countries surveyed, Gallup determined that there was a substantial decrease since 2007−08 in the number of Americans and Europeans who viewed global warming as a serious threat. In the US, just a little over half the population (53%) now viewed it as a serious concern for either themselves or their families; this was 10 points below the 2008 poll (63%). Latin America had the biggest rise in concern: 73% said global warming is a serious threat to their families. This global poll also found that people are more likely to attribute global warming to human activities than to natural causes, except in the US where nearly half (47%) of the population attributed global warming to natural causes.

A March−May 2013 survey by Pew Research Center for the People & the Press polled 39 countries about global threats. According to 54% of those questioned, global warming featured top of the perceived global threats. In a January 2013 survey, Pew found that 69% of Americans say there is solid evidence that the Earth's average temperature has got warmer over the past few decades, up six points since November 2011 and 12 points since 2009.

A 2010 survey of 14 industrialized countries found that skepticism about the danger of global warming was highest in Australia, Norway, New Zealand and the United States, in that order, correlating positively with per capita emissions of carbon dioxide.

Etymology

In the 1950s, research suggested increasing temperatures, and a 1952 newspaper reported "climate change". This phrase next appeared in a November 1957 report in *The Hammond Times* which described Roger Revelle's research into the effects of increasing human-caused CO_2 emissions on the greenhouse effect, "a large scale global warming, with radical climate changes may result". Both phrases were only used occasionally until 1975, when Wallace Smith Broecker published a scientific paper on the topic; "Climatic Change: Are We on the Brink of a Pronounced Global Warming?" The phrase began to come into common use, and in 1976 Mikhail Budyko's statement that "a global warming up has started" was widely reported. Other studies, such as a 1971 MIT report, referred to the human impact as "inadvertent climate modification", but an influential 1979 National Academy of Sciences study headed by Jule Charney followed Broecker in using *global warming* for rising surface temperatures, while describing the wider effects of increased CO_2 as *climate change*.

In 1986 and November 1987, NASA climate scientist James Hansen gave testimony to Congress on global warming. There were increasing heatwaves and drought problems in the summer of 1988, and when Hansen testified in the Senate on 23 June he sparked worldwide interest. He said: "global warming has reached a level such that we can ascribe with a high degree of confidence a cause and effect relationship between the greenhouse effect and the observed warming." Public attention increased over the summer, and *global warming* became the dominant popular term, commonly used both by the press and in public discourse.

In a 2008 NASA article on usage, Erik M. Conway defined *Global warming* as "the increase in Earth's average surface temperature due to rising levels of greenhouse gases", while *Climate change* was "a long-term change in the Earth's climate, or of a region on Earth." As effects such as changing patterns of rainfall and rising sea levels would probably have more impact than temperatures alone, he considered *global climate change* a more scientifically accurate term, and like the Intergovernmental Panel on Climate Change, the NASA website would emphasize this wider context.

Business Action on Climate Change

Business action on climate change includes a range of activities relating to global warming, and to influencing political decisions on global-warming-related regulation, such as the Kyoto Protocol. Major multinationals have played and to some extent continue to play a significant role in the politics of global warming, especially in the United States, through lobbying of government and funding of global warming skeptics. Business also plays a key role in the mitigation of global warming, through decisions to invest in researching and implementing new energy technologies and energy efficiency measures.

Overview

In 1989 in the US, the petroleum and automotive industries and the National Association of Manufacturers created the Global Climate Coalition (GCC) to oppose mandatory actions to address global warming. In 1997, when the US Senate overwhelmingly passed a resolution against ratifying the Kyoto Protocol, the industry funded a $13 million industry advertising blitz in the run-up to the vote.

In 1998 the *New York Times* published an American Petroleum Institute (API) memo outlining a strategy aiming to make "recognition of uncertainty … part of the 'conventional wisdom.'" The memo has been compared to a late 1960s memo by tobacco company Brown and Williamson, which observed: "Doubt is our product since it is the best means of competing with the 'body of fact' that exists in the mind of the general public. It is also the means of establishing a controversy." Those involved in the memo included Jeffrey Salmon, then executive director of the George C. Marshall Institute, Steven Milloy, a prominent skeptic commentator, and the Competitive Enterprise Institute's Myron Ebell. In June 2005 a former API lawyer, Philip Cooney, resigned his White House post after accusations of politically motivated tampering with scientific reports.

In 2002 the GCC considered its work in the US against regulation on global warming to have been so successful that it "deactivated" itself, although the loss of some leading members may also have been a factor.

At the same time, since 1989 many previously skeptical petroleum and automobile industry corporations have changed their position as the political and scientific consensus has grown, with the creation of the Kyoto Protocol and the publication of the International Panel on Climate Change's Second and Third Assessment Reports. These corporations include major petroleum companies like Royal Dutch Shell, Texaco, and BP, as well as automobile manufacturers like Ford, General Motors, and DaimlerChrysler. Some of these have joined with the Center for Climate and Energy Solutions (formerly the Pew Center on Global Climate Change), a non-profit organization aiming to support efforts to address global climate change.

Since 2000, the Carbon Disclosure Project has been working with major corporations and investors to disclose the emissions of the largest companies. By 2007, the CDP published the emissions data for 2400 of the largest corporations in the world, and represented major institutional investors with $41 trillion combined assets under management. The pressure from these investors had had some success in working with companies to reduce emissions.

The World Business Council for Sustainable Development, a CEO-led association of some 200 multinational companies, has called on governments to agree on a global targets, and suggests that it is necessary to cut emissions by 60-80 percent from current levels by 2050.

Global Climate Coalition

A central organization in climate skepticism was the Global Climate Coalition (1989–2002), a group of mainly United States businesses opposing immediate action to reduce greenhouse gas emissions. The coalition funded skeptical scientists to be public spokespeople, provided industry a voice on climate change, and fought the Kyoto Protocol. The *New York Times* reported that "even as the coalition worked to sway opinion [towards skepticism], its own scientific and technical experts were advising that the science backing the role of greenhouse gases in global warming could not be refuted."

In the year 2000, the rate of corporate members leaving accelerated when they became the target of a national divestiture campaign run by John Passacantando and Phil Radford with the organization Ozone Action. According to the New York Times, when Ford Motor Company was the first company to leave the coalition, it was "the latest sign of divisions within heavy industry over how to respond to global warming." After that, between December, 1999 and early March, 2000, the GCC was deserted by Daimler-Chrysler, Texaco, the Southern Company and General Motors.

The organization closed in 2002, or in their own words, 'deactivated'.

U.S. Climate Action Partnership

The U.S. Climate Action Partnership (USCAP) was formed in January 2007 with the primary goal of influencing the US government's regulation of greenhouse gas emissions. Original members included General Electric, Alcoa, Natural Resources Defense Council, etc., but they were joined in April, 2007 by ConocoPhilips and AIG.

Energy Industry

Exxonmobil

ExxonMobil has been a leading figure in the business world's position on climate change, providing substantial funding to a range of global-warming-skeptical organizations. *Mother Jones* counted some 40 ExxonMobil-funded organization that "either have sought to undermine mainstream scientific findings on global climate change or have maintained affiliations with a small group of "skeptic" scientists who continue to do so." Between 2000 and 2003 these organizations received more than $8m in funding.

It has also had a key influence in the Bush administration's energy policy, including on the Kyoto Protocol, supported by both $55m spent on lobbying since 1999, and direct contacts between the company and leading politicians. It was a leading member of the Global Climate Coalition. It encouraged (and may have been instrumental in) the replacement in 2002 of the head of the IPCC, Robert Watson. It has also invested $100m into the Global Climate and Energy Project, with Stanford University, and other programs at institutions such as the Massachusetts Institute of Technology, Carnegie Mellon University and the International Energy Agency Greenhouse Gas Research and Development Program.

Some of Exxon's activities on climate change produced strong criticism from environmental groups, including reactions such as a leaflet produced by the Stop Esso campaign, saying 'Don't buy E$$o', and featuring a tiger hand setting fire to the Earth. The company's carbon dioxide emissions are more than 50% higher than those of British rival BP, despite the US firm's oil and gas production being only slightly larger.

According to a 2004 study commissioned by Friends of the Earth, ExxonMobil and its predecessors caused 4.7 to 5.3 percent of the world's man-made carbon dioxide emissions between 1882 and 2002. The group suggested that such studies could form the basis for eventual legal action.

BP

BP left the Global Climate Coalition in 1997 and said that global warming was a problem that had to be dealt with, although it subsequently joined others in lobbying the Australian government not to sign the Kyoto Protocol unless the US did. In March 2002 BP's chief executive, Lord Browne, declared in a speech that global warming was real and that urgent action was needed, saying that "Companies composed of highly skilled and trained people can't live in denial of mounting evidence gathered by hundreds of the most reputable scientists in the world.". In 2005 BP was considering testing carbon sequestration in one of its North Sea oil fields, by pumping carbon dioxide into them (and thereby also increasing yields). Throughout 2006 BP, led by their CEO Lord John Browne, has continued to take a leadership stance on climate change. It has cut its own operational emissions of CO_2 by 10%. It is investing $8 billion in renewable energy over the next 10 years. And most recently it has launched a 'target zero' campaign in the UK to encourage its customers to offset their vehicle emissions when they fill up at the petrol station.

BP's American division is a member of the U.S. Climate Action Partnership (USCAP).

Koch Industries

From 2005 to 2008, Koch Industries donated $5.7 million on political campaigns and $37 million on direct lobbying to support fossil fuel industries. Between 1997 and 2008, Koch Industries donated a total of nearly $48 million to climate opposition groups. According to Greenpeace, Koch Industries is the major source of funds of what Greenpeace calls "climate denial". Koch Industries and its subsidiaries spent more than $20 million on lobbying in 2008 and $12.3 million in 2009, according to the Center for Responsive Politics, a nonpartisan research group.

Others

American Electric Power, the world's largest private producer of carbon dioxide, said in 2005 that targets for carbon reduction "represent a common-sense approach that can begin the process of lowering emissions along a gradual, cost-effective path." The company complained that "uncertainties over the cost of carbon" made it very difficult to make decisions about capital investment.

DuPont has cut its greenhouse gas emissions by 65% since 1990, saving hundreds of millions of dollars in the process. "Give us a date, tell us how much we need to cut, give us the flexibility to meet the goals, and we'll get it done", Xcel Energy CEO Wayne Brunetti told *Business Week* in 2004.

Duke Energy, FPL Group, and PG&E Corporation are members of the U.S. Climate Action Partnership (USCAP).

Transportation

A large proportion of carbon dioxide emissions occur because of transportation. Several companies have formed or invested in electric substitutes for standard automobiles. The Tesla Roadster is an all-electric sports car, available now (production ended). Tesla produces a sedan, Tesla Model S now. Vectrix produces and sells an electric scooter rated for 100 km/h (60 MPH).

There has also been greatly increased interest in personal rapid transit, which applies system engineering principles to reduce energy use, eliminate traffic jams, and produce an acceptable substitute to replace cars, all at the same time. Most systems fully meet Kyoto Treaty carbon emission goals now, 60 years ahead of schedule. Korean steel maker POSCO and its partner Vectus Ltd. have produced a working safety case, including test track and vehicles, that remains fully functional in Swedish winters. Vectus and Suncheon S. Korea signed a memorandum of understanding to install a system. Advanced Transportation Systems' ULTra passed safety certification by the UK Rail Inspectorate in 2003, and won a demonstration project at Heathrow Airport due to be in service in early 2010. ATS Ltd. estimates its ULTra PRT will consume 839 BTU per passenger mile (0.55 MJ per passenger km). By comparison, automobiles consume 3,496 BTU, and personal trucks consume 4,329 BTU per passenger mile. 2getthere Inc. sells automated electric freight handling and transit vehicles designed to share existing rights of way with normal traffic. The company recently won the personal rapid transit competition for Masdar.

Insurance Industry

In 2004 Swiss Re, the world's second largest reinsurance company, warned that the economic costs of climate-related disasters threatened to reach $150 billion a year within ten years.

In 2006 Lloyd's of London, published a report highlighting the latest science and implications for the insurance industry.

Swiss Re, has said that if the shore communities of four Gulf Coast states choose not to implement adaptation strategies, they could see annual climate-change related damages jump 65 percent a year to $23 billion by 2030. "Society needs to reduce its vulnerability to climate risks, and as long as they remain manageable, they remain insurable, which is our interest as well," said Mark D. Way, head of Swiss Re's sustainable development for the Americas.

AIG is a member of the U.S. Climate Action Partnership (USCAP).

Media

In the UK, some newspapers (*Daily Mail, Daily Telegraph*) are significantly skeptical, while most others (with varying enthusiasm, *The Independent* giving it most prominence) support action on global warming. Overall, British newspapers have given the issue three times more coverage than US newspapers. In 2006 (*British Sky Broadcasting* (Sky) became the world's first media company to go 'climate neutral' by purchasing enough carbon offsets. The CEO of the company James Murdoch (son of Rupert Murdoch and heir apparent for the News International empire) is a strong advocate of action on climate change and is thought to be influential on the issue within the wider group of companies, *The Sun* announced it was "going green" and now covers the global warming issue extensively. In June 2006, to much industry interest, Rupert Murdoch invited Al Gore to make his climate change presentation at the annual News Corp (including the Fox Network) gathering at the Pebble Beach golf resort, (USA). In August 2007, Rupert Murdoch announced plans for News Corp. to be carbon neutral by 2010.

More on Business Action

Businesses take action on climate change for several reasons. Action improves corporate image and better aligns corporate actions with the environmental interests of owners, employees, suppliers, and customers. Action also occurs to reduce costs, increase return on investments, and to reduce dependency on uncontrollable costs.

Increased Energy Efficiency

For many companies, looking at more efficient energy use can pay off in the medium to long term; unfortunately, shareholders need to be satisfied in the short term, so regulatory intervention is often required, to encourage prudent conservation measures. However, as carbon intensity starts to show up on balance books through organizations such as the Carbon Disclosure Project, voluntary action is starting to take place.

Recently there has been a spate of companies acting to improve their energy efficiency. Possibly the most prominent of these companies is Wal-Mart. Wal-Mart, the largest retailer in the US, has announced specific environmental goals to reduce energy use in its stores and pressure its 60,000 suppliers in its worldwide supply chain to follow its lead. On energy efficiency, Wal-Mart wants to increase the fuel efficiency of its truck fleet by 25% over the next three years and double it within ten years, moving from 6.5 mpg. This seems an attainable goal, and by 2020, it is expected to save the company $494 million a year. The company also wants to build a store that is at least 25% more energy efficient within four years.

Use of Renewable Energies

In August 2002, the largest gathering of ministers in the history of the world met at the World Summit on Sustainable Development WSSD in Johannesburg. The global environmental community discussed the role of renewables and energy efficiency in lowering carbon emissions, mitigating poverty reduction (energy access) and improving energy security. One result from WSSD was the formation of Partnerships for Sustainable Development to carry forward the international dialogue on sustainable energy and its role in the energy mix.

Partnerships formed include the Renewable Energy and Energy Efficiency Partnership REEEP, the Global Village Energy Partnership GVEP, the Johannesburg Renewable Energy Coalition (JREC), and the Global Network on Energy for Sustainable Development GNESD.

Renewable energies and renewable energy technologies have many advantages over their fossil fuel counterparts. These advantages include the absence of local pollution such as particulates, sulphur oxides (SOX's) and nitrous oxides (NOX's). For the business community, the economic advantages are also becoming clearer. Numerous studies have shown that the working environment has a significant effect on workforce morale. Renewable energy solutions are a part of this, wind turbines in particular being seen by many as a potent symbol of a new modernity, where environmental considerations are taken seriously. A workforce seeing a forward-looking and responsible company is more likely to feel good about working for such a company. A happier workforce is a more productive workforce.

More directly, the high petroleum (oil) and gas prices of 2005 have only added to the attraction of renewable energy sources. Although most renewable energies are more expensive at current fuel prices, the difference is narrowing, and uncertainty in oil and gas markets is a factor worth considering for highly energy-intensive businesses.

Another factor affecting the uptake of renewable energies in Europe is the EU Energy Trading Scheme (ETS or EUTS). Many large businesses are fined for increases in emissions, but can sell any "excess" reductions they make.

Companies with high-profile renewable energy portfolios include an aluminium smelter (Alcan), a cement company (Lafarge), and a microchip manufacturer (Intel). Many examples of corporate leadership in this area can be found on the website of The Climate Group, an independent organization set up for promoting such action by business and government.

Carbon Offsets

The principle of carbon offset is fairly simple: a business decides that it doesn't want to contribute further to global warming, and it has already made efforts to reduce its carbon (dioxide) emissions, so it decides to pay someone else to further reduce its net emissions by planting trees or by taking up low-carbon technologies. Every unit of carbon that is absorbed by trees—or not emitted due to funding of renewable energy deployment—offsets the emissions from fossil fuel use. In many cases, funding of renewable energy, energy efficiency, or tree planting—particularly in developing nations—can be a relatively cheap way of making an event, project, or business "carbon neutral". Many carbon offset providers—some as inexpensive as $0.10 per ton of carbon dioxide—are referenced in the Carbon Offset article of this encyclopedia.

Many businesses are now looking to carbon offset all their work. An example of a business going carbon neutral is FIFA: their 2006 World Cup Final will be carbon neutral. FIFA estimate they are offsetting one hundred thousand tons of carbon dioxide created by the event, largely as a result of people travelling there. Other carbon neutral companies include the bank HSBC, the consumer staples manufacturer Annie's Homegrown, world leading society publisher Blackwell Publishing, and the publishing house New Society Publishers. The Guardian newspaper also offsets its carbon emissions resulting from international air travel.

Land Surface Effects on Climate

Land surface effects on climate are wide-ranging and vary by region. Deforestation and exploitation of natural landscapes play a significant role. Some of these environmental changes are similar to those caused by the effects of global warming.

Deforestation Effects

Major land surface changes affecting climate include deforestation (especially in tropical areas), and destruction of grasslands and xeric woodlands by overgrazing, or lack of grazing. These changes in the natural landscape reduce evapotranspiration, and thus water vapor, in the atmosphere, limiting clouds and precipitation (this may contribute to the retreat of glaciers). It has been proposed, in the journal Atmospheric Chemistry and Physics, that evaporation rates from forested areas may exceed that of the oceans, creating zones of low pressure, which enhance the development of storms and rainfall through atmospheric moisture recycling. The American Institute of Biological Sciences published a similar paper in support of this concept in 2009. In addition, with deforestation and/or destruction of grasslands, the amount of dew harvested (or condensed) by plants is greatly diminished. All of this helps lead to desertification in these regions.

This concept of land-atmosphere feedback is common among permaculturists, such as Masanobu Fukuoka, who, in his book, *The One Straw Revolution*, said "rain comes from the ground, not the sky."

Deforestation, and conversion of grasslands to desert, may also lead to cooling of the regional climate. This is because of the albedo effect (sunlight reflected by bare ground) during the day, and rapid radiation of heat into space at night, due to the lack of vegetation and atmospheric moisture.

Reforestation, conservation grazing, holistic land management, and, in drylands, water harvesting and keyline design, are examples of methods that might help prevent or lessen these drying effects.

Deforestation and Climate Change

Deforestation is one of the main causes of climate change. It is the second largest anthropogenic source of carbon dioxide to the atmosphere, after fossil fuel combustion. Deforestation and forest degradation contribute to atmospheric greenhouse gas emissions through combustion of forest biomass and decomposition of remaining plant material and soil carbon. It used to account for more than 20% of carbon dioxide emissions, but it's currently somewhere around the 10% mark. By 2008, deforestation was 12% of total CO_2, or 15% if peatlands are included. These proportions are likely to have fallen since given the continued rise of fossil fuel use.

Averaged over all land and ocean surfaces, temperatures warmed roughly 1.53 °F (0.85 °C) between 1880 and 2012, according to the Intergovernmental Panel on Climate Change. In the Northern Hemisphere, 1983 to 2012 were the warmest 30-year period of the last 1400 years.

Effect on Climate Change

Decrease in Biodiversity

A 2007 study conducted by the National Science Foundation found that biodiversity and genetic diversity are codependent—that diversity among species requires diversity within a species, and vice versa. "If any one type is removed from the system, the cycle can break down, and the community becomes dominated by a single species."

Counteracting Climate Change

Reforestation

Reforestation is the natural or intentional restocking of existing forests and woodlands that have been depleted, usually through deforestation. It is the reestablishment of forest cover either naturally or artificially. Similar to the other methods of forestation, reforestation can be very effective because a single tree can absorb as much as 48 pounds of carbon dioxide per year and can sequester 1 ton of carbon dioxide by the time it reaches 40 years old.

Afforestation

Afforestation is the establishment of a forest or stand of trees in an area where there was no forest.

China

Although China has set official goals for reforestation, these goals were set for an 80-year time horizon and were not significantly met by 2008. China is trying to correct these problems with projects such as the Green Wall of China, which aims to replant forests and halt the expansion of the Gobi Desert. A law promulgated in 1981 requires that every school student over the age of 11 plant at least one tree per year. But average success rates, especially in state-sponsored plantings, remains relatively low. And even the properly planted trees have had great difficulty surviving the combined impacts of prolonged droughts, pest infestation and fires. Nonetheless, China currently has the highest afforestation rate of any country or region in the world, with 4.77 million hectares (47,000 square kilometers) of afforestation in 2008.

Japan

The primary goal of afforestation projects in Japan is to develop the forest structure of the nation and to maintain the biodiversity found in the Japanese wilderness. The Japanese temperate rainforest is scattered throughout the Japanese archipelago and is home to many endemic species that are not naturally found anywhere else. As development of the country's caused a decline in forest cover, a reduction in biodiversity was seen in those areas.

Agroforestry

Agroforestry or agro-sylviculture is a land use management system in which trees or shrubs are grown around or among crops or pastureland. It combines agricultural and forestry technologies to create more diverse, productive, profitable, healthy, and sustainable land-use systems.

Projects and foundations

Arbor Day Foundation

Founded in 1972, the centennial of the first Arbor Day observance in the 19th century, the Foundation has grown to become the largest nonprofit membership organization dedicated to planting trees, with over one million members, supporters, and valued partners. They work on projects focused on planting trees around campuses, low-income communities, and communities that have been affected by natural disasters among other places.

Billion Tree Campaign

The Billion Tree Campaign was launched in 2006 by the United Nations Environment Programme (UNEP) as a response to the challenges of global warming, as well as to a wider array of sustainability challenges, from water supply to biodiversity loss. Its initial target was the planting of one billion trees in 2007. Only one year later in 2008, the campaign's objective was raised to 7 billion trees—a target to be met by the climate change conference that was held in Copenhagen, Denmark in December 2009. Three months before the conference, the 7 billion planted trees mark had been surpassed. In December 2011, after more than 12 billion trees had been planted, UNEP formally handed management of the program over to the not-for-profit Plant-for-the-Planet initiative, based in Munich, Germany.

The Amazon Fund (Brazil)

Considered the largest reserve of biological diversity in the world, the Amazon Basin is also the largest Brazilian biome, taking up almost half the nation's territory. The Amazon Basin corresponds to two fifths of South America's territory. Its area of approximately seven million square kilometers covers the largest hydrographic network on the planet, through which runs about one fifth of the fresh water on the world's surface. Deforestation in the Amazon rainforest is a major cause to climate change due to the decreasing number of trees available to capture increasing carbon dioxide levels in the atmosphere.

The Amazon Fund is aimed at raising donations for non-reimbursable investments in efforts to prevent, monitor and combat deforestation, as well as to promote the preservation and sustainable use of forests in the Amazon Biome, under the terms of Decree N.º 6,527, dated August 1, 2008. The Amazon Fund supports the following areas: management of public forests and protected areas, environmental control, monitoring and inspection, sustainable forest management, economic activities created with sustainable use of forests, ecological and economic zoning, territorial arrangement and agricultural regulation, preservation and sustainable use of biodiversity, and recovery of deforested areas. Besides those, the Amazon Fund may use up to 20% of its donations to support the development of systems to monitor and control deforestation in other Brazilian biomes and in biomes of other tropical countries.

Air Pollution

Air pollution is the introduction of particulates, biological molecules, or other harmful substances into Earth's atmosphere, causing diseases, allergies, death to humans, damage to other living organisms such as animals and food crops, or the natural or built environment. Air pollution may come from anthropogenic or natural sources.

The atmosphere is a complex natural gaseous system that is essential to support life on planet Earth.

Indoor air pollution and urban air quality are listed as two of the world's worst toxic pollution problems in the 2008 Blacksmith Institute World's Worst Polluted Places report. According to the 2014 WHO report, air pollution in 2012 caused the deaths of around 7 million people worldwide, an estimate roughly matched by the International Energy Agency.

Pollutants

An air pollutant is a substance in the air that can have adverse effects on humans and the ecosystem. The substance can be solid particles, liquid droplets, or gases. A pollutant can be of natural origin or man-made. Pollutants are classified as primary or secondary. Primary pollutants are usually produced from a process, such as ash from a volcanic eruption. Other examples include carbon monoxide gas from motor vehicle exhaust, or the sulfur dioxide released from factories. Secondary pollutants are not emitted directly. Rather, they form in the air when primary pollutants react or interact. Ground level ozone is a prominent example of a secondary pollutant. Some pollutants may be both primary and secondary: they are both emitted directly and formed from other primary pollutants.

Major primary pollutants produced by human activity include:

1. Sulfur oxides (SO_x) - particularly sulfur dioxide, a chemical compound with the formula SO_2. SO_2 is produced by volcanoes and in various industrial processes. Coal and petroleum often contain sulfur compounds, and their combustion generates sulfur dioxide. Further oxidation of SO_2, usually in the presence of a catalyst such as NO_2, forms H_2SO_4, and thus acid rain. This is one of the causes for concern over the environmental impact of the use of these fuels as power sources.

2. Nitrogen oxides (NO_x) - Nitrogen oxides, particularly nitrogen dioxide, are expelled from high temperature combustion, and are also produced during thunderstorms by electric discharge. They can be seen as a brown haze dome above or a plume downwind of cities. Nitrogen dioxide is a chemical compound with the formula NO_2. It is one of several nitrogen oxides. One of the most prominent air pollutants, this reddish-brown toxic gas has a characteristic sharp, biting odor.

3. Carbon monoxide (CO) - CO is a colorless, odorless, toxic yet non-irritating gas. It is a product by incomplete combustion of fuel such as natural gas, coal or wood. Vehicular exhaust is a major source of carbon monoxide.

4. Volatile organic compounds (VOC) - VOCs are a well-known outdoor air pollutant. They are categorized as either methane (CH_4) or non-methane (NMVOCs). Methane is an extremely efficient greenhouse gas which contributes to enhanced global warming. Other hydrocarbon VOCs are also significant greenhouse gases because of their role in creating ozone and prolonging the life of methane in the atmosphere. This effect varies depending on local air quality. The aromatic NMVOCs benzene, toluene and xylene are suspected carcinogens and may lead to leukemia with prolonged exposure. 1,3-butadiene is another dangerous compound often associated with industrial use.

5. Particulates, alternatively referred to as particulate matter (PM), atmospheric particulate matter, or fine particles, are tiny particles of solid or liquid suspended in a gas. In contrast, aero-

sol refers to combined particles and gas. Some particulates occur naturally, originating from volcanoes, dust storms, forest and grassland fires, living vegetation, and sea spray. Human activities, such as the burning of fossil fuels in vehicles, power plants and various industrial processes also generate significant amounts of aerosols. Averaged worldwide, anthropogenic aerosols—those made by human activities—currently account for approximately 10 percent of our atmosphere. Increased levels of fine particles in the air are linked to health hazards such as heart disease, altered lung function and lung cancer.

6. Persistent free radicals connected to airborne fine particles are linked to cardiopulmonary disease.

7. Toxic metals, such as lead and mercury, especially their compounds.

8. Chlorofluorocarbons (CFCs) - harmful to the ozone layer; emitted from products are currently banned from use. These are gases which are released from air conditioners, refrigerators, aerosol sprays, etc. CFC's on being released into the air rises to stratosphere. Here they come in contact with other gases and damage the ozone layer. This allows harmful ultraviolet rays to reach the earth's surface. This can lead to skin cancer, disease to eye and can even cause damage to plants.

9. Ammonia (NH_3) - emitted from agricultural processes. Ammonia is a compound with the formula NH_3. It is normally encountered as a gas with a characteristic pungent odor. Ammonia contributes significantly to the nutritional needs of terrestrial organisms by serving as a precursor to foodstuffs and fertilizers. Ammonia, either directly or indirectly, is also a building block for the synthesis of many pharmaceuticals. Although in wide use, ammonia is both caustic and hazardous. In the atmosphere, ammonia reacts with oxides of nitrogen and sulfur to form secondary particles.

10. Odours — such as from garbage, sewage, and industrial processes

11. Radioactive pollutants - produced by nuclear explosions, nuclear events, war explosives, and natural processes such as the radioactive decay of radon.

Secondary pollutants include:

1. Particulates created from gaseous primary pollutants and compounds in photochemical smog. Smog is a kind of air pollution. Classic smog results from large amounts of coal burning in an area caused by a mixture of smoke and sulfur dioxide. Modern smog does not usually come from coal but from vehicular and industrial emissions that are acted on in the atmosphere by ultraviolet light from the sun to form secondary pollutants that also combine with the primary emissions to form photochemical smog.

2. Ground level ozone (O_3) formed from NO_x and VOCs. Ozone (O_3) is a key constituent of the troposphere. It is also an important constituent of certain regions of the stratosphere commonly known as the Ozone layer. Photochemical and chemical reactions involving it drive many of the chemical processes that occur in the atmosphere by day and by night. At abnormally high concentrations brought about by human activities (largely the combustion of fossil fuel), it is a pollutant, and a constituent of smog.

3. Peroxyacetyl nitrate (PAN) - similarly formed from NO_x and VOCs.

Minor air pollutants include:

1. A large number of minor hazardous air pollutants. Some of these are regulated in USA under the Clean Air Act and in Europe under the Air Framework Directive

2. A variety of persistent organic pollutants, which can attach to particulates

Persistent organic pollutants (POPs) are organic compounds that are resistant to environmental degradation through chemical, biological, and photolytic processes. Because of this, they have been observed to persist in the environment, to be capable of long-range transport, bioaccumulate in human and animal tissue, biomagnify in food chains, and to have potentially significant impacts on human health and the environment.

Sources

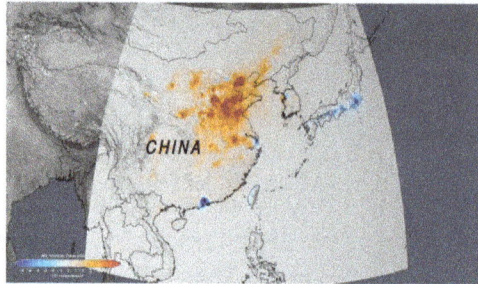

This video provides an overview of a NASA study on the human fingerprint on global air quality.

There are various locations, activities or factors which are responsible for releasing pollutants into the atmosphere. These sources can be classified into two major categories.

Anthropogenic (man-made) Sources:

Controlled burning of a field outside of Statesboro, Georgia in preparation for spring planting.

These are mostly related to the burning of multiple types of fuel.

1. Stationary sources include smoke stacks of power plants, manufacturing facilities (factories) and waste incinerators, as well as furnaces and other types of fuel-burning heating devices. In developing and poor countries, traditional biomass burning is the major source of air pollutants; traditional biomass includes wood, crop waste and dung.

2. Mobile sources include motor vehicles, marine vessels, and aircraft.

3. Controlled burn practices in agriculture and forest management. Controlled or prescribed

burning is a technique sometimes used in forest management, farming, prairie restoration or greenhouse gas abatement. Fire is a natural part of both forest and grassland ecology and controlled fire can be a tool for foresters. Controlled burning stimulates the germination of some desirable forest trees, thus renewing the forest.

4. Fumes from paint, hair spray, varnish, aerosol sprays and other solvents

5. Waste deposition in landfills, which generate methane. Methane is highly flammable and may form explosive mixtures with air. Methane is also an asphyxiant and may displace oxygen in an enclosed space. Asphyxia or suffocation may result if the oxygen concentration is reduced to below 19.5% by displacement.

6. Military resources, such as nuclear weapons, toxic gases, germ warfare and rocketry

Natural Sources:

Dust storm approaching Stratford, Texas.

1. Dust from natural sources, usually large areas of land with little or no vegetation

2. Methane, emitted by the digestion of food by animals, for example cattle

3. Radon gas from radioactive decay within the Earth's crust. Radon is a colorless, odorless, naturally occurring, radioactive noble gas that is formed from the decay of radium. It is considered to be a health hazard. Radon gas from natural sources can accumulate in buildings, especially in confined areas such as the basement and it is the second most frequent cause of lung cancer, after cigarette smoking.

4. Smoke and carbon monoxide from wildfires

5. Vegetation, in some regions, emits environmentally significant amounts of Volatile organic compounds (VOCs) on warmer days. These VOCs react with primary anthropogenic pollutants—specifically, NO_x, SO_2, and anthropogenic organic carbon compounds — to produce a seasonal haze of secondary pollutants. Black gum, poplar, oak and willow are some examples of vegetation that can produce abundant VOCs. The VOC production from these species result in ozone levels up to eight times higher than the low-impact tree species.

6. Volcanic activity, which produces sulfur, chlorine, and ash particulates

Emission Factors

Air pollutant emission factors are reported representative values that attempt to relate the quantity of a pollutant released to the ambient air with an activity associated with the release of that pollutant. These factors are usually expressed as the weight of pollutant divided by a unit weight, volume, distance, or duration of the activity emitting the pollutant (e.g., kilograms of particulate emitted per tonne of coal burned). Such factors facilitate estimation of emissions from various sources of air pollution. In most cases, these factors are simply averages of all available data of acceptable quality, and are generally assumed to be representative of long-term averages.

Beijing air on a 2005-day after rain (left) and a smoggy day (right)

There are 12 compounds in the list of persistent organic pollutants. Dioxins and furans are two of them and intentionally created by combustion of organics, like open burning of plastics. These compounds are also endocrine disruptors and can mutate the human genes.

The United States Environmental Protection Agency has published a compilation of air pollutant emission factors for a multitude of industrial sources. The United Kingdom, Australia, Canada and many other countries have published similar compilations, as well as the European Environment Agency.

Exposure

Air pollution risk is a function of the hazard of the pollutant and the exposure to that pollutant. Air pollution exposure can be expressed for an individual, for certain groups (e.g. neighborhoods or children living in a country), or for entire populations. For example, one may want to calculate the exposure to a hazardous air pollutant for a geographic area, which includes the various microenvironments and age groups. This can be calculated as an inhalation exposure. This would account for daily exposure in various settings (e.g. different indoor micro-environments and outdoor locations). The exposure needs to include different age and other demographic groups, especially infants, children, pregnant women and other sensitive subpopulations. The exposure to an air pollutant must integrate the concentrations of the air pollutant with respect to the time spent in each setting and the respective inhalation rates for each subgroup for each specific time that the subgroup is in the setting and engaged in particular activities (playing, cooking, reading, working, etc.). For example, a small child's inhalation rate will be less than that of an adult. A child engaged in vigorous exercise will have a higher respiration rate than the same child in a sedentary activity. The daily exposure, then, needs to reflect the time spent in each micro-environmental setting and the type of activities in these settings. The air pollutant concentration in each microactivity/micro-environmental setting is summed to indicate the exposure.

Indoor air Quality (IAQ)

A lack of ventilation indoors concentrates air pollution where people often spend the majority of their time. Radon (Rn) gas, a carcinogen, is exuded from the Earth in certain locations and trapped inside houses. Building materials including carpeting and plywood emit formaldehyde (H_2CO)

gas. Paint and solvents give off volatile organic compounds (VOCs) as they dry. Lead paint can degenerate into dust and be inhaled. Intentional air pollution is introduced with the use of air fresheners, incense, and other scented items. Controlled wood fires in stoves and fireplaces can add significant amounts of smoke particulates into the air, inside and out. Indoor pollution fatalities may be caused by using pesticides and other chemical sprays indoors without proper ventilation.

Air quality monitoring, New Delhi, India.

Carbon monoxide poisoning and fatalities are often caused by faulty vents and chimneys, or by the burning of charcoal indoors or in a confined space, such as a tent. Chronic carbon monoxide poisoning can result even from poorly-adjusted pilot lights. Traps are built into all domestic plumbing to keep sewer gas and hydrogen sulfide, out of interiors. Clothing emits tetrachloroethylene, or other dry cleaning fluids, for days after dry cleaning.

Though its use has now been banned in many countries, the extensive use of asbestos in industrial and domestic environments in the past has left a potentially very dangerous material in many localities. Asbestosis is a chronic inflammatory medical condition affecting the tissue of the lungs. It occurs after long-term, heavy exposure to asbestos from asbestos-containing materials in structures. Sufferers have severe dyspnea (shortness of breath) and are at an increased risk regarding several different types of lung cancer. As clear explanations are not always stressed in non-technical literature, care should be taken to distinguish between several forms of relevant diseases. According to the World Health Organisation (WHO), these may defined as; asbestosis, *lung cancer*, and *Peritoneal Mesothelioma* (generally a very rare form of cancer, when more widespread it is almost always associated with prolonged exposure to asbestos).

Biological sources of air pollution are also found indoors, as gases and airborne particulates. Pets produce dander, people produce dust from minute skin flakes and decomposed hair, dust mites in bedding, carpeting and furniture produce enzymes and micrometre-sized fecal droppings, inhabitants emit methane, mold forms on walls and generates mycotoxins and spores, air conditioning systems can incubate Legionnaires' disease and mold, and houseplants, soil and surrounding gardens can produce pollen, dust, and mold. Indoors, the lack of air circulation allows these airborne pollutants to accumulate more than they would otherwise occur in nature.

Health Effects

Air pollution is a significant risk factor for a number of pollution-related diseases and health conditions including respiratory infections, heart disease, COPD, stroke and lung cancer. The health effects caused by air pollution may include difficulty in breathing, wheezing, coughing, asthma and worsening of existing respiratory and cardiac conditions. These effects can result in increased medication use, increased doctor or emergency room visits, more hospital admissions and premature death. The human health effects of poor air quality are far reaching, but principally affect the body's respiratory system and the cardiovascular system. Individual reactions to air pollutants depend on the type of pollutant a person is exposed to, the degree of exposure, and the individual's health status and genetics. The most common sources of air pollution include particulates, ozone, nitrogen dioxide, and sulphur dioxide. Children aged less than five years that live in developing countries are the most vulnerable population in terms of total deaths attributable to indoor and outdoor air pollution.

Mortality

The World Health Organization estimated in 2014 that every year air pollution causes the premature death of some 7 million people worldwide. India has the highest death rate due to air pollution. India also has more deaths from asthma than any other nation according to the World Health Organization. In December 2013 air pollution was estimated to kill 500,000 people in China each year. There is a positive correlation between pneumonia-related deaths and air pollution from motor vehicle emissions.

Annual premature European deaths caused by air pollution are estimated at 430,000. An important cause of these deaths is nitrogen dioxide and other nitrogen oxides (NOx) emitted by road vehicles. Across the European Union, air pollution is estimated to reduce life expectancy by almost nine months. Causes of deaths include strokes, heart disease, COPD, lung cancer, and lung infections.

The US EPA estimates that a proposed set of changes in diesel engine technology (*Tier 2*) could result in 12,000 fewer *premature mortalities*, 15,000 fewer heart attacks, 6,000 fewer emergency room visits by children with asthma, and 8,900 fewer respiratory-related hospital admissions each year in the United States.

The US EPA has estimated that limiting ground-level ozone concentration to 65 parts per billion, would avert 1,700 to 5,100 premature deaths nationwide in 2020 compared with the 75-ppb standard. The agency projected the more protective standard would also prevent an additional 26,000 cases of aggravated asthma, and more than a million cases of missed work or school. Following this assessment, the EPA acted to protect public health by lowering the National Ambient Air Quality Standards (NAAQS) for ground-level ozone to 70 parts per billion (ppb).

A new economic study of the health impacts and associated costs of air pollution in the Los Angeles Basin and San Joaquin Valley of Southern California shows that more than 3,800 people die prematurely (approximately 14 years earlier than normal) each year because air pollution levels violate federal standards. The number of annual premature deaths is considerably higher than the fatalities related to auto collisions in the same area, which average fewer than 2,000 per year.

Diesel exhaust (DE) is a major contributor to combustion-derived particulate matter air pollution. In several human experimental studies, using a well-validated exposure chamber setup, DE has been linked to acute vascular dysfunction and increased thrombus formation.

The mechanisms linking air pollution to increased cardiovascular mortality are uncertain, but probably include pulmonary and systemic inflammation.

Cardiovascular Disease

A 2007 review of evidence found ambient air pollution exposure is a risk factor correlating with increased total mortality from cardiovascular events (range: 12% to 14% per 10 microg/m^3 increase).

Air pollution is also emerging as a risk factor for stroke, particularly in developing countries where pollutant levels are highest. A 2007 study found that in women, air pollution is not associated with hemorrhagic but with ischemic stroke. Air pollution was also found to be associated with increased incidence and mortality from coronary stroke in a cohort study in 2011. Associations are believed to be causal and effects may be mediated by vasoconstriction, low-grade inflammation and atherosclerosis Other mechanisms such as autonomic nervous system imbalance have also been suggested.

Lung Disease

Chronic obstructive pulmonary disease (COPD) includes diseases such as chronic bronchitis and emphysema.

Research has demonstrated increased risk of developing asthma and COPD from increased exposure to traffic-related air pollution. Additionally, air pollution has been associated with increased hospitalization and mortality from asthma and COPD.

A study conducted in 1960-1961 in the wake of the Great Smog of 1952 compared 293 London residents with 477 residents of Gloucester, Peterborough, and Norwich, three towns with low reported death rates from chronic bronchitis. All subjects were male postal truck drivers aged 40 to 59. Compared to the subjects from the outlying towns, the London subjects exhibited more severe respiratory symptoms (including cough, phlegm, and dyspnea), reduced lung function (FEV$_1$ and peak flow rate), and increased sputum production and purulence. The differences were more pronounced for subjects aged 50 to 59. The study controlled for age and smoking habits, so concluded that air pollution was the most likely cause of the observed differences.

It is believed that much like cystic fibrosis, by living in a more urban environment serious health hazards become more apparent. Studies have shown that in urban areas patients suffer mucus hypersecretion, lower levels of lung function, and more self-diagnosis of chronic bronchitis and emphysema.

Cancer

Cancer mainly the result of environmental factors.

A review of evidence regarding whether ambient air pollution exposure is a risk factor for cancer in 2007 found solid data to conclude that long-term exposure to PM2.5 (fine particulates) increases the overall risk of non-accidental mortality by 6% per a 10 microg/m³ increase. Exposure to PM2.5 was also associated with an increased risk of mortality from lung cancer (range: 15% to 21% per 10 microg/m³ increase) and total cardiovascular mortality (range: 12% to 14% per a 10 microg/m³ increase). The review further noted that living close to busy traffic appears to be associated with elevated risks of these three outcomes --- increase in lung cancer deaths, cardiovascular deaths, and overall non-accidental deaths. The reviewers also found suggestive evidence that exposure to PM2.5 is positively associated with mortality from coronary heart diseases and exposure to SO_2 increases mortality from lung cancer, but the data was insufficient to provide solid conclusions. Another investigation showed that higher activity level increases deposition fraction of aerosol particles in human lung and recommended avoiding heavy activities like running in outdoor space at polluted areas.

In 2011, a large Danish epidemiological study found an increased risk of lung cancer for patients who lived in areas with high nitrogen oxide concentrations. In this study, the association was higher for non-smokers than smokers. An additional Danish study, also in 2011, likewise noted evidence of possible associations between air pollution and other forms of cancer, including cervical cancer and brain cancer.

In December 2015, medical scientists reported that cancer is overwhelmingly a result of environmental factors, and not largely down to bad luck. Maintaining a healthy weight, eating a healthy diet, minimizing alcohol and eliminating smoking reduces the risk of developing the disease, according to the researchers.

Children

In the United States, despite the passage of the Clean Air Act in 1970, in 2002 at least 146 million Americans were living in non-attainment areas—regions in which the concentration of certain air pollutants exceeded federal standards. These dangerous pollutants are known as the criteria pollutants, and include ozone, particulate matter, sulfur dioxide, nitrogen dioxide, carbon monoxide, and lead. Protective measures to ensure children's health are being taken in cities such as New Delhi, India where buses now use compressed natural gas to help eliminate the "pea-soup" smog. A recent study in Europe has found that exposure to ultrafine particles can increase blood pressure in children.

"Clean" Areas

Even in the areas with relatively low levels of air pollution, public health effects can be significant and costly, since a large number of people breathe in such pollutants. A 2005 scientific study for the British Columbia Lung Association showed that a small improvement in air quality (1% reduction of ambient PM2.5 and ozone concentrations) would produce $29 million in annual savings in the Metro Vancouver region in 2010. This finding is based on health valuation of lethal (death) and sub-lethal (illness) affects.

Central Nervous System

Data is accumulating that air pollution exposure also affects the central nervous system.

In a June 2014 study conducted by researchers at the University of Rochester Medical Center, published in the journal Environmental Health Perspectives, it was discovered that early exposure to air pollution causes the same damaging changes in the brain as autism and schizophrenia. The study also shows that air pollution also affected short-term memory, learning ability, and impulsivity. Lead researcher Professor Deborah Cory-Slechta said that "When we looked closely at the ventricles, we could see that the white matter that normally surrounds them hadn't fully developed. It appears that inflammation had damaged those brain cells and prevented that region of the brain from developing, and the ventricles simply expanded to fill the space. Our findings add to the growing body of evidence that air pollution may play a role in autism, as well as in other neurodevelopmental disorders." Air pollution has a more significant negative effect on males than on females.

In 2015, experimental studies reported the detection of significant episodic (situational) cognitive impairment from impurities in indoor air breathed by test subjects who were not informed about changes in the air quality. Researchers at the Harvard University and SUNY Upstate Medical University and Syracuse University measured the cognitive performance of 24 participants in three different controlled laboratory atmospheres that simulated those found in "conventional" and "green" buildings, as well as green buildings with enhanced ventilation. Performance was evaluated objectively using the widely used Strategic Management Simulation software simulation tool, which is a well-validated assessment test for executive decision-making in an unconstrained situation allowing initiative and improvisation. Significant deficits were observed in the performance scores achieved in increasing concentrations of either volatile organic compounds (VOCs) or carbon dioxide, while keeping other factors constant. The highest impurity levels reached are not uncommon in some classroom or office environments.

Agricultural Effects

In India in 2014, it was reported that air pollution by black carbon and ground level ozone had cut crop yields in the most affected areas by almost half in 2010 when compared to 1980 levels.

Economic Effects

Air pollution costs the world economy $5 trillion per year as a result of productivity losses and degraded quality of life, according to a joint study by the World Bank and the Institute for Health Metrics and Evaluation (IHME) at the University of Washington These productivity losses are caused by deaths due to diseases caused by air pollution. One out of ten deaths in 2013 was caused by diseases associated with air pollution and the problem is getting worse. The problem is even more acute in the developing world. "Children under age 5 in lower-income countries are more than 60 times as likely to die from exposure to air pollution as children in high-income countries." The report states that additional economic losses caused by air pollution, including health costs and the adverse effect on agricultural and other productivity were not calculated in the report, and thus the actual costs to the world economy are far higher than $5 trillion.

Historical Disasters

The world's worst short-term civilian pollution crisis was the 1984 Bhopal Disaster in India. Leaked industrial vapours from the Union Carbide factory, belonging to Union Carbide, Inc., U.S.A. (later bought by Dow Chemical Company), killed at least 3787 people and injured anywhere from 150,000 to 600,000. The United Kingdom suffered its worst air pollution event when the December 4 Great Smog of 1952 formed over London. In six days more than 4,000 died and more recent estimates put the figure at nearer 12,000. An accidental leak of anthrax spores from a biological warfare laboratory in the former USSR in 1979 near Sverdlovsk is believed to have caused at least 64 deaths. The worst single incident of air pollution to occur in the US occurred in Donora, Pennsylvania in late October, 1948, when 20 people died and over 7,000 were injured.

Alternatives to Pollution

There are now practical alternatives to the three principal causes of air pollution.

1. Combustion of fossil fuels for space heating can be replaced by using ground source heat pumps and seasonal thermal energy storage.

2. Electric power generation from burning fossil fuels can be replaced by power generation from nuclear and renewables.

3. Motor vehicles driven by fossil fuels, a key factor in urban air pollution, can be replaced by electric vehicles.

Reduction Efforts

There are various air pollution control technologies and strategies available to reduce air pollution. At its most basic level, land-use planning is likely to involve zoning and transport infrastructure planning. In most developed countries, land-use planning is an important part of social policy, ensuring that land is used efficiently for the benefit of the wider economy and population, as well as to protect the environment.

Because a large share of air pollution is caused by combustion of fossil fuels such as coal and oil, the reduction of these fuels can reduce air pollution drastically. Most effective is the switch to clean power sources such as wind power, solar power, hydro power which don't cause air pollution. Efforts to reduce pollution from mobile sources includes primary regulation (many developing countries have permissive regulations), expanding regulation to new sources (such as cruise and transport ships, farm equipment, and small gas-powered equipment such as string trimmers, chainsaws, and snowmobiles), increased fuel efficiency (such as through the use of hybrid vehicles), conversion to cleaner fuels or conversion to electric vehicles.

Titanium dioxide has been researched for its ability to reduce air pollution. Ultraviolet light will release free electrons from material, thereby creating free radicals, which break up VOCs and NOx gases. One form is superhydrophilic.

In 2014, Prof. Tony Ryan and Prof. Simon Armitage of University of Sheffield prepared a 10 meter by 20 meter-sized poster coated with microscopic, pollution-eating nanoparticles of titanium dioxide. Placed on a building, this giant poster can absorb the toxic emission from around 20 cars each day.

A very effective means to reduce air pollution is the transition to renewable energy. According to a study published in Energy and Environmental Science in 2015 the switch to 100% renewable energy in the United States would eliminate about 62,000 premature mortalities per year and about 42,000 in 2050, if no biomass were used. This would save about $600 billion in health costs a year due to reduced air pollution in 2050, or about 3.6% of the 2014 U.S. gross domestic product.

Control Devices

The following items are commonly used as pollution control devices in industry and transportation. They can either destroy contaminants or remove them from an exhaust stream before it is emitted into the atmosphere.

1. Particulate Control

    1. Mechanical collectors (dust cyclones, multicyclones)

    2. Electrostatic precipitators An electrostatic precipitator (ESP), or electrostatic air cleaner is a particulate collection device that removes particles from a flowing gas (such as air), using the force of an induced electrostatic charge. Electrostatic precipitators are highly efficient filtration devices that minimally impede the flow of gases through the device, and can easily remove fine particulates such as dust and smoke from the air stream.

    3. Baghouses Designed to handle heavy dust loads, a dust collector consists of a blower, dust filter, a filter-cleaning system, and a dust receptacle or dust removal system (distinguished from air cleaners which utilize disposable filters to remove the dust).

    4. Particulate scrubbers Wet scrubber is a form of pollution control technology. The term describes a variety of devices that use pollutants from a furnace flue gas or from other gas streams. In a wet scrubber, the polluted gas stream is brought into contact with the scrubbing liquid, by spraying it with the liquid, by forcing it through a pool of liquid, or by some other contact method, so as to remove the pollutants.

2. Scrubbers

    1. Baffle spray scrubber

    2. Cyclonic spray scrubber

    3. Ejector venturi scrubber

    4. Mechanically aided scrubber

    5. Spray tower

    6. Wet scrubber

3. NOx control

    1. Low NOx burners

    2. Selective catalytic reduction (SCR)

    3. Selective non-catalytic reduction (SNCR)

    4. NOx scrubbers

    5. Exhaust gas recirculation

    6. Catalytic converter (also for VOC control)

4. VOC abatement

    1. Adsorption systems, using activated carbon, such as Fluidized Bed Concentrator

    2. Flares

    3. Thermal oxidizers

    4. Catalytic converters

    5. Biofilters

    6. Absorption (scrubbing)

    7. Cryogenic condensers

    8. Vapor recovery systems

5. Acid Gas/SO_2 control

    1. Wet scrubbers

    2. Dry scrubbers

    3. Flue-gas desulfurization

6. Mercury control

    1. Sorbent Injection Technology

    2. Electro-Catalytic Oxidation (ECO)

    3. K-Fuel

7. Dioxin and furan control

8. Miscellaneous associated equipment

    1. Source capturing systems

    2. Continuous emissions monitoring systems (CEMS)

Regulations

In general, there are two types of air quality standards. The first class of standards (such as the U.S. National Ambient Air Quality Standards and E.U. Air Quality Directive) set maximum atmospheric concentrations for specific pollutants. Environmental agencies enact regulations which are intended to result in attainment of these target levels. The second class (such as the North American

Air Quality Index) take the form of a scale with various thresholds, which is used to communicate to the public the relative risk of outdoor activity. The scale may or may not distinguish between different pollutants.

Smog in Cairo

Canada

In Canada, air pollution and associated health risks are measured with the Air Quality Health Index or (AQHI). It is a health protection tool used to make decisions to reduce short-term exposure to air pollution by adjusting activity levels during increased levels of air pollution.

The Air Quality Health Index or "AQHI" is a federal program jointly coordinated by Health Canada and Environment Canada. However, the AQHI program would not be possible without the commitment and support of the provinces, municipalities and NGOs. From air quality monitoring to health risk communication and community engagement, local partners are responsible for the vast majority of work related to AQHI implementation. The AQHI provides a number from 1 to 10+ to indicate the level of health risk associated with local air quality. Occasionally, when the amount of air pollution is abnormally high, the number may exceed 10. The AQHI provides a local air quality current value as well as a local air quality maximums forecast for today, tonight and tomorrow and provides associated health advice.

Risk: Low (1-3) Moderate (4-6) High (7-10) Very high (above 10)

As it is now known that even low levels of air pollution can trigger discomfort for the sensitive population, the index has been developed as a continuum: The higher the number, the greater the health risk and need to take precautions. The index describes the level of health risk associated with this number as 'low', 'moderate', 'high' or 'very high', and suggests steps that can be taken to reduce exposure.

Health Risk	Air Quality Health Index	Health Messages	
		At Risk population	General Population
Low	1-3	Enjoy your usual outdoor activities.	Ideal air quality for outdoor activities

Health Risk	Air Quality Health Index	Health Messages	
Moderate	4-6	Consider reducing or rescheduling strenuous activities outdoors if you are experiencing symptoms.	No need to modify your usual outdoor activities unless you experience symptoms such as coughing and throat irritation.
High	7-10	Reduce or reschedule strenuous activities outdoors. Children and the elderly should also take it easy.	Consider reducing or rescheduling strenuous activities outdoors if you experience symptoms such as coughing and throat irritation.
Very high	Above 10	Avoid strenuous activities outdoors. Children and the elderly should also avoid outdoor physical exertion and should stay indoors.	Reduce or reschedule strenuous activities outdoors, especially if you experience symptoms such as coughing and throat irritation.

The measurement is based on the observed relationship of Nitrogen Dioxide (NO_2), ground-level Ozone (O_3) and particulates ($PM_{2.5}$) with mortality, from an analysis of several Canadian cities. Significantly, all three of these pollutants can pose health risks, even at low levels of exposure, especially among those with pre-existing health problems.

When developing the AQHI, Health Canada's original analysis of health effects included five major air pollutants: particulates, ozone, and nitrogen dioxide (NO2), as well as sulfur dioxide (SO_2), and carbon monoxide (CO). The latter two pollutants provided little information in predicting health effects and were removed from the AQHI formulation.

The AQHI does not measure the effects of odour, pollen, dust, heat or humidity.

Germany

TA Luft is the German air quality regulation.

Hotspots

Air pollution hotspots are areas where air pollution emissions expose individuals to increased negative health effects. They are particularly common in highly populated, urban areas, where there may be a combination of stationary sources (e.g. industrial facilities) and mobile sources (e.g. cars and trucks) of pollution. Emissions from these sources can cause respiratory disease, childhood asthma, cancer, and other health problems. Fine particulate matter such as diesel soot, which contributes to more than 3.2 million premature deaths around the world each year, is a significant problem. It is very small and can lodge itself within the lungs and enter the bloodstream. Diesel soot is concentrated in densely populated areas, and one in six people in the U.S. live near a diesel pollution hot spot.

While air pollution hotspots affect a variety of populations, some groups are more likely to be located in hotspots. Previous studies have shown disparities in exposure to pollution by race and/or income. Hazardous land uses (toxic storage and disposal facilities, manufacturing facilities, major roadways) tend to be located where property values and income levels are low. Low socioeconomic status can be a proxy for other kinds of social vulnerability, including race, a lack of ability to influence regulation and a lack of ability to move to neighborhoods with less environmental pollution. These communities bear a disproportionate burden of environmental pollution and are more likely to face health risks such as cancer or asthma.

Studies show that patterns in race and income disparities not only indicate a higher exposure to pollution but also higher risk of adverse health outcomes. Communities characterized by low socioeconomic status and racial minorities can be more vulnerable to cumulative adverse health impacts resulting from elevated exposure to pollutants than more privileged communities. Blacks and Latinos generally face more pollution than whites and Asians, and low-income communities bear a higher burden of risk than affluent ones. Racial discrepancies are particularly distinct in suburban areas of the US South and metropolitan areas of the US West. Residents in public housing, who are generally low-income and cannot move to healthier neighborhoods, are highly affected by nearby refineries and chemical plants.

Cities

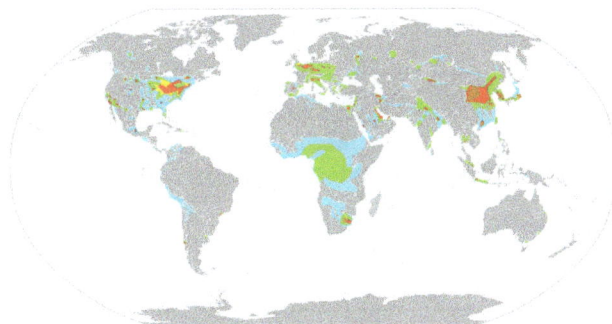

Nitrogen dioxide concentrations as measured from satellite 2002-2004

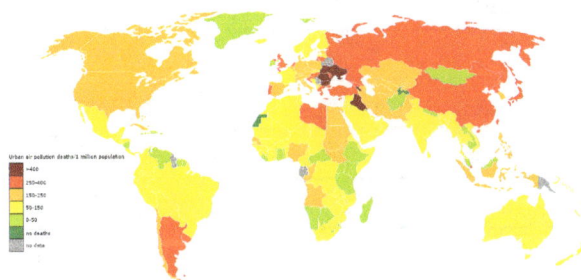

Deaths from air pollution in 2004

Air pollution is usually concentrated in densely populated metropolitan areas, especially in developing countries where environmental regulations are relatively lax or nonexistent. However, even populated areas in developed countries attain unhealthy levels of pollution, with Los Angeles and Rome being two examples. Between 2002 and 2011 the incidence of lung cancer in Beijing near doubled. While smoking remains the leading cause of lung cancer in China, the number of smokers is falling while lung cancer rates are rising. Another project focusing on the effects on pollution in vegetation has been researched by the local university in Sheffield, UK.

National-scale Air Toxics Assessments 1995-2005

The national-scale air toxics assessment(NATA) is an evaluation of air toxics by the U.S. EPA. EPA has furnished four assessments that characterize nationwide chronic cancer risk estimates and noncancer hazards from inhaling air toxics. The lates was from 2005, and made publicly available in early 2011.

"EPA developed the NATA as a state-of-the-science screening tool for State/Local/Tribal Agencies to prioritize pollutants, emission sources and locations of interest for further study, in order to gain a better understanding of the risks. NATA assessments do not incorporate refined information about emission sources, but rather, use general information about sources to develop estimates of risks which are more likely to overestimate impacts than underestimate them. NATA provides estimates of the risk of cancer and other serious health effects from breathing (inhaling) air toxics in order to inform both national and more localized efforts to identify and prioritize air toxics, emission source types and locations which are of greatest potential concern in terms of contributing to population risk. This in turn helps air pollution experts focus limited analytical resources on areas and or populations where the potential for health risks are highest. Assessments include estimates of cancer and non-cancer health effects based on chronic exposure from outdoor sources, including assessments of non-cancer health effects for Diesel Particulate Matter. Assessments provide a snapshot of the outdoor air quality and the risks to human health that would result if air toxic emissions levels remained unchanged."

Most polluted cities by PM	
Particulate matter, μg/m³ (2004)	City
168	Cairo, Egypt
150	Delhi, India
128	Kolkata, India (Calcutta)
125	Tianjin, China
123	Chongqing, China
109	Kanpur, India
109	Lucknow, India
104	Jakarta, Indonesia
101	Shenyang, China

Governing Urban Air Pollution

In Europe, Council Directive 96/62/EC on ambient air quality assessment and management provides a common strategy against which member states can "set objectives for ambient air quality in order to avoid, prevent or reduce harmful effects on human health and the environment . . . and improve air quality where it is unsatisfactory".

On 25 July 2008 in the case Dieter Janecek v Freistaat Bayern CURIA, the European Court of Justice ruled that under this directive citizens have the right to require national authorities to implement a short term action plan that aims to maintain or achieve compliance to air quality limit values.

This important case law appears to confirm the role of the EC as centralised regulator to European nation-states as regards air pollution control. It places a supranational legal obligation on the UK to protect its citizens from dangerous levels of air pollution, furthermore superseding national interests with those of the citizen.

In 2010, the European Commission (EC) threatened the UK with legal action against the successive breaching of PM10 limit values. The UK government has identified that if fines are imposed, they could cost the nation upwards of £300 million per year.

In March 2011, the Greater London Built-up Area remains the only UK region in breach of the EC's limit values, and has been given 3 months to implement an emergency action plan aimed at meeting the EU Air Quality Directive. The City of London has dangerous levels of PM10 concentrations, estimated to cause 3000 deaths per year within the city. As well as the threat of EU fines, in 2010 it was threatened with legal action for scrapping the western congestion charge zone, which is claimed to have led to an increase in air pollution levels.

In response to these charges, Boris Johnson, Mayor of London, has criticised the current need for European cities to communicate with Europe through their nation state's central government, arguing that in future "A great city like London" should be permitted to bypass its government and deal directly with the European Commission regarding its air quality action plan.

This can be interpreted as recognition that cities can transcend the traditional national government organisational hierarchy and develop solutions to air pollution using global governance networks, for example through transnational relations. Transnational relations include but are not exclusive to national governments and intergovernmental organisations, allowing sub-national actors including cities and regions to partake in air pollution control as independent actors.

Particularly promising at present are global city partnerships. These can be built into networks, for example the C40 Cities Climate Leadership Group, of which London is a member. The C40 is a public 'non-state' network of the world's leading cities that aims to curb their greenhouse emissions. The C40 has been identified as 'governance from the middle' and is an alternative to intergovernmental policy. It has the potential to improve urban air quality as participating cities "exchange information, learn from best practices and consequently mitigate carbon dioxide emissions independently from national government decisions". A criticism of the C40 network is that its exclusive nature limits influence to participating cities and risks drawing resources away from less powerful city and regional actors.

Atmospheric Dispersion

The basic technology for analyzing air pollution is through the use of a variety of mathematical models for predicting the transport of air pollutants in the lower atmosphere. The principal methodologies are:

1. Point source dispersion, used for industrial sources

2. Line source dispersion, used for airport and roadway air dispersion modeling

3. Area source dispersion, used for forest fires or duststorms

4. Photochemical models, used to analyze reactive pollutants that form smog

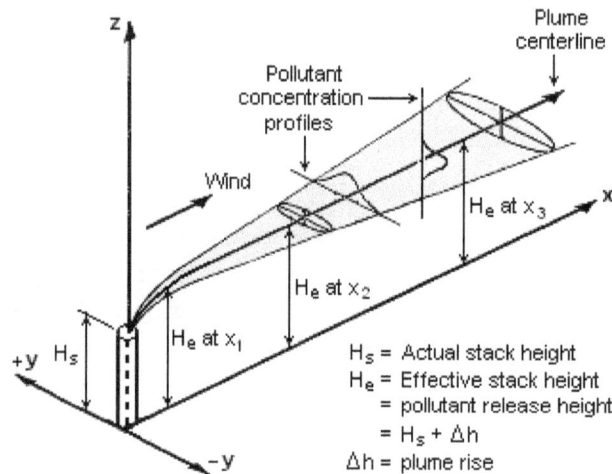

Visualization of a buoyant Gaussian air pollution dispersion plume as used in many atmospheric dispersion models.

The point source problem is the best understood, since it involves simpler mathematics and has been studied for a long period of time, dating back to about the year 1900. It uses a Gaussian dispersion model for continuous buoyant pollution plumes to predict the air pollution isopleths, with consideration given to wind velocity, stack height, emission rate and stability class (a measure of atmospheric turbulence). This model has been extensively validated and calibrated with experimental data for all sorts of atmospheric conditions.

The roadway air dispersion model was developed starting in the late 1950s and early 1960s in response to requirements of the National Environmental Policy Act and the U.S. Department of Transportation (then known as the Federal Highway Administration) to understand impacts of proposed new highways upon air quality, especially in urban areas. Several research groups were active in this model development, among which were: the Environmental Research and Technology (ERT) group in Lexington, Massachusetts, the ESL Inc. group in Sunnyvale, California and the California Air Resources Board group in Sacramento, California. The research of the ESL group received a boost with a contract award from the United States Environmental Protection Agency to validate a line source model using sulfur hexafluoride as a tracer gas. This program was successful in validating the line source model developed by ESL Inc. Some of the earliest uses of the model were in court cases involving highway air pollution; the Arlington, Virginia portion of Interstate 66 and the New Jersey Turnpike widening project through East Brunswick, New Jersey.

Area source models were developed in 1971 through 1974 by the ERT and ESL groups, but addressed a smaller fraction of total air pollution emissions, so that their use and need was not as widespread as the line source model, which enjoyed hundreds of different applications as early as the 1970s. Similarly photochemical models were developed primarily in the 1960s and 70s, but their use was more specialized and for regional needs, such as understanding smog formation in Los Angeles, California.

References

- National Research Council (2010). America's Climate Choices: Panel on Advancing the Science of Climate Change;. Washington, D.C.: The National Academies Press. ISBN 0-309-14588-0.

- Ross Gelbspan, Boiling Point: How Politicians, Big Oil and Coal, Journalists and Activists Are Fueling the Climate Crisis—And What We Can Do to Avert Disaster, Basic Books, (August 1, 2004) ISBN 0-465-02761-X

- Davis, Devra (2002). When Smoke Ran Like Water: Tales of Environmental Deception and the Battle Against Pollution. Basic Books. ISBN 0-465-01521-2.

- Turner, D.B. (1994). Workbook of atmospheric dispersion estimates: an introduction to dispersion modeling (2nd ed.). CRC Press. ISBN 1-56670-023-X.

- "Tropical drying trends in global warming models and observations". UCLA Atmospheric and Oceanic Sciences. Retrieved May 13, 2016.

- "Huge parts of world are drying up: Land 'evapotranspiration' taking unexpected turn". ScienceDaily. October 11, 2010. Retrieved May 13, 2016.

- St. Fleur, Nicholas (10 November 2015). "Atmospheric Greenhouse Gas Levels Hit Record, Report Says". New York Times. Retrieved 11 November 2015.

- Ritter, Karl (9 November 2015). "UK: In 1st, global temps average could be 1 degree C higher". AP News. Retrieved 11 November 2015.

- Cole, Steve; Gray, Ellen (14 December 2015). "New NASA Satellite Maps Show Human Fingerprint on Global Air Quality". NASA. Retrieved 14 December 2015.

- "Bucknell tent death: Hannah Thomas-Jones died from carbon monoxide poisoning". BBC News. 17 January 2013. Retrieved 22 September 2015.

- Gallagher, James (17 December 2015). "Cancer is not just 'bad luck' but down to environment, study suggests". BBC. Retrieved 17 December 2015.

- "New Study Demonstrates Indoor Building Environment Has Significant, Positive Impact on Cognitive Function". New York Times. 26 October 2015.

- European Court of Justice, CURIA (2008). "PRESS RELEASE No 58/08 Judgment of the Court of Justice in Case C-237/07" (PDF). Retrieved 24 January 2015.

- House of Commons Environmental Audit Committee (2010). "Environmental Audit Committee - Fifth Report Air Quality". Retrieved 24 January 2015.

Climate Change Mitigation

Climate change mitigation is a number of precautions taken to limit the extent or rate of long-term climate change. It usually involves reduction of greenhouse gas effects. Examples for climate change mitigation include discontinuation of fossil fuels by using low carbon energy sources and by growing forests to remove greater amounts of carbon dioxide. The topics discussed in the section are of great importance to broaden the knowledge on climate change mitigation.

Climate Change Mitigation

Climate change mitigation consists of actions to limit the magnitude or rate of long-term climate change. Climate change mitigation generally involves reductions in human (anthropogenic) emissions of greenhouse gases (GHGs). Mitigation may also be achieved by increasing the capacity of carbon sinks, e.g., through reforestation. Mitigation policies can substantially reduce the risks associated with human-induced global warming.

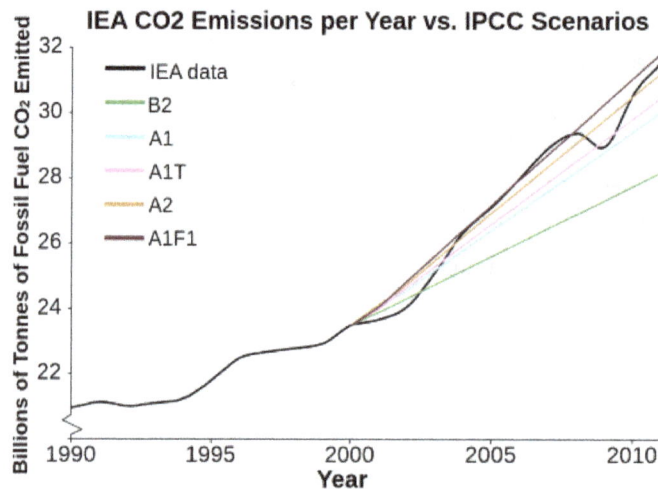

Fossil fuel related CO2 emissions compared to five of IPCC's emissions scenarios. The dips are related to global recessions. Data from IPCC SRES scenarios; Data spreadsheet included with International Energy Agency's "CO2 Emissions from Fuel Combustion 2010 – Highlights"; and Supplemental IEA data. Image source: Skeptical Science

"Mitigation is a public good; climate change is a case of the 'tragedy of the commons'". Effective climate change mitigation will not be achieved if each agent (individual, institution or country) acts independently in its own selfish interest, suggesting the need for collective action. Some adaptation actions, on the other hand, have characteristics of a private good as benefits of actions may accrue more directly to the individuals, regions, or countries that undertake them, at least in the short term. Nevertheless, financing such adaptive activities remains an issue, particularly for poor individuals and countries."

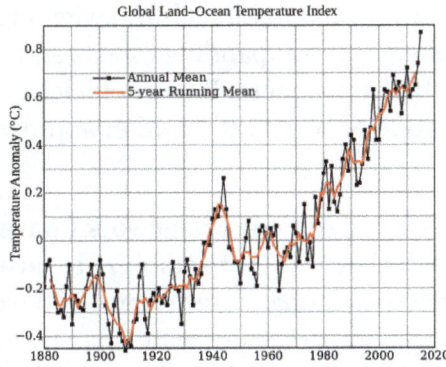

Global mean surface temperature change since 1880, relative to the 1951–1980 mean. The black line is the annual mean and the red line is the 5-year running mean. Source: NASA GISS. Global dimming, from sulfate aerosol air pollution, between 1950 and 1980 is believed to have mitigated global warming somewhat.

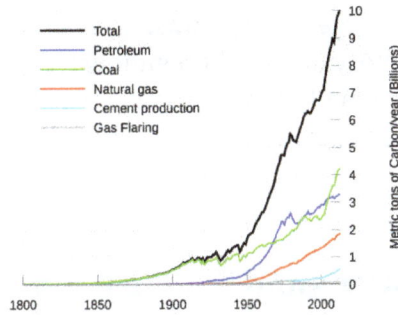

Global carbon dioxide emissions from human activities 1800–2007.

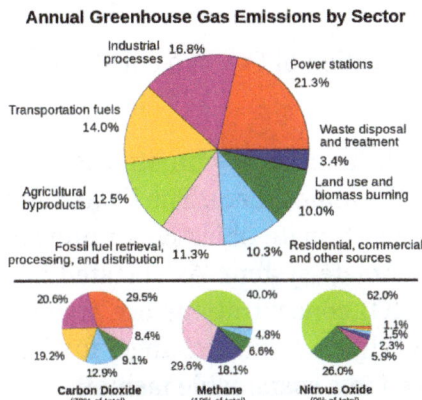

Greenhouse gas emissions by sector.

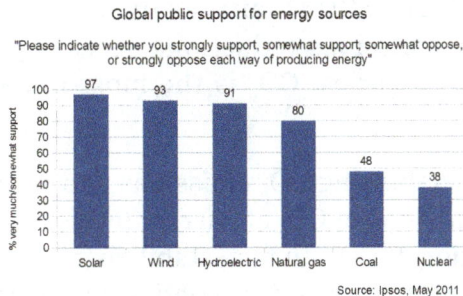

Global public support for energy sources, based on a survey by Ipsos (2011).

Examples of mitigation include phasing out fossil fuels by switching to low-carbon energy sources, such as renewable and nuclear energy, and expanding forests and other "sinks" to remove greater amounts of carbon dioxide from the atmosphere. Energy efficiency may also play a role, for example, through improving the insulation of buildings. Another approach to climate change mitigation is climate engineering.

Most countries are parties to the United Nations Framework Convention on Climate Change (UN-FCCC). The ultimate objective of the UNFCCC is to stabilize atmospheric concentrations of GHGs at a level that would prevent dangerous human interference of the climate system. Scientific analysis can provide information on the impacts of climate change, but deciding which impacts are dangerous requires value judgments.

In 2010, Parties to the UNFCCC agreed that future global warming should be limited to below 2.0 °C (3.6 °F) relative to the pre-industrial level. With the Paris Agreement of 2015 this was confirmed, but was revised with a new target laying down "parties will do the best" to achieve warming below 1.5 °C. The current trajectory of global greenhouse gas emissions does not appear to be consistent with limiting global warming to below 1.5 or 2 °C. Other mitigation policies have been proposed, some of which are more stringent or modest than the 2 °C limit.

Background

Greenhouse Gas Concentrations and Stabilization

Stabilizing CO_2 emissions at their present level would not stabilize its concentration in the atmosphere.

Stabilizing the atmospheric concentration of CO_2 at a constant level would require emissions to be effectively eliminated.

One of the issues often discussed in relation to climate change mitigation is the stabilization of greenhouse gas concentrations in the atmosphere. The United Nations Framework Convention on Climate Change (UNFCCC) has the ultimate objective of preventing "dangerous" anthropogenic (i.e., human) interference of the climate system. As is stated in Article 2 of the Convention, this requires that greenhouse gas (GHG) concentrations are stabilized in the atmosphere at a level where ecosystems can adapt naturally to climate change, food production is not threatened, and economic development can proceed in a sustainable fashion.

There are a number of anthropogenic greenhouse gases. These include carbon dioxide (chemical formula: CO_2), methane (CH4), nitrous oxide (N2O), and a group of gases referred to as halocarbons. The emissions reductions necessary to stabilize the atmospheric concentrations of these gases varies. CO_2 is the most important of the anthropogenic greenhouse gases.

There is a difference between stabilizing CO_2 emissions and stabilizing atmospheric concentrations of CO_2. Stabilizing emissions of CO_2 at current levels would not lead to a stabilization in the atmospheric concentration of CO_2. In fact, stabilizing emissions at current levels would result in the atmospheric concentration of CO_2 continuing to rise over the 21st century and beyond.

The reason for this is that human activities are adding CO_2 to the atmosphere far faster than natural processes can remove it. This is analogous to a flow of water into a bathtub. So long as the tap runs water (analogous to the emission of carbon dioxide) into the tub faster than water escapes through the plughole (the natural removal of carbon dioxide from the atmosphere), then the level of water in the tub (analogous to the concentration of carbon dioxide in the atmosphere) will continue to rise.

According to some studies, stabilizing atmospheric CO_2 concentrations would require anthropogenic CO_2 emissions to be reduced by 80% relative to the peak emissions level. An 80% reduction in emissions would stabilize CO_2 concentrations for around a century, but even greater reductions would be required beyond this. Other research has found that, after leaving room for emissions for food production for 9 billion people and to keep the global temperature rise below 2 °C, emissions from energy production and transport will have to peak almost immediately in the developed world and decline at ca. 10% per annum until zero emissions are reached around 2030. In developing countries energy and transport emissions would have to peak by 2025 and then decline similarly.

Stabilizing the atmospheric concentration of the other greenhouse gases humans emit also depends on how fast their emissions are added to the atmosphere, and how fast the GHGs are removed. Stabilization for these gases is described in the later section on non-CO_2 GHGs.

Projections

Projections of future greenhouse gas emissions are highly uncertain. In the absence of policies to mitigate climate change, GHG emissions could rise significantly over the 21st century.

Numerous assessments have considered how atmospheric GHG concentrations could be stabilized. The lower the desired stabilization level, the sooner global GHG emissions must peak and decline. GHG concentrations are unlikely to stabilize this century without major policy changes.

Projected carbon dioxide emissions and atmospheric concentrations over the 21st century for reference and mitigation scenarios.	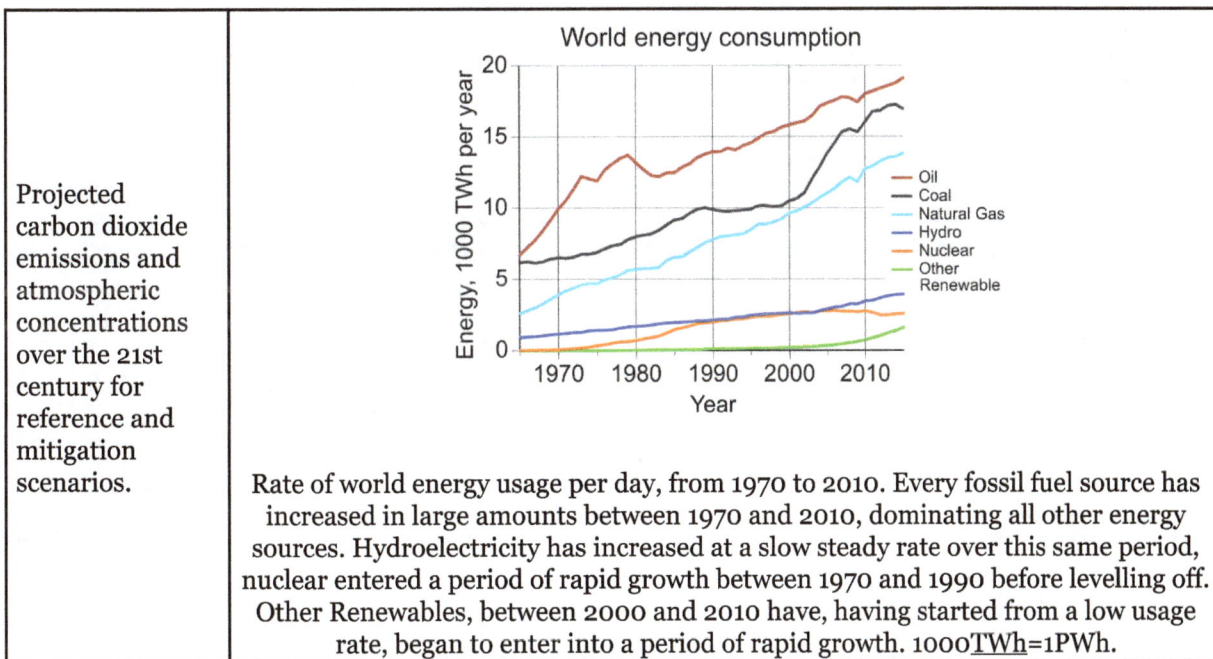
	Rate of world energy usage per day, from 1970 to 2010. Every fossil fuel source has increased in large amounts between 1970 and 2010, dominating all other energy sources. Hydroelectricity has increased at a slow steady rate over this same period, nuclear entered a period of rapid growth between 1970 and 1990 before levelling off. Other Renewables, between 2000 and 2010 have, having started from a low usage rate, began to enter into a period of rapid growth. 1000TWh=1PWh.

Energy Consumption by Power Source

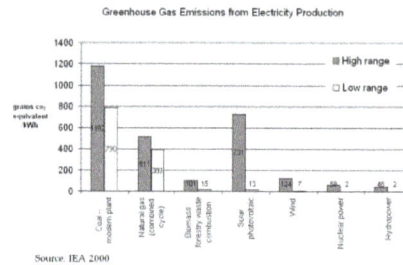

Source: IEA 2000

"Hydropower-Internalised Costs and Externalised Benefits"; Frans H. Koch; International Energy Agency (IEA)-Implementing Agreement for Hydropower Technologies and Programmes; 2000.

To create lasting climate change mitigation, the replacement of high carbon emission intensity power sources, such as conventional fossil fuels—oil, coal and natural gas—with low-carbon power sources is required. Fossil fuels supply humanity with the vast majority of our energy demands, and at a growing rate. In 2012 the IEA noted that coal accounted for half the increased energy use of the prior decade, growing faster than all renewable energy sources. Both hydroelectricity and nuclear power together provide the majority of the generated low-carbon power fraction of global total power consumption.

Fuel type	Average total global power consumption in TW		
	1980	2004	2006
Oil	4.38	5.58	5.74
Gas	1.80	3.45	3.61
Coal	2.34	3.87	4.27
Hydroelectric	0.60	0.93	1.00
Nuclear power	0.25	0.91	0.93
Geothermal, wind, solar energy, wood	0.02	0.13	0.16
Total	9.48	15.0	15.8
Source: The USA Energy Information Administration			

Change and use of energy, by source, in units of (PWh) in that year.				
	Fossil	Nuclear	All renewables	Total
1990	83.374	6.113	13.082	102.569
2000	94.493	7.857	15.337	117.687
2008	117.076	8.283	18.492	143.851
Change 2000–2008	22.583	0.426	3.155	26.164

Methods and Means

This graph shows the projected contribution of various energy sources to world primary electricity consumption (PEC). It is based on a climate change mitigation scenario, in which GHG emissions are substantially reduced over the 21st century. In the scenario, emission reductions are achieved using a portfolio of energy sources, as well as reductions in energy demand. *Also available in greyscale.*

Assessments often suggest that GHG emissions can be reduced using a portfolio of low-carbon technologies. At the core of most proposals is the reduction of greenhouse gas (GHG) emissions through reducing energy waste and switching to low-carbon power sources of energy. As the cost of reducing GHG emissions in the electricity sector appears to be lower than in other sectors, such as in the transportation sector, the electricity sector may deliver the largest proportional carbon reductions under an economically efficient climate policy.

"Economic tools can be useful in designing climate change mitigation policies." "While the limitations of economics and social welfare analysis, including cost–benefit analysis, are widely documented, economics nevertheless provides useful tools for assessing the pros and cons of taking, or not taking, action on climate change mitigation, as well as of adaptation measures, in achieving competing societal goals. Understanding these pros and cons can help in making policy decisions on climate change mitigation and can influence the actions taken by countries, institutions and individuals."

Other frequently discussed means include energy conservation, increasing fuel economy in automobiles (which includes the use of electric hybrids), charging plug-in hybrids and electric cars by low-carbon electricity, making individual-lifestyle changes (e.g., cycling instead of driving), and changing business practices. Many fossil fuel driven vehicles can be converted to use electricity, the U.S. has an estimated capacity of supporting 73% light duty vehicles (LDV). In terms of transportation, the net result would be a 27% total reduction in emissions of the greenhouse gases carbon dioxide, methane, and nitrous oxide, a 31% total reduction in nitrogen oxides, a slight reduction in nitrous oxide emissions, an increase in particulate matter emissions, the same sulfur dioxide emissions, and the near elimination of carbon monoxide and volatile organic compound emissions (a 98% decrease in carbon monoxide and a 93% decrease in volatile organic compounds). The emissions would be displaced away from street level, where they have "high human-health implications."

A range of energy technologies may contribute to climate change mitigation. These include nuclear power and renewable energy sources such as biomass, hydroelectricity, wind power, solar power, geothermal power, ocean energy, and; the use of carbon sinks, and carbon capture and storage. For example, Pacala and Socolow of Princeton have proposed a 15 part program to reduce CO_2 emissions by 1 billion metric tons per year – or 25 billion tons over the 50-year period using today's technologies as a type of Global warming game.

Another consideration is how future socio-economic development proceeds. Development choices (or "pathways") can lead differences in GHG emissions. Political and social attitudes may affect how easy or difficult it is to implement effective policies to reduce emissions.

Alternative Energy Sources

Renewable Energy

Renewable energy flows involve natural phenomena such as sunlight, wind, rain, tides, plant growth, and geothermal heat, as the International Energy Agency explains:

The 22,500 MW nameplate capacity Three Gorges Dam in the Peoples Republic of China, the largest hydroelectric power station in the world.

The Shepherds Flat Wind Farm is an 845 megawatt (MW) nameplate capacity, wind farm in the U.S. state of Oregon, each turbine is a nameplate 2 or 2.5 MW electricity generator.

The 150 MW Andasol solar power station is a commercial parabolic trough solar thermal power plant, located in Spain. The Andasol plant uses tanks of molten salt to store solar energy so that it can continue generating electricity for 7.5 hours after the sun has stopped shining.

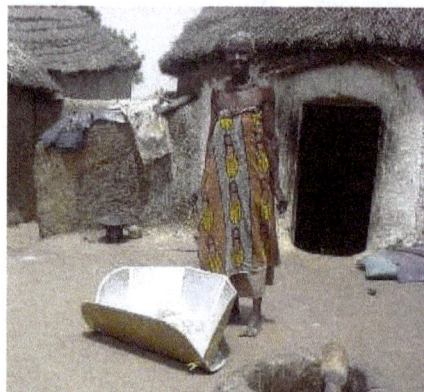

Solar cookers use sunlight as energy source for outdoor cooking.

Renewable energy is derived from natural processes that are replenished constantly. In its various forms, it derives directly from the sun, or from heat generated deep within the earth. Included in the definition is electricity and heat generated from solar, wind, ocean, hydropower, biomass, geothermal resources, and biofuels and hydrogen derived from renewable resources.

Climate change concerns and the need to reduce carbon emissions are driving increasing growth in the renewable energy industries. Low-carbon renewable energy replaces conventional fossil fuels in three main areas: power generation, hot water/ space heating, and transport fuels. In 2011, the share of renewables in electricity generation worldwide grew for the fourth year in a row to 20.2%. Based on REN21's 2014 report, renewables contributed 19% to supply global energy consumption. This energy consumption is divided as 9% coming from burning biomass, 4.2% as heat energy (non-biomass), 3.8% hydro electricity and 2% as electricity from wind, solar, geothermal, and biomass thermal power plants.

Renewable energy use has grown much faster than anyone anticipated. The Intergovernmental Panel on Climate Change (IPCC) has said that there are few fundamental technological limits to integrating a portfolio of renewable energy technologies to meet most of total global energy demand. At the national level, at least 30 nations around the world already have renewable energy contributing more than 20% of energy supply.

As of 2012, renewable energy accounts for almost half of new electricity capacity installed and costs are continuing to fall. Public policy and political leadership helps to "level the playing field" and drive the wider acceptance of renewable energy technologies. As of 2011, 118 countries have targets for their own renewable energy futures, and have enacted wide-ranging public policies to promote renewables. Leading renewable energy companies include BrightSource Energy, First Solar, Gamesa, GE Energy, Goldwind, Sinovel, Suntech, Trina Solar, Vestas and Yingli.

The incentive to use 100% renewable energy has been created by global warming and other ecological as well as economic concerns. Mark Z. Jacobson says producing all new energy with wind power, solar power, and hydropower by 2030 is feasible and existing energy supply arrangements could be replaced by 2050. Barriers to implementing the renewable energy plan are seen to be "primarily social and political, not technological or economic". Jacobson says that energy costs with a wind, solar, water system should be similar to today's energy costs. According to a 2011 projection by the (IEA)Intewwwrnational Energy Agency, solar power generators may produce most of the world's electricity within 50 years, dramatically reducing harmful greenhouse gas emissions. Critics of the "100% renewable energy" approach include Vaclav Smil and James E. Hansen. Smil and Hansen are concerned about the variable output of solar and wind power, NIMBYism, and a lack of infrastructure.

Economic analysts expect market gains for renewable energy (and efficient energy use) following the 2011 Japanese nuclear accidents. In his 2012 State of the Union address, President Barack Obama restated his commitment to renewable energy and mentioned the long-standing Interior Department commitment to permit 10,000 MW of renewable energy projects on public land in 2012. Globally, there are an estimated 3 million direct jobs in renewable energy industries, with about half of them in the biofuels industry.

Some countries, with favorable geography, geology and weather well suited to an economical exploitation of renewable energy sources, already get most of their electricity from renewables, including from geothermal energy in Iceland (100 percent), and Hydroelectric power in Brazil (85 percent), Austria (62 percent), New Zealand (65 percent), and Sweden (54 percent). Renewable power generators are spread across many countries, with wind power providing a significant share of electricity in some regional areas: for example, 14 percent in the U.S. state of Iowa, 40 percent in the northern German state of Schleswig-Holstein, and 20 percent in Denmark. Solar water heating makes an important and growing contribution in many countries, most notably in China, which now has 70 percent of the global total (180 GWth). Worldwide, total installed solar water heating systems meet a portion of the water heating needs of over 70 million households. The use of biomass for heating continues to grow as well. In Sweden, national use of biomass energy has surpassed that of oil. Direct geothermal heating is also growing rapidly. Renewable biofuels for transportation, such as ethanol fuel and biodiesel, have contributed to a significant decline in oil consumption in the United States since 2006. The 93 billion liters of biofuels produced worldwide in 2009 displaced the equivalent of an estimated 68 billion liters of gasoline, equal to about 5 percent of world gasoline production.

Some of the world's largest solar power stations: Ivanpah (CSP) and Topaz (PV)

Nuclear Power

Blue Cherenkov light being produced near the core of the Fission powered Advanced Test Reactor

Since about 2001 the term "nuclear renaissance" has been used to refer to a possible nuclear power industry revival, driven by rising fossil fuel prices and new concerns about meeting greenhouse gas emission limits. However, in March 2011 the Fukushima nuclear disaster in Japan and associated shutdowns at other nuclear facilities raised questions among some commentators over the future of nuclear power. Platts has reported that "the crisis at Japan's Fukushima nuclear plants has prompted leading energy-consuming countries to review the safety of their existing reactors and cast doubt on the speed and scale of planned expansions around the world".

The World Nuclear Association has reported that nuclear electricity generation in 2012 was at its lowest level since 1999. Several previous international studies and assessments, suggested that as part of the portfolio of other low-carbon energy technologies, nuclear power will continue to play a role in reducing greenhouse gas emissions. Historically, nuclear power usage is estimated to have prevented the atmospheric emission of 64 gigatonnes of CO2-equivalent as of 2013. Public concerns about nuclear power include the fate of spent nuclear fuel, nuclear accidents, security risks, nuclear proliferation, and a concern that nuclear power plants are very expensive. Of these concerns, nuclear accidents and disposal of long-lived radioactive fuel/"waste" have probably had the greatest public impact worldwide. Although generally unaware of it, both of these glaring public concerns are greatly diminished by present passive safety designs, the experimentally proven, "melt-down proof" EBR-II, future molten salt reactors, and the use of conventional and more advanced fuel/"waste" pyroprocessing, with the latter recycling or reprocessing not presently being commonplace as it is often considered to be cheaper to use a once-through nuclear fuel cycle in many countries, depending on the varying levels of intrinsic value given by a society in reducing the long-lived waste in their country, with France doing a considerable amount of reprocessing when compared to the US.

Nuclear power, with a 10.6% share of world electricity production as of 2013, is second only to hydroelectricity as the largest source of low-carbon power. Over 400 reactors generate electricity in 31 countries.

A Yale University review published in the Journal of Industrial Ecology analyzing CO_2 life cycle assessment(LCA) emissions from nuclear power(Light water reactors) determined that: "The collective LCA literature indicates that life cycle GHG emissions from nuclear power are only a fraction of traditional fossil sources and comparable to renewable technologies." While some have raised uncertainty surrounding the future GHG emissions of nuclear power as a result of an extreme potential decline in uranium ore grade without a corresponding increase in the efficiency of enrichment methods. In a scenario analysis of future global nuclear development, as it could be effected by a decreasing global uranium market of average ore grade, the analysis determined that depending on conditions, median life cycle nuclear power GHG emissions could be between 9 and 110 g CO_2-eq/kWh by 2050, with the latter high figure being derived from a "worst-case scenario" that is not "considered very robust" by the authors of the paper, as the "ore grade" in the scenario is lower than the uranium concentration in many lignite coal ashes.

Although this future analyses primarily deals with extrapolations for present Generation II reactor technology, the same paper also summarizes the literature on "FBRs"/Fast Breeder Reactors, of which two are in operation as of 2014 with the newest being the BN-800, for these reactors it states that the "median life cycle GHG emissions ... [are] similar to or lower than [present light water reactors] LWRs and purports to consume little or no uranium ore.

In their 2014 report, the IPCC comparison of energy sources global warming potential per unit of electricity generated, which notably included albedo effects, mirror the median emission value derived from the Warner and Heath Yale meta-analysis for the more common non-breeding Light water reactors, a CO2-equivalent value of 12 g CO2-eq/kWh, which is the lowest global warming forcing of all baseload power sources, with comparable low carbon power baseload sources, such as hydropower and biomass, producing substantially more global warming forcing 24 and 230 g CO2-eq/kWh respectively.

In 2014, Brookings Institution published *The Net Benefits of Low and No-Carbon Electricity Technologies* which states, after performing an energy and emissions cost analysis, that "The net benefits of new nuclear, hydro, and natural gas combined cycle plants far outweigh the net benefits of new wind or solar plants", with the most cost effective low carbon power technology being determined to be nuclear power.

During his presidential campaign, Barack Obama stated, "Nuclear power represents more than 70% of our noncarbon generated electricity. It is unlikely that we can meet our aggressive climate goals if we eliminate nuclear power as an option."

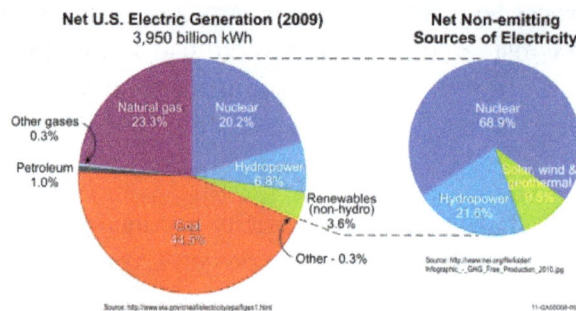

This graph illustrates nuclear power is the United States's largest contributor of non-greenhouse-gas-emitting electric power generation, comprising nearly three-quarters of the non-emitting sources.

Analysis in 2015 by Professor and Chair of Environmental Sustainability Barry W. Brook and his colleagues on the topic of replacing fossil fuels entirely, from the electric grid of the world, has determined that at the historically modest and proven-rate at which nuclear energy was added to and replaced fossil fuels in France and Sweden during each nation's building programs in the 1980s, within 10 years nuclear energy could displace or remove fossil fuels from the electric grid completely, "allow[ing] the world to meet the most stringent greenhouse-gas mitigation targets.". In a similar analysis, Brook had earlier determined that 50% of all global energy, that is not solely electricity, but transportation synfuels etc. could be generated within approximately 30 years, if the global nuclear fission build rate was identical to each of these nation's already proven decadal rates(in units of installed nameplate capacity, GW per year, per unit of global GDP(GW/year/$).

This is in contrast to the completely conceptual paper-studies for a *100% renewable energy* world, which would require an orders of magnitude more costly global investment per year, an investment rate that has no historical precedent, having never been attempted due to its prohibitive cost, and with far greater land area that would be required to be devoted to the wind, wave and solar projects, along with the inherent assumption that humanity will use less, and not more, energy in the future. As Brook notes the "principal limitations on nuclear fission are not technical, economic or fuel-related, but are instead linked to complex issues of societal acceptance, fiscal and political inertia, and inadequate critical evaluation of the real-world constraints facing [the other] low-carbon alternatives."

Nuclear power may be uncompetitive compared with fossil fuel energy sources in countries without a carbon tax program, and in comparison to a fossil fuel plant of the same power output, nuclear power plants take a longer amount of time to construct.

Two new, first of their kind, EPR reactors under construction in Finland and France have been delayed and are running over-budget. However learning from experience, two further EPR reactors under construction in China are on, and ahead, of schedule respectively. As of 2013, according to the IAEA and the European Nuclear Society, worldwide there were 68 civil nuclear power reactors under construction in 15 countries. China has 29 of these nuclear power reactors under construction, as of 2013, with plans to build many more, while in the US the licenses of almost half its reactors have been extended to 60 years, and plans to build another dozen are under serious consideration. There are also a considerable number of new reactors being built in South Korea, India, and Russia. At least 100 older and smaller reactors will "most probably be closed over the next 10–15 years". This is probable only if one does not factor in the ongoing Light Water Reactor Sustainability Program, created to permit the extension of the life span of the USA's 104 nuclear reactors to 60 years. The licenses of almost half of the USA's reactors have been extended to 60 years as of 2008. Two new "passive safety" AP1000 reactors are, as of 2013, being constructed at Vogtle Electric Generating Plant.

Public opinion about nuclear power varies widely between countries. A poll by Gallup International (2011) assessed public opinion in 47 countries. The poll was conducted following a tsunami and earthquake which caused an accident at the Fukushima nuclear power plant in Japan. 49% stated that they held favourable views about nuclear energy, while 43% held an unfavourable view. Another global survey by Ipsos (2011) assessed public opinion in 24 countries. Respondents to this survey showed a clear preference for renewable energy sources over coal and nuclear energy (refer to graph opposite). Ipsos (2012) found that solar and wind were viewed by the public as being more environmentally friendly and more viable long-term energy sources relative to nuclear power and natural gas. However, solar and wind were viewed as being less reliable relative to nuclear power and natural gas. In 2012 a poll done in the UK found that 63% of those surveyed support nuclear power, and with opposition to nuclear power at 11%. In Germany, strong anti-nuclear sentiment led to eight of the seventeen operating reactors being permanently shut down following the March 2011 Fukushima nuclear disaster.

Nuclear fusion research, in the form of the International Thermonuclear Experimental Reactor is underway. Fusion powered electricity generation was initially believed to be readily achievable, as fission power had been. However, the extreme requirements for continuous reactions and plasma containment led to projections being extended by several decades. In 2010, more than 60 years after the first attempts, commercial power production was still believed to be unlikely before 2050. Although rather than an either, or, issue economical fusion-fission hybrid reactors could be built before any attempt at this more demanding commercial "pure-fusion reactor"/DEMO reactor takes place.

Coal to Gas Fuel Switching

Most mitigation proposals imply—rather than directly state—an eventual reduction in global fossil fuel production. Also proposed are direct quotas on global fossil fuel production.

Natural gas emits far fewer greenhouse gases (i.e. CO_2 and methane—CH_4) than coal when burned at power plants, but evidence has been emerging that this benefit could be completely negated by methane leakage at gas drilling fields and other points in the supply chain.

A study performed by the Environmental Protection Agency (EPA) and the Gas Research Institute (GRI) in 1997 sought to discover whether the reduction in carbon dioxide emissions from increased natural gas (predominantly methane) use would be offset by a possible increased level of methane emissions from sources such as leaks and emissions. The study concluded that the reduction in emissions from increased natural gas use outweighs the detrimental effects of increased methane emissions. More recent peer-reviewed studies have challenged the findings of this study, with researchers from the National Oceanic and Atmospheric Administration (NOAA) reconfirming findings of high rates of methane (CH4) leakage from natural gas fields.

A 2011 study by noted climate research scientist, Tom Wigley, found that while carbon dioxide (CO_2) emissions from fossil fuel combustion may be reduced by using natural gas rather than coal to produce energy, it also found that additional methane (CH4) from leakage adds to the radiative forcing of the climate system, offsetting the reduction in CO_2 forcing that accompanies the transition from coal to gas. The study looked at methane leakage from coal mining; changes in radiative forcing due to changes in the emissions of sulfur dioxide and carbonaceous aerosols; and differences in the efficiency of electricity production between coal- and gas-fired power generation. On balance, these factors more than offset the reduction in warming due to reduced CO_2 emissions. When gas replaces coal there is additional warming out to 2,050 with an assumed leakage rate of 0%, and out to 2,140 if the leakage rate is as high as 10%. The overall effects on global-mean temperature over the 21st century, however, are small. Petron et al. (2013) and Alvarez et al. (2012) note that estimated that leakage from gas infrastructure is likely to be underestimated. These studies indicate that the exploitation of natural gas as a "cleaner" fuel is questionable. A 2014 meta-study of 20 years of natural gas technical literature shows that methane emissions are consistently underestimated but on a 100-year scale, the climate benefits of coal to gas fuel switching are likely larger than the negative effects of natural gas leakage.

Heat Pump

Outside unit of an air-source heat pump.

A heat pump is a device that provides heat energy from a source of heat to a destination called a "heat sink". Heat pumps are designed to move thermal energy opposite to the direction of spontaneous heat flow by absorbing heat from a cold space and releasing it to a warmer one. A heat pump uses some amount of external power to accomplish the work of transferring energy from the heat source to the heat sink.

While air conditioners and freezers are familiar examples of heat pumps, the term "heat pump" is more general and applies to many HVAC (heating, ventilating, and air conditioning) devices used for space heating or space cooling. When a heat pump is used for heating, it employs the same basic refrigeration-type cycle used by an air conditioner or a refrigerator, but in the opposite direction—releasing heat into the conditioned space rather than the surrounding environment. In this use, heat pumps generally draw heat from the cooler external air or from the ground. In heating mode, heat pumps are three to four times more efficient in their use of electric power than simple electrical resistance heaters.

It has been concluded that heat pumps are the single technology that could reduce the greenhouse gas emissions of households better than every other technology that is available on the market. With a market share of 30% and (potentially) clean electricity, heat pumps could reduce global CO_2 emissions by 8% annually. Using ground source heat pumps could reduce around 60% of the primary energy demand and 90% of CO_2 emissions in Europe in 2050 and make handling high shares of renewable energy easier. Using surplus renewable energy in heat pumps is regarded as the most effective household means to reduce global warming and fossil fuel depletion.

With significant amounts of fossil fuel used in electricity production, demands on the electrical grid also generate greenhouse gases. Without a high share of low-carbon electricity, a domestic heat pump will produce more carbon emissions than using natural gas.

Fossil fuel Phase-out: Carbon Neutral and Negative Fuels

3,500-4,000 environmental activists blocking a coal mine in Germany to limit climate change (Ende Gelände 2016).

Fossil fuel may be phased-out with carbon neutral and carbon negative pipeline and transportation fuels created with power to gas and gas to liquids technologies. Carbon dioxide from fossil fuel flue gas can be used to produce plastic lumber allowing carbon negative reforestation.

Demand Side Management

Energy Efficiency and Conservation

Efficient energy use, sometimes simply called "energy efficiency", is the goal of efforts to reduce the amount of energy required to provide products and services. For example, insulating a home allows a building to use less heating and cooling energy to achieve and maintain a comfortable temperature. Installing fluorescent lights or natural skylights reduces the amount of energy required to attain the same level of illumination compared to using traditional incandescent light bulbs. Compact fluorescent lights use two-thirds less energy and may last 6 to 10 times longer than incandescent lights.

A spiral-type integrated compact fluorescent lamp, use has grown among North American consumers since its introduction in the mid-1990s.

Energy efficiency has proved to be a cost-effective strategy for building economies without necessarily growing energy consumption. For example, the state of California began implementing energy-efficiency measures in the mid-1970s, including building code and appliance standards with strict efficiency requirements. During the following years, California's energy consumption has remained approximately flat on a per capita basis while national U.S. consumption doubled. As part of its strategy, California implemented a "loading order" for new energy resources that puts energy efficiency first, renewable electricity supplies second, and new fossil-fired power plants last.

Energy conservation is broader than energy efficiency in that it encompasses using less energy to achieve a lesser energy service, for example through behavioural change, as well as encompassing energy efficiency. Examples of conservation without efficiency improvements would be heating a room less in winter, driving less, or working in a less brightly lit room. As with other definitions, the boundary between efficient energy use and energy conservation can be fuzzy, but both are important in environmental and economic terms. This is especially the case when actions are directed at the saving of fossil fuels.

Reducing energy use is seen as a key solution to the problem of reducing greenhouse gas emissions. According to the International Energy Agency, improved energy efficiency in buildings, industrial processes and transportation could reduce the world's energy needs in 2050 by one third, and help control global emissions of greenhouse gases.

Demand Side Switching Sources

Fuel switching on the demand side refers to changing the type of fuel used to satisfy a need for an energy service. To meet deep decarbonization goals, like the 80% reduction by 2050 goal being discussed in California and the European Union, many primary energy changes are needed. Energy efficiency alone may not be sufficient to meet these goals, switching fuels used on the demand side will help lower carbon emissions. Progressively coal, oil and eventually natural gas for space and water heating in buildings will need to be reduced. For an equivalent amount of heat, burning natural gas produces about 45 per cent less carbon dioxide than burning coal. There are various ways in which this could happen, and different strategies will likely make sense in different locations. While the system efficiency of a gas furnace may be higher than the combination of natural gas power plant and electric heat, the combination of the same natural gas power plant and an electric heat pump has lower emissions per unit of heat delivered in all but the coldest climates. This is possible because of the very efficient coefficient of performance of heat pumps.

At the beginning of this century 70% of all electricity was generated by fossil fuels, and as carbon free sources eventually make up half of the generation mix, replacing gas or oil furnaces and water heaters with electric ones will have a climate benefit. In areas like Norway, Brazil and Quebec that have abundant hydroelectricity, electric heat and hot water is common.

The economics of switching the demand side from fossil fuels to electricity for heating, will depend on the price of fuels vs electricity and the relative prices of the equipment. The EIA Annual Energy Outlook 2014 suggests that domestic gas prices will rise faster than electricity prices which will encourage electrification in the coming decades. Electrifying heating loads may also provide a flexible resource that can participate in demand response. Since thermostatically controlled loads have inherent energy storage, electrification of heating could provide a valuable resource to integrate variable renewable resources into the grid.

Alternatives to electrification, include decarbonizing pipeline gas through power to gas, biogas, or other carbon neutral fuels. A 2015 study by Energy+Environmental Economics shows that a hybrid approach of decarbonizing pipeline gas, electrification, and energy efficiency can meet carbon reduction goals at a similar cost as only electrification and energy efficiency in Southern California.

Demand Side Grid Management

Expanding intermittent electrical sources such as wind power, creates a growing problem balancing grid fluctuations. Some of the plans include building pumped storage or continental super grids costing billions of dollars. However instead of building for more power,there are a variety of ways to affect the size and timing of electricity demand on the consumer side. Designing for reduced demands on a smaller power grid is more efficient and economic than having extra generation and transmission for intermittentcy, power failures and peak demands. Having these abilities is one of the chief aims of a smart grid.

Time of use metering is a common way to motivate electricity users to reduce their peak load consumption. For instance, running dishwashers and laundry at night after the peak has passed, reduces electricity costs.

Dynamic demand plans have devices passively shut off when stress is sensed on the electrical grid. This method may work very well with thermostats, when power on the grid sags a small amount, a low power temperature setting is automatically selected reducing the load on the grid. For instance millions of refrigerators reduce their consumption when clouds pass over solar installations. Consumers would need to have a smart meter in order for the utility to calculate credits.

Demand response devices could receive all sorts of messages from the grid. The message could be a request to use a low power mode similar to dynamic demand, to shut off entirely during a sudden failure on the grid, or notifications about the current and expected prices for power. This would allow electric cars to recharge at the least expensive rates independent of the time of day. The vehicle-to-grid suggestion would use a car's battery or fuel cell to supply the grid temporarily.

Lifestyle and Behavior

The IPCC Fifth Assessment Report emphasises that behaviour, lifestyle and cultural change have a high mitigation potential in some sectors, particularly when complementing technological and structural change. In general, higher consumption lifestyles have a greater environmental impact. Overall, food accounts for the largest share of consumption-based GHG emissions with nearly 20% of the global carbon footprint, followed by housing, mobility, services, manufactured products, and construction. Food and services are more significant in poor countries, while mobility and manufactured goods are more significant in rich countries.

Dietary Change

A 2014 study into the real-life diets of British people estimates their greenhouse gas contributions (CO_2eq) to be: 7.19 kg/day for high meat-eaters through to 3.81 kg/day for vegetarians and 2.89 kg/day for vegans. The widespread adoption of a vegetarian diet could cut food-related greenhouse gas emissions by 63% by 2050. China introduced new dietary guidelines in 2016 which aim to cut meat consumption by 50% and thereby reduce greenhouse gas emissions by 1 billion tonnes by 2030.

Sinks and Negative Emissions

A carbon sink is a natural or artificial reservoir that accumulates and stores some carbon-containing chemical compound for an indefinite period, such as a growing forest. A negative carbon dioxide emission on the other hand is a permanent removal of carbon dioxide out of the atmosphere, such as directly capturing carbon dioxide in the atmosphere and storing it in geologic formations underground.

The Antarctic Climate and Ecosystems Cooperative Research Centre (ACE-CRC) notes that one third of humankind's annual emissions of CO_2 are absorbed by the oceans. However, this also leads to ocean acidification, with potentially significant impacts on marine life. Acidification lowers the level of carbonate ions available for calcifying organisms to form their shells. These organisms include plankton species that contribute to the foundation of the Southern Ocean food web. However acidification may impact on a broad range of other physiological and ecological processes, such as fish respiration, larval development and changes in the solubility of both nutrients and toxins.

Reforestation and Afforestation

Almost 20 percent (8 $GtCO_2$/year) of total greenhouse-gas emissions were from deforestation in 2007. It is estimated that avoided deforestation reduces CO_2 emissions at a rate of 1 ton of CO_2 per \$1-\$5 in opportunity costs from lost agriculture. Reforestation and afforestation, where there was previously no forest, could save at least another 1$GtCO_2$/year, at an estimated cost of \$5-\$15/tCO_2.

Transferring land rights to indigenous inhabitants is argued to efficiently conserve forests. Regrowth of forests on abandoned farmland restores more forest than that lost to deforestation.

Transferring rights over land from public domain to its indigenous inhabitants is argued to be a cost effective strategy to conserve forests. This includes the protection of such rights entitled in existing laws, such as India's Forest Rights Act. The transferring of such rights in China, perhaps the largest land reform in modern times, has been argued to have increased forest cover. In Brazil, forested areas given tenure to indigenous groups have even lower rates of clearing than national parks.

With increased intensive agriculture and urbanization, there is an increase in the amount of abandoned farmland. By some estimates, for every half a hectare of original old-growth forest cut down, more than 20 hectares of new secondary forests are growing, even though they do not have the same biodiversity as the original forests and original forests store 60% more carbon than these new secondary forests. According to a study in Science, promoting regrowth on abandoned farmland could offset years of carbon emissions.

Avoided Desertification

Managed grazing methods are argued to be able to restore grasslands, thereby significantly decreasing atmospheric CO_2 levels.

Restoring grasslands store CO_2 from the air into plant material. Grazing livestock, usually not left to wander, would eat the grass and would minimize any grass growth. However, grass left alone would eventually grow to cover its own growing buds, preventing them from photosynthesizing and the dying plant would stay in place. A method proposed to restore grasslands uses fences with many small paddocks and moving herds from one paddock to another after a day a two in order to mimick natural grazers and allowing the grass to grow optimally. Additionally, when part of leaf matter is consumed by a herding animal, a corresponding amount of root matter is sloughed off too as it would not be able to sustain the previous amount of root matter and while most of the lost root matter would rot and enter the atmosphere, part of the carbon is sequestered into the soil. It is estimated that increasing the carbon content of the soils in the world's 3.5 billion hectares of agricultural grassland by 1% would offset nearly 12 years of CO_2 emissions. Allan Savory, as part of holistic management, claims that while large herds are often blamed for desertification, prehistoric lands supported large or larger herds and areas where herds were removed in the United States are still desertifying.

Carbon Capture and Storage

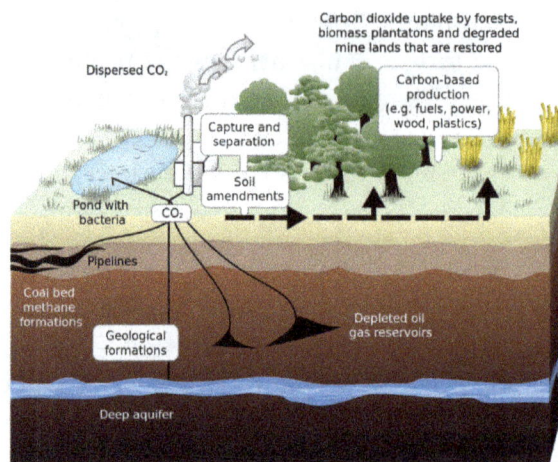

Schematic showing both terrestrial and geological sequestration of carbon dioxide emissions from a coal-fired plant.

Carbon capture and storage (CCS) is a method to mitigate climate change by capturing carbon dioxide (CO_2) from large point sources such as power plants and subsequently storing it away safely instead of releasing it into the atmosphere. The Intergovernmental Panel on Climate Change says CCS could contribute between 10% and 55% of the cumulative worldwide carbon-mitigation effort over the next 90 years. The International Energy Agency says CCS is "the most important single new technology for CO_2 savings" in power generation and industry. Though it requires up to 40% more energy to run a CCS coal power plant than a regular coal plant, CCS could potentially capture about 90% of all the carbon emitted by the plant. Norway, which first began storing CO_2, has cut its emissions by almost a million tons a year, or about 3% of the country's 1990 levels. As of late 2011, the total CO_2 storage capacity of all 14 projects in operation or under construction is over 33 million tonnes a year. This is broadly equivalent to preventing the emissions from more than six million cars from entering the atmosphere each year.

Negative Carbon Dioxide Emissions

Creating negative carbon dioxide emissions literally removes carbon from the atmosphere. Examples are direct air capture, biochar, bio-energy with carbon capture and storage and enhanced weathering technologies. These processes are sometimes considered as variations of sinks or mitigation, and sometimes as geoengineering.

In combination with other mitigation measures, sinks in combination with negative carbon emissions are considered crucial for meeting the 350 ppm target, and even the less conservative 450 ppm target.

Geoengineering

Geoengineering is seen by some as an alternative to mitigation and adaptation, but by others as an entirely separate response to climate change. In a literature assessment, Barker *et al.* (2007) described geoengineering as a type of mitigation policy. IPCC (2007) concluded that geoengineering options, such as ocean fertilization to remove CO_2 from the atmosphere, remained largely unproven. It was judged that reliable cost estimates for geoengineering had not yet been published.

Chapter 28 of the National Academy of Sciences report *Policy Implications of Greenhouse Warming: Mitigation, Adaptation, and the Science Base* (1992) defined geoengineering as "options that would involve large-scale engineering of our environment in order to combat or counteract the effects of changes in atmospheric chemistry." They evaluated a range of options to try to give preliminary answers to two questions: can these options work and could they be carried out with a reasonable cost. They also sought to encourage discussion of a third question — what adverse side effects might there be. The following types of option were examined: reforestation, increasing ocean absorption of carbon dioxide (carbon sequestration) and screening out some sunlight. NAS also argued "Engineered countermeasures need to be evaluated but should not be implemented without broad understanding of the direct effects and the potential side effects, the ethical issues, and the risks.". In July 2011 a report by the United States Government Accountability Office on geoengineering found that "[c]limate engineering technologies do not now offer a viable response to global climate change."

Carbon Dioxide Removal

Carbon dioxide removal has been proposed as a method of reducing the amount of radiative forcing. A variety of means of artificially capturing and storing carbon, as well as of enhancing natural sequestration processes, are being explored. The main natural process is photosynthesis by plants and single-celled organisms. Artificial processes vary, and concerns have been expressed about the long-term effects of some of these processes.

It is notable that the availability of cheap energy and appropriate sites for geological storage of carbon may make carbon dioxide air capture viable commercially. It is, however, generally expected that carbon dioxide air capture may be uneconomic when compared to carbon capture and storage from major sources — in particular, fossil fuel powered power stations, refineries, etc. In such cases, costs of energy produced will grow significantly. However, captured CO_2 can be used to force more crude oil out of oil fields, as Statoil and Shell have made plans to do. CO_2 can also be used in commercial greenhouses, giving an opportunity to kick-start the technology. Some attempts have been made to use algae to capture smokestack emissions, notably the GreenFuel Technologies Corporation, who have now shut down operations.

Solar Radiation Management

The main purpose of solar radiation management seek to reflect sunlight and thus reduce global warming. The ability of stratospheric sulfate aerosols to create a global dimming effect has made them a possible candidate for use in climate engineering projects.

Non-Co$_2$ Greenhouse Gases

CO_2 is not the only GHG relevant to mitigation, and governments have acted to regulate the emissions of other GHGs emitted by human activities (anthropogenic GHGs). The emissions caps agreed to by most developed countries under the Kyoto Protocol regulate the emissions of almost all the anthropogenic GHGs. These gases are CO_2, methane (CH_4), nitrous oxide (N_2O), the hydrofluorocarbons (HFC), perfluorocarbons (PFC), and sulfur hexafluoride (SF_6).

Stabilizing the atmospheric concentrations of the different anthropogenic GHGs requires an understanding of their different physical properties. Stabilization depends both on how quickly GHGs are added to the atmosphere and how fast they are removed. The rate of removal is measured by the atmospheric lifetime of the GHG in question. Here, the lifetime is defined as the time required for a given perturbation of the GHG in the atmosphere to be reduced to 37% of its initial amount. Methane has a relatively short atmospheric lifetime of about 12 years, while N_2O's lifetime is about 110 years. For methane, a reduction of about 30% below current emission levels would lead to a stabilization in its atmospheric concentration, while for N_2O, an emissions reduction of more than 50% would be required.

Methane is a significantly more potent greenhouse gas than carbon dioxide in the amount of heat it can trap, especially in the short term. Burning one molecule of methane generates one molecule of carbon dioxide, indicating there may be no net benefit in using gas as a fuel source. Reducing the amount of waste methane produced in the first place and moving away from use of gas as a fuel source will have a greater beneficial impact, as might other approaches to productive use of otherwise-wasted methane. In terms of prevention, vaccines are being developed in Australia to reduce the significant global warming contributions from methane released by livestock via flatulence and eructation.

Another physical property of the anthropogenic GHGs relevant to mitigation is the different abilities of the gases to trap heat (in the form of infrared radiation). Some gases are more effective at trapping heat than others, e.g., SF_6 is 22,200 times more effective a GHG than CO_2 on a per-kilogram basis. A measure for this physical property is the global warming potential (GWP), and is used in the Kyoto Protocol.

Although not designed for this purpose, the Montreal Protocol has probably benefited climate change mitigation efforts. The Montreal Protocol is an international treaty that has successfully reduced emissions of ozone-depleting substances (for example, CFCs), which are also greenhouse gases.

By Sector

Transport

Modern energy-efficient technologies, such as plug-in hybrid electric vehicles, and development of new technologies, such as carbon-neutral synthetic gasoline & Jet fuel, may reduce the consumption of petroleum, land use changes and emissions of carbon dioxide. A shift from air transport and truck transport to electric rail transport would reduce emissions significantly. For electric vehicles, the reduction of carbon emissions will improve further if the way the required electricity is generated is low-carbon power in origin.

The Tesla Roadster emits no tailpipe emissions, uses lithium ion batteries to achieve 220 mi (350 km) per charge, while also capable of going 0–60 in under 4 seconds.

Bicycles have almost no carbon footprint compared to cars, and canal transport may represent a positive option for certain types of freight in the 21st century

Urban Planning

Effective urban planning to reduce sprawl would decrease Vehicle Miles Travelled (VMT), lowering emissions from transportation. Increased use of public transport can also reduce greenhouse gas emissions per passenger kilometer. Between 1982 and 1997, the amount of land consumed for urban development in the United States increased by 47 percent while the nation's population grew by only 17 percent. Inefficient land use development practices have increased infrastructure costs as well as the amount of energy needed for transportation, community services, and buildings.

At the same time, a growing number of citizens and government officials have begun advocating a smarter approach to land use planning. These smart growth practices include compact community development, multiple transportation choices, mixed land uses, and practices to conserve green space. These programs offer environmental, economic, and quality-of-life benefits; and they also serve to reduce energy usage and greenhouse gas emissions.

Approaches such as New Urbanism and Transit-oriented development seek to reduce distances travelled, especially by private vehicles, encourage public transit and make walking and cycling more attractive options. This is achieved through "medium-density", mixed-use planning and the concentration of housing within walking distance of town centers and transport nodes.

Smarter growth land use policies have both a direct and indirect effect on energy consuming behavior. For example, transportation energy usage, the number one user of petroleum fuels, could be significantly reduced through more compact and mixed use land development patterns, which in turn could be served by a greater variety of non-automotive based transportation choices.

Building Design

Emissions from housing are substantial, and government-supported energy efficiency programmes can make a difference.

For institutions of higher learning in the United States, greenhouse gas emissions depend primarily on total area of buildings and secondarily on climate. If climate is not taken into account, annual greenhouse gas emissions due to energy consumed on campuses plus purchased electricity can be estimated with the formula, $E=aS^b$, where a =0.001621 metric tonnes of CO_2 equivalent/square foot or 0.0241 metric tonnes of CO_2 equivalent/square meter and b = 1.1354.

New buildings can be constructed using passive solar building design, low-energy building, or zero-energy building techniques, using renewable heat sources. Existing buildings can be made more efficient through the use of insulation, high-efficiency appliances (particularly hot water heaters and furnaces), double- or triple-glazed gas-filled windows, external window shades, and building orientation and siting. Renewable heat sources such as shallow geothermal and passive solar energy reduce the amount of greenhouse gasses emitted. In addition to designing buildings which are more energy-efficient to heat, it is possible to design buildings that are more energy-efficient to cool by using lighter-coloured, more reflective materials in the development of urban areas (e.g. by painting roofs white) and planting trees. This saves energy because it cools buildings and reduces the urban heat island effect thus reducing the use of air conditioning.

Agriculture

According to the EPA, agricultural soil management practices can lead to production and emission of nitrous oxide (N2O), a major greenhouse gas and air pollutant. Activities that can contribute to N2O emissions include fertilizer usage, irrigation and tillage. The management of soils accounts for over half of the emissions from the Agriculture sector. Cattle livestocks account for one third of emissions, through methane emissions. Manure management and rice cultivation also produce gaseous emissions.

Methods that significantly enhance carbon sequestration in soil include no-till farming, residue mulching, cover cropping, and crop rotation, all of which are more widely used in organic farming than in conventional farming. Because only 5% of US farmland currently uses no-till and residue mulching, there is a large potential for carbon sequestration.

A 2015 study found that farming can deplete soil carbon and render soil incapable of supporting life. Instead the study showed that conservation farming can protect carbon in soils, and repair damage over time.

The farming practise of cover crops has been recognized as climate-smart agriculture by the White House.

Societal Controls

Another method being examined is to make carbon a new currency by introducing tradeable "personal carbon credits". The idea being it will encourage and motivate individuals to reduce their 'carbon footprint' by the way they live. Each citizen will receive a free annual quota of carbon that they can use to travel, buy food, and go about their business. It has been suggested that by using this concept it could actually solve two problems; pollution and poverty, old age pensioners will actually be better off because they fly less often, so they can cash in their quota at the end of the year to pay heating bills and so forth.

Population

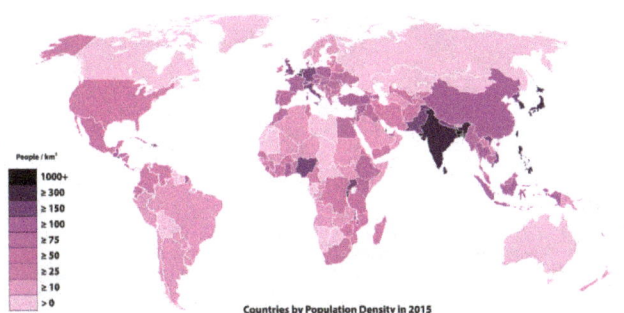

Population density by country

Various organizations promote population control as a means for mitigating global warming. Proposed measures include improving access to family planning and reproductive health care and information, reducing natalistic politics, public education about the consequences of continued population growth, and improving access of women to education and economic opportunities.

Population control efforts are impeded by there being somewhat of a taboo in some countries against considering any such efforts. Also, various religions discourage or prohibit some or all forms of birth control.

Population size has a different per capita effect on global warming in different countries, since the per capita production of anthropogenic greenhouse gases varies greatly by country.

Costs and Benefits

Costs

The Stern Review proposes stabilising the concentration of greenhouse-gas emissions in the atmosphere at a maximum of 550ppm CO_2e by 2050. The Review estimates that this would mean cutting total greenhouse-gas emissions to three quarters of 2007 levels. The Review further estimates that the cost of these cuts would be in the range −1.0 to +3.5% of World GDP, (i.e. GWP), with an average estimate of approximately 1%. Stern has since revised his estimate to 2% of GWP. For comparison, the Gross World Product (GWP) at PPP was estimated at $74.5

trillion in 2010, thus 2% is approximately $1.5 trillion. The Review emphasises that these costs are contingent on steady reductions in the cost of low-carbon technologies. Mitigation costs will also vary according to how and when emissions are cut: early, well-planned action will minimise the costs.

One way of estimating the cost of reducing emissions is by considering the likely costs of potential technological and output changes. Policy makers can compare the marginal abatement costs of different methods to assess the cost and amount of possible abatement over time. The marginal abatement costs of the various measures will differ by country, by sector, and over time.

Benefits

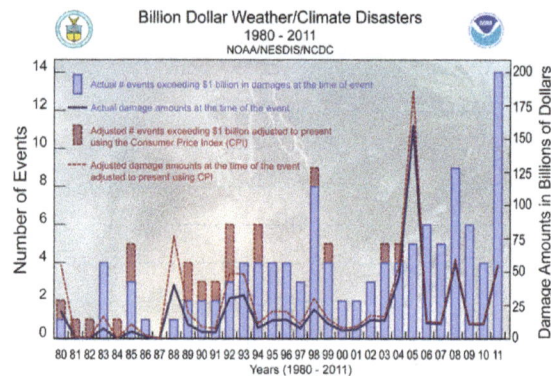

Total extreme weather cost and number of events costing more than $1 billion in the United States from 1980 to 2011.

Yohe *et al.* (2007) assessed the literature on sustainability and climate change. With high confidence, they suggested that up to the year 2050, an effort to cap greenhouse gas (GHG) emissions at 550 ppm would benefit developing countries significantly. This was judged to be especially the case when combined with enhanced adaptation. By 2100, however, it was still judged likely that there would be significant effects of global warming. This was judged to be the case even with aggressive mitigation and significantly enhanced adaptive capacity.

Sharing

One of the aspects of mitigation is how to share the costs and benefits of mitigation policies. There is no scientific consensus over how to share these costs and benefits (Toth *et al.*, 2001). In terms of the politics of mitigation, the UNFCCC's ultimate objective is to stabilize concentrations of GHG in the atmosphere at a level that would prevent "dangerous" climate change (Rogner *et al.*, 2007).

GHG emissions are an important correlate of wealth, at least at present (Banuri *et al.*, 1996, pp. 91–92). Wealth, as measured by per capita income (i.e., income per head of population), varies widely between different countries. Activities of the poor that involve emissions of GHGs are often associated with basic needs, such as heating to stay tolerably warm. In richer countries, emissions tend to be associated with things like cars, central heating, etc. The impacts of cutting emissions could therefore have different impacts on human welfare according to wealth.

Distributing Emissions Abatement Costs

There have been different proposals on how to allocate responsibility for cutting emissions (Banuri *et al.*, 1996, pp. 103–105):

1. Egalitarianism: this system interprets the problem as one where each person has equal rights to a global resource, i.e., polluting the atmosphere.

2. Basic needs: this system would have emissions allocated according to basic needs, as defined according to a minimum level of consumption. Consumption above basic needs would require countries to buy more emission rights. From this viewpoint, developing countries would need to be at least as well off under an emissions control regime as they would be outside the regime.

3. Proportionality and polluter-pays principle: Proportionality reflects the ancient Aristotelian principle that people should receive in proportion to what they put in, and pay in proportion to the damages they cause. This has a potential relationship with the "polluter-pays principle", which can be interpreted in a number of ways:

   1. *Historical responsibilities*: this asserts that allocation of emission rights should be based on patterns of past emissions. Two-thirds of the stock of GHGs in the atmosphere at present is due to the past actions of developed countries (Goldemberg *et al.*, 1996, p. 29).

   2. *Comparable burdens and ability to pay*: with this approach, countries would reduce emissions based on comparable burdens and their ability to take on the costs of reduction. Ways to assess burdens include monetary costs per head of population, as well as other, more complex measures, like the UNDP's Human Development Index.

   3. *Willingness to pay*: with this approach, countries take on emission reductions based on their ability to pay along with how much they benefit from reducing their emissions.

Specific Proposals

1. Ad hoc: Lashof (1992) and Cline (1992) (referred to by Banuri *et al.*, 1996, p. 106), for example, suggested that allocations based partly on GNP could be a way of sharing the burdens of emission reductions. This is because GNP and economic activity are partially tied to carbon emissions.

2. Equal per capita entitlements: this is the most widely cited method of distributing abatement costs, and is derived from egalitarianism (Banuri *et al.*, 1996, pp. 106–107). This approach can be divided into two categories. In the first category, emissions are allocated according to national population. In the second category, emissions are allocated in a way that attempts to account for historical (cumulative) emissions.

3. Status quo: with this approach, historical emissions are ignored, and current emission levels are taken as a status quo right to emit (Banuri *et al.*, 1996, p. 107). An analogy for this approach can be made with fisheries, which is a common, limited resource. The analogy would be with the atmosphere, which can be viewed as an exhaustible natural resource (Goldemberg *et al.*, 1996, p. 27). In international law, one state recognized the long-established use of another state's use of the fisheries resource. It was also recognized by the state that part of the other state's economy was dependent on that resource.

Governmental and Intergovernmental Action

" Bringing down emissions of greenhouse gases asks a good deal of people, not least that they accept the science of climate change. It requires them to make sacrifices today so that future generations will suffer less, and to weigh the needs of people who are living far away. "

— *The Economist, 28 November 2015*

Many countries, both developing and developed, are aiming to use cleaner technologies (World Bank, 2010, p. 192). Use of these technologies aids mitigation and could result in substantial reductions in CO_2 emissions. Policies include targets for emissions reductions, increased use of renewable energy, and increased energy efficiency. It is often argued that the results of climate change are more damaging in poor nations, where infrastructures are weak and few social services exist. The Commitment to Development Index is one attempt to analyze rich country policies taken to reduce their disproportionate use of the global commons. Countries do well if their greenhouse gas emissions are falling, if their gas taxes are high, if they do not subsidize the fishing industry, if they have a low fossil fuel rate per capita, and if they control imports of illegally cut tropical timber.

Kyoto Protocol

The main current international agreement on combating climate change is the Kyoto Protocol, which came into force on 16 February 2005. The Kyoto Protocol is an amendment to the United Nations Framework Convention on Climate Change (UNFCCC). Countries that have ratified this protocol have committed to reduce their emissions of carbon dioxide and five other greenhouse gases, or engage in emissions trading if they maintain or increase emissions of these gases.

Temperature Targets

The graph on the right shows three "pathways" to meet the UNFCCC's 2 °C target, labelled "global technology", "decentralised solutions", and "consumption change". Each pathway shows how various measures (e.g., improved energy efficiency, increased use of renewable energy) could contribute to emissions reductions. Image credit: PBL Netherlands Environmental Assessment Agency.

Actions to mitigate climate change are sometimes based on the goal of achieving a particular temperature target. One of the targets that has been suggested is to limit the future increase in global mean temperature (global warming) to below 2 °C, relative to the pre-industrial level. The 2 °C target was adopted in 2010 by Parties to the United Nations Framework Convention on Climate Change. Most countries of the world are Parties to the UNFCCC. The target had been adopted in 1996 by the European Union Council.

Feasibility of 2 °C

Temperatures have increased by 0.8 °C compared to the pre-industrial level, and another 0.5–0.7 °C is already committed. The 2 °C rise is typically associated in climate models with a carbon dioxide equivalent concentration of 400–500 ppm by volume; the current (January 2015) level of carbon dioxide alone is 400 ppm by volume, and rising at 1–3 ppm annually. Hence, to avoid a very likely breach of the 2 °C target, CO_2 levels would have to be stabilised very soon; this is generally regarded as unlikely, based on current programs in place to date. The importance of change is illustrated by the fact that world economic energy efficiency is improving at only half the rate of world economic growth.

Views in the literature

There is disagreement among experts over whether or not the 2 °C target can be met. For example, according to Anderson and Bows (2011), "there is little to no chance" of meeting the target. On the other hand, according to Alcamo *et al.* (2013):

1. Policies adopted by parties to the UNFCCC are too weak to meet a 2 or 1.5 °C target. However, these targets might still be achievable if more stringent mitigation policies are adopted immediately.

2. Cost-effective 2 °C scenarios project annual global greenhouse gas emissions to peak before the year 2020, with deep cuts in emissions thereafter, leading to a reduction in 2050 of 41% compared to 1990 levels.

Discussion on other targets

Scientific analysis can provide information on the impacts of climate change and associated policies, such as reducing GHG emissions. However, deciding what policies are best requires value judgements. For example, limiting global warming to 1 °C relative to pre-industrial levels may help to reduce climate change damages more than a 2 °C limit. However, a 1 °C limit may be more costly to achieve than a 2 °C limit.

According to some analysts, the 2 °C "guardrail" is inadequate for the needed degree and timeliness of mitigation. On the other hand, some economic studies suggest more modest mitigation policies. For example, the emissions reductions proposed by Nordhaus (2010) might lead to global warming (in the year 2100) of around 3 °C, relative to pre-industrial levels.

Official long-term target of 1.5 °C

In 2015, two official UNFCCC scientific expert bodies came to the conclusion that, "in some regions and vulnerable ecosystems, high risks are projected even for warming above 1.5°C". This expert position was, together with the strong diplomatic voice of the poorest countries and the island nations in the Pacific, the driving force leading to the decision of the Paris Conference 2015, to lay down this 1.5 °C long-term target on top of the existing 2 °C goal.

Encouraging Use Changes

Emissions tax

An emissions tax on greenhouse gas emissions requires individual emitters to pay a fee, charge or tax for every tonne of greenhouse gas released into the atmosphere. Most environmentally related taxes with implications for greenhouse gas emissions in OECD countries are levied on energy products and motor vehicles, rather than on CO_2 emissions directly.

Emission taxes can be both cost-effective and environmentally effective. Difficulties with emission taxes include their potential unpopularity, and the fact that they cannot guarantee a particular level of emissions reduction. Emissions or energy taxes also often fall disproportionately on lower income classes. In developing countries, institutions may be insufficiently developed for the collection of emissions fees from a wide variety of sources.

Subsidies

According to Mark Z. Jacobson, a program of subsidization balanced against expected flood costs could pay for conversion to 100% renewable power by 2030. Jacobson, and his colleague Mark Delucchi, suggest that the cost to generate and transmit power in 2020 will be less than 4 cents per kilowatt hour (in 2007 dollars) for wind, about 4 cents for wave and hydroelectric, from 4 to 7 cents for geothermal, and 8 cents per kWh for solar, fossil, and nuclear power.

Investment

Another indirect method of encouraging uses of renewable energy, and pursue sustainability and environmental protection, is that of prompting investment in this area through legal means, something that is already being done at national level as well as in the field of international investment.

Carbon Emissions Trading

With the creation of a market for trading carbon dioxide emissions within the Kyoto Protocol, it is likely that London financial markets will be the centre for this potentially highly lucrative business; the New York and Chicago stock markets may have a lower trade volume than expected as long as the US maintains its rejection of the Kyoto.

However, emissions trading may delay the phase-out of fossil fuels.

In the north-east United States, a successful cap and trade program has shown potential for this solution.

The European Union Emission Trading Scheme (EU ETS) is the largest multi-national, greenhouse gas emissions trading scheme in the world. It commenced operation on 1 January 2005, and all 28 member states of the European Union participate in the scheme which has created a new market in carbon dioxide allowances estimated at 35 billion Euros (US$43 billion) per year. The Chicago Climate Exchange was the first (voluntary) emissions market, and is soon to be followed by Asia's first market (Asia Carbon Exchange). A total of 107 million metric tonnes of carbon dioxide equivalent have been exchanged through projects in 2004, a 38% increase relative to 2003 (78 Mt CO_2e).

Twenty three multinational corporations have come together in the G8 Climate Change Roundtable, a business group formed at the January 2005 World Economic Forum. The group includes Ford, Toyota, British Airways and BP. On 9 June 2005 the Group published a statement stating that there was a need to act on climate change and claiming that market-based solutions can help. It called on governments to establish "clear, transparent, and consistent price signals" through "creation of a long-term policy framework" that would include all major producers of greenhouse gases.

The Regional Greenhouse Gas Initiative is a proposed carbon trading scheme being created by nine North-eastern and Mid-Atlantic American states; Connecticut, Delaware, Maine, Massachusetts, New Hampshire, New Jersey, New York, Rhode Island and Vermont. The scheme was due to be developed by April 2005 but has not yet been completed.

Implementation

Implementation puts into effect climate change mitigation strategies and targets. These can be targets set by international bodies or voluntary action by individuals or institutions. This is the most important, expensive and least appealing aspect of environmental governance.

Funding

Implementation requires funding sources but is often beset by disputes over who should provide funds and under what conditions. A lack of funding can be a barrier to successful strategies as there are no formal arrangements to finance climate change development and implementation. Funding is often provided by nations, groups of nations and increasingly NGO and private sources. These funds are often channelled through the Global Environmental Facility (GEF). This is an environmental funding mechanism in the World Bank which is designed to deal with global environmental issues. The GEF was originally designed to tackle four main areas: biological diversity, climate change, international waters and ozone layer depletion, to which land degradation and persistent organic pollutant were added. The GEF funds projects that are agreed to achieve global environmental benefits that are endorsed by governments and screened by one of the GEF's implementing agencies.

Problems

There are numerous issues which result in a current perceived lack of implementation. It has been suggested that the main barriers to implementation are, Uncertainty, Fragmentation, Institutional void, Short time horizon of policies and politicians and Missing motives and willingness to start adapting. The relationships between many climatic processes can cause large levels of uncertainty as they are not fully understood and can be a barrier to implementation. When information on climate change is held between the large numbers of actors involved it can be highly dispersed, context specific or difficult to access causing fragmentation to be a barrier. Institutional void is the lack of commonly accepted rules and norms for policy processes to take place, calling into question the legitimacy and efficacy of policy processes. The Short time horizon of policies and politicians often means that climate change policies are not implemented in favour of socially favoured societal issues. Statements are often posed to keep the illusion of political action to prevent or postpone decisions being made. Missing motives and willingness to start adapting is a large barrier as it prevents any implementation.

The issues that arise with a system which involves international government cooperation, such as Cap and Trade, could potentially be improved with a polycentric approach where the rules are enforced by many small sections of authority as apposed to one overall enforcement agency.

Occurrence

Despite a perceived lack of occurrence, evidence of implementation is emerging internationally. Some examples of this are the initiation of NAPA's and of joint implementation. Many developing nations have made National Adaptation Programs of Action (NAPAs) which are frameworks to prioritize adaption needs. The implementation of many of these is supported by GEF agencies. Many developed countries are implementing 'first generation' institutional adaption plans particularly at the state and local government scale. There has also been a push towards joint implementation between countries by the UNFCC as this has been suggested as a cost-effective way for objectives to be achieved.

Territorial Policies

United States

Efforts to reduce greenhouse gas emissions by the United States include energy policies which encourage efficiency through programs like Energy Star, Commercial Building Integration, and the Industrial Technologies Program. On 12 November 1998, Vice President Al Gore symbolically signed the Kyoto Protocol, but he indicated participation by the developing nations was necessary prior its being submitted for ratification by the United States Senate.

In 2007, Transportation Secretary Mary Peters, with White House approval, urged governors and dozens of members of the House of Representatives to block California's first-in-the-nation limits on greenhouse gases from cars and trucks, according to e-mails obtained by Congress. The U.S. Climate Change Science Program is a group of about twenty federal agencies and US Cabinet Departments, all working together to address global warming.

The Bush administration pressured American scientists to suppress discussion of global warming, according to the testimony of the Union of Concerned Scientists to the Oversight and Government Reform Committee of the U.S. House of Representatives. "High-quality science" was "struggling to get out," as the Bush administration pressured scientists to tailor their writings on global warming to fit the Bush administration's skepticism, in some cases at the behest of an ex-oil industry lobbyist. "Nearly half of all respondents perceived or personally experienced pressure to eliminate the words 'climate change,' 'global warming' or other similar terms from a variety of communications." Similarly, according to the testimony of senior officers of the Government Accountability Project, the White House attempted to bury the report "National Assessment of the Potential Consequences of Climate Variability and Change," produced by U.S. scientists pursuant to U.S. law. Some U.S. scientists resigned their jobs rather than give in to White House pressure to underreport global warming.

In the absence of substantial federal action, state governments have adopted emissions-control laws such as the Regional Greenhouse Gas Initiative in the Northeast and the Global Warming Solutions Act of 2006 in California.

Developing Countries

In order to reconcile economic development with mitigating carbon emissions, developing countries need particular support, both financial and technical. One of the means of achieving this is the Kyoto Protocol's Clean Development Mechanism (CDM). The World Bank's Prototype Carbon Fund is a public private partnership that operates within the CDM.

An important point of contention, however, is how overseas development assistance not directly related to climate change mitigation is affected by funds provided to climate change mitigation. One of the outcomes of the UNFCC Copenhagen Climate Conference was the Copenhagen Accord, in which developed countries promised to provide US $30 million between 2010 and 2012 of new and additional resources. Yet it remains unclear what exactly the definition of additional is and the European Commission has requested its member states to define what they understand to be additional, and researchers at the Overseas Development Institute have found four main understandings:

1. Climate finance classified as aid, but additional to (over and above) the '0.7%' ODA target;

2. Increase on previous year's Official Development Assistance (ODA) spent on climate change mitigation;

3. Rising ODA levels that include climate change finance but where it is limited to a specified percentage; and

4. Increase in climate finance not connected to ODA.

The main point being that there is a conflict between the OECD states budget deficit cuts, the need to help developing countries adapt to develop sustainably and the need to ensure that funding does not come from cutting aid to other important Millennium Development Goals.

However, none of these initiatives suggest a quantitative cap on the emissions from developing countries. This is considered as a particularly difficult policy proposal as the economic growth of developing countries are proportionally reflected in the growth of greenhouse emissions. Critics of mitigation often argue that, the developing countries' drive to attain a comparable living standard to the developed countries would doom the attempt at mitigation of global warming. Critics also argue that holding down emissions would shift the human cost of global warming from a general one to one that was borne most heavily by the poorest populations on the planet.

In an attempt to provide more opportunities for developing countries to adapt clean technologies, UNEP and WTO urged the international community to reduce trade barriers and to conclude the Doha trade round "which includes opening trade in environmental goods and services".

Non-governmental Approaches

While many of the proposed methods of mitigating global warming require governmental funding, legislation and regulatory action, individuals and businesses can also play a part in the mitigation effort.

Choices in Personal Actions And Business Operations

Environmental groups encourage individual action against global warming, often aimed at the consumer. Common recommendations include lowering home heating and cooling usage, burning less gasoline, supporting renewable energy sources, buying local products to reduce transportation, turning off unused devices, and various others.

A geophysicist at Utrecht University has urged similar institutions to hold the vanguard in voluntary mitigation, suggesting the use of communications technologies such as videoconferencing to reduce their dependence on long-haul flights.

Air Travel and Shipment

In 2008, climate scientist Kevin Anderson raised concern about the growing effect of rapidly increasing global air transport on the climate in a paper, and a presentation, suggesting that reversing this trend is necessary to reduce emissions.

Part of the difficulty is that when aviation emissions are made at high altitude, the climate impacts are much greater than otherwise. Others have been raising the related concerns of the increasing hypermobility of individuals, whether traveling for business or pleasure, involving frequent and often long distance air travel, as well as air shipment of goods.

Business Opportunities and Risks

On 9 May 2005 Jeff Immelt, the chief executive of General Electric (GE), announced plans to reduce GE's global warming related emissions by one percent by 2012. "GE said that given its projected growth, those emissions would have risen by 40 percent without such action."

On 21 June 2005 a group of leading airlines, airports and aerospace manufacturers pledged to work together to reduce the negative environmental impact of aviation, including limiting the impact of air travel on climate change by improving fuel efficiency and reducing carbon dioxide emissions of new aircraft by fifty percent per seat kilometre by 2020 from 2000 levels. The group aims to develop a common reporting system for carbon dioxide emissions per aircraft by the end of 2005, and pressed for the early inclusion of aviation in the European Union's carbon emission trading scheme.

Investor Response

Climate change is also a concern for large institutional investors who have a long term time horizon and potentially large exposure to the negative impacts of global warming because of the large geographic footprint of their multi-national holdings. SRI (Socially responsible investing) Funds allow investors to invest in funds that meet high ESG (environmental, social, governance) standards as such funds invest in companies that are aligned with these goals. Proxy firms can be used to draft guidelines for investment managers that take these concerns into account.

Legal Action

In some countries, those affected by climate change may be able to sue major producers. Attempts at litigation have been initiated by entire peoples such as Palau and the Inuit, as well as non-governmental organizations such as the Sierra Club. Although proving that particular weather events are due specifically to global warming may never be possible, methodologies have been developed to show the increased risk of such events caused by global warming.

For a legal action for negligence (or similar) to succeed, "Plaintiffs ... must show that, more probably than not, their individual injuries were caused by the risk factor in question, as opposed to any other cause. This has sometimes been translated to a requirement of a relative risk of at least two." Another route (though with little legal bite) is the World Heritage Convention, if it can be shown that climate change is affecting World Heritage Sites like Mount Everest.

Besides countries suing one another, there are also cases where people in a country have taken legal steps against their own government. Legal action for instance has been taken to try to force the U.S. Environmental Protection Agency to regulate greenhouse gas emissions under the Clean Air Act, and against the Export-Import Bank and OPIC for failing to assess environmental impacts (including global warming impacts) under NEPA.

In the Netherlands and Belgium, organisations as Urgenda and the vzw Klimaatzaak in Belgium have also sued their governments as they believe their governments aren't meeting the emission reductions they agreed to. Urgenda has all ready won their case against the Dutch government.

According to a 2004 study commissioned by Friends of the Earth, ExxonMobil and its predecessors caused 4.7 to 5.3 percent of the world's man-made carbon dioxide emissions between 1882 and 2002. The group suggested that such studies could form the basis for eventual legal action.

In 2015, Exxon, received a subpoena. According to the Washington Post and confirmed by the company, the attorney general of New York, Eric Schneiderman, opened an investigation into the possibility that the company had mislead the public and investors about the risks of climate change.

Climate Change Mitigation Strategies

Carbon Accounting

Carbon accounting refers generally to processes undertaken to "measure" amounts of carbon dioxide equivalents emitted by an entity. It is used inter alia by nation states, corporations, individuals – to create the carbon credit commodity traded on carbon markets (or to establish the demand for carbon credits). Correspondingly, examples for products based upon forms of carbon accounting can be found in national inventories, corporate environmental reports or carbon footprint calculators. Likening sustainability measurement, as an instance of ecological modernisation discourses and policy, carbon accounting is hoped to provide a factual ground for carbon-related decision-making. However, social scientific studies of accounting challenge this hope, pointing to the socially constructed character of carbon conversion factors or of the accountants' work practice which cannot implement abstract accounting schemes into reality.

While natural sciences claim to know and measure carbon, for organisations it is usually easier to employ forms of carbon accounting to represent carbon. The trustworthiness of accounts of carbon emissions can easily be contested. Thus, how well carbon accounting represents carbon is difficult to exactly know. Science and Technology Studies scholar Donna Haraway's pluralised concept of knowledge, i.e. knowledges, can well be used to understand better the status of knowledge produced by carbon accounting: carbon accounting produced a version of understanding of carbon emissions. Other carbon accountants would produce other results.

Carbon Accounting in Corporations

Carbon accounting can be used as part of sustainability accounting by for-profit and non-profit organisations. A corporate or organisational "carbon" or greenhouse gas (GHG) emissions assessment promises to quantify the greenhouse gases produced directly and indirectly from a business or organisation's activities within a set of boundaries. Also known as a carbon footprint, it is a business tool that constructs information that may (or may not) be useful for understanding and managing climate change impacts.

The drivers for corporate carbon accounting include mandatory GHG reporting in directors' reports, investment due diligence, shareholder and stakeholder communication, staff engagement, green messaging, and tender requirements for business and government contracts. Accounting for greenhouse gas emissions is increasingly framed as a standard requirement for business. As of June 2011, 60% of UK FTSE 100 companies had published environmental targets, with 53% of these 240+ targets relating to carbon, greenhouse gas emissions or energy reductions (representing 59% of the FTSE 100). In June 2012, the UK coalition government announced the introduction of mandatory carbon reporting, requiring around 1,100 of the UK's largest listed companies to report their greenhouse gas emissions every year. Deputy Prime Minister Nick Clegg confirmed that emission reporting rules would come into effect from April 2013 in his piece for The Guardian.

Carbon Accounting of Avoided Emissions

A special case of carbon accounting is the accounting process undertaken to measure the amount of carbon dioxide equivalents that will not be released into the atmosphere as a result of flexible mechanisms projects under the Kyoto Protocol. These projects thus include (but are not limited to) renewable energy projects and biomass, forage and tree plantations.

Carbon Accounting Software

A number of programs are created in order to assist with carbon accounting.

Carbon Dioxide Removal

Carbon dioxide removal (CDR) methods refers to a number of technologies which reduce the levels of carbon dioxide in the atmosphere. Among such technologies are bio-energy with carbon capture and storage, biochar, direct air capture, ocean fertilization and enhanced weathering. CDR is a different approach than removing CO_2 from the stack emissions of large fossil fuel point sources, such as power stations. The latter reduces emission to the atmosphere but cannot reduce the amount of carbon dioxide already in the atmosphere. As CDR removes carbon dioxide from the atmosphere, it creates negative emissions, offsetting emissions from small and dispersed point sources such as domestic heating systems, airplanes and vehicle exhausts. It is regarded by some as a form of climate engineering, while other commentators describe it as a form of carbon capture and storage or extreme mitigation. Whether CDR would satisfy common definitions of "climate engineering" or "geoengineering" usually depends upon the scale on which it would be undertaken.

The likely need for CDR has been publicly expressed by a range of individuals and organizations involved with climate change issues, including IPCC chief Rajendra Pachauri, the UNFCCC executive secretary Christiana Figueres, and the World Watch Institute. Institutions with major programs focusing on CDR include the Lenfest Center for Sustainable Energy at the Earth Institute, Columbia University, and the Climate Decision Making Center, an international collaboration operated out of Carnegie-Mellon University's Department of Engineering and Public Policy.

The mitigation effectiveness of air capture is limited by societal investment, land use, availability of geologic reservoirs, and leakage. The reservoirs are estimated to be sufficient to for storing at least 545 GtC. Storing 771 GtC would cause an 186 ppm atmospheric reduction. In order to return the atmospheric CO_2 content to 350 ppm we need atmospheric reduction of 50 ppm plus an additional 2 ppm per year of current emissions.

Methods

Bio-energy with carbon Capture & Storage

Bio-energy with carbon capture and storage, or BECCS, uses biomass to extract carbon dioxide from the atmosphere, and carbon capture and storage technologies to concentrate and permanently store it in deep geological formations.

BECCS is currently (as of October 2012) the only CDR technology deployed at full industrial scale, with 550 000 tonnes CO_2/year in total capacity operating, divided between three different facilities (as of January 2012).

The Imperial College London, the UK Met Office Hadley Centre for Climate Prediction and Research, the Tyndall Centre for Climate Change Research, the Walker Institute for Climate System Research, and the Grantham Institute for Climate Change issued a joint report on carbon dioxide removal technologies as part of the *AVOID: Avoiding dangerous climate change* research program, stating that "Overall, of the technologies studied in this report, BECCS has the greatest maturity and there are no major practical barriers to its introduction into today's energy system. The presence of a primary product will support early deployment."

According to the OECD, "Achieving lower concentration targets (450 ppm) depends significantly on the use of BECCS".

Biochar

Biochar is created by the pyrolysis of biomass, and is under investigation as a method of carbon sequestration. Biochar is a charcoal that is used for agricultural purposes which also aids in carbon sequestration, the capture or hold of carbon. It is created using a process called pyrolysis, which is basically the act of high temperature heating biomass in an environment with low oxygen levels. What remains is a material known as char, similar to charcoal but is made through a sustainable process, thus the use of biomass. Biomass is organic matter produced by living organisms or recently living organisms, most commonly plants or plant based material. The offset of GHG emission, if biochar were to be implemented, would be a maximum of 12%. This equates to about 106 metric tons of CO_2 equivalents. On a medium conservative level, it would be 23% less than that, at 82 metric tons. A study done by the UK Biochar Research Center has stated that, on a conservative level, biochar can store 1 gigaton of carbon per year. With greater effort in marketing and acceptance of biochar, the benefit would the storage of 5-9 gigatons per year of carbon in biochar soils.

Enhanced Weathering

Enhanced weathering refers to chemical approach to remove carbon dioxide involving land or ocean based techniques. Examples of land based enhanced weathering techniques are in-situ carbonation of silicates. Ultramafic rock, for example, has the potential to store thousands of years worth of CO_2 emissions according to one estimate. Ocean based techniques involve alkalinity enhancement, such as, grinding, dispersing and dissolving olivine, limestone, silicates, or calcium hydroxide to address ocean acidification and CO_2 sequestration. Enhanced weathering is considered as one of the least expensive of geoengineering options. One example of a research project on the feasibility of enhanced weathering is the CarbFix project in Iceland.

Direct Air Capture

Carbon dioxide can be removed from ambient air through chemical processes, sequestered, and stored. One proposed method is by so-called *artificial trees*. This concept, proposed by climate scientist Wallace S. Broecker and science writer Robert Kunzig, imagines huge numbers of artificial trees around the world to remove ambient CO_2. The technology is now being pioneered by Klaus Lackner, a researcher at the Earth Institute, Columbia University, whose artificial tree technology can suck up to 1,000 times more CO_2 from the air than real trees can, at a rate of about one ton of carbon per day if the artificial tree is approximately the size of an actual tree. The CO_2 would be captured in a filter and then removed from the filter and stored.

The chemistry used is a variant of that described below, as it is based on sodium hydroxide. However, in a more recent design proposed by Klaus Lackner, the process can be carried out at only 40 °C by using a polymer-based ion exchange resin, which takes advantage of changes in humidity to prompt the release of captured CO_2, instead of using a kiln. This reduces the energy required to operate the process.

Another substance which can be used are Metal-organic frameworks (or MOF's). A special MOF has been made specifically for locking CO_2 by Joeri Denayer.

In 2008, the Discovery Channel covered the work of David Keith, of University of Calgary, who built a tower, 4 feet wide and 20 feet tall (1.2×6.1 meters), with a fan at the bottom that sucks air in, which comes out again at the top. In the process, about half the CO_2 is removed from the air.

This device uses the chemical process described in detail below. The system demonstrated on the Discovery Channel was a 1/90,000th scale test system of the capture section; the reagents are regenerated in a separate facility. The main costs of a full plant will be the cost to build it, and the energy input to regenerate the chemicals and produce a pure stream of CO_2.

To put this into perspective, people in the U.S. emit about 20 tonnes of CO_2 per person annually. In other words, each person in the U.S. would require a tower like the one featured by the Discovery Channel to remove this amount of CO_2 from the air, requiring an annual 2 megawatt-hours of electricity to operate it. By comparison, a refrigerator consumes about 1.2 megawatt-hours annually (2001 figures). But, by combining many small systems such as this into one large system, the construction costs and energy use can be reduced.

It has been proposed that the Solar updraft tower to generate electricity from thermal air currents also be used at the same time for amine gravity scrubbing of CO_2. Some heat would be required to regenerate the amine.

Finally, a similar CO_2 scrubber has also been build by Carbon Engineering. Besides simply focusing on capturing the CO_2, the company also puts emphasis on reuse of the CO_2, for example in the production of fuels, which would thus be carbon-neutral.

Direct air capture has been proposed as a way of generating carbon-neutral organic chemicals, by harvesting the atmospheric compounds and then using them in the production and synthesis of polymers and fuels.

Ocean Fertilization

Ocean fertilization or ocean nourishment is a type of climate engineering based on the purposeful introduction of nutrients to the upper ocean to increase marine food production and to remove carbon dioxide from the atmosphere. A number of techniques, including fertilization by iron, urea and phosphorus have been proposed.

Example Co$_2$ Scrubbing Chemistry

Calcium Oxide

Calcium oxide (quicklime) will absorb CO_2 from atmospheric air mixed with steam at 400 °C (forming calcium carbonate) and release it at 1,000 °C. This process, proposed by A. Steinfeld, can be performed using renewable energy from thermal concentrated solar power. Quicklime is made by heating limestone to release the CO_2 within it. Quicklime is mixed with sand for brick building as mortar, where it hardens by absorption of CO_2.

Potassium Hydroxide

Zeman and Lackner outlined a specific method of air capture using sodium hydroxide. Carbon Engineering, a Calgary, Alberta firm founded in 2009 and partially funded by Bill Gates, is developing a process to capture carbon dioxide using a solution of potassium hydroxide mixed with some water at their pilot plant . They hope to create and sell synthetic fuels at a cost of $100 a ton.

Economic Issues

A crucial issue for CDR methods is their cost, which differs substantially among the different technologies: some of these are not sufficiently developed to perform cost assessments. The American Physical Society estimates the costs for direct air capture to be $600/tonne with optimistic assumptions. The IEA Greenhouse Gas R&D Programme and Ecofys provides an estimate that 3.5 billion tonnes could be removed annually from the atmosphere with BECCS (Bio-Energy with Carbon Capture and Storage) at carbon prices as low as €50, whereas a report from Biorecro and the Global Carbon Capture and Storage Institute estimates costs "below €100" per tonne for large scale BECCS deployment.

Risks, Problems and Criticisms

CDR is slow to act, and requires a long-term political and engineering program to effect. CDR is even slower to take effect on acidified oceans. In a *Business as usual* concentration pathway, the deep ocean will remain acidified for centuries, and as a consequence many marine species are in danger of extinction.

Greenhouse Gas Removal

Greenhouse gas removal projects are a type of climate engineering that seek to remove greenhouse gases from the atmosphere, and thus they tackle the root cause of global warming. These techniques either directly remove greenhouse gases, or alternatively seek to influence natural processes to remove greenhouse gases indirectly. The discipline overlaps with carbon capture and storage and carbon sequestration, and some projects listed may not be considered to be geoengineering by all commentators, instead being described as mitigation.

Carbon Sequestration

A wide range of techniques for carbon sequestration exist. These range from ideas to remove CO_2 from the atmosphere (carbon dioxide air capture), flue gases (carbon capture and storage) and by preventing carbon in biomass from re-entering the atmosphere, such as with Bio-energy with carbon capture and storage (BECCS).

Chlorofluorocarbon Photochemistry

Atmospheric chlorofluorocarbon (CFC) removal is an idea which suggests using lasers to break up CFCs, an important family of greenhouse gases, in the atmosphere.

Methane Removal

Methane potentially poses major challenges for remediation. It is around 20 times as powerful a greenhouse gas as CO_2. Large quantities may be outgassed from permafrost and clathrates as a result of global warming, notably in the Arctic.

There are existing climate engineering proposals. Methane is removed by several natural processes, which can be enhanced.

Chemical decomposition — reaction with hydroxyl radicals produced from photochemical decomposition of ozone in the stratosphere.

Biological decomposition — by methanotrophs in soils and water.

Solar Radiation Management

Proposed solar radiation management using a tethered balloon to inject sulfate aerosols into the stratosphere.

Solar radiation management (SRM) projects (proposed and theoretical) are a type of climate engineering which seek to reflect sunlight and thus reduce global warming. Proposed examples include the creation of stratospheric sulfate aerosols. Their principal advantages as an approach to climate engineering is the speed with which they can be deployed and become fully active, their potential low financial cost, and the reversibility of their direct climatic effects. Solar radiation management projects could, for example, be used as a temporary response while levels of greenhouse gases can be brought under control by greenhouse gas remediation techniques. They would not reduce greenhouse gas concentrations in the atmosphere, and thus do not address problems such as ocean acidification caused by excess carbon dioxide (CO_2). By comparison, other climate engineering techniques based on greenhouse gas remediation, such as ocean iron fertilization, need to sequester the anthropogenic carbon excess before any reversal of global warming would occur.

Background

Climate engineering projects have been proposed in order to reduce global warming. The effect of rising greenhouse gas concentrations in the atmosphere on global climate is a warming effect on the planet. By modifying the albedo of the Earth's surface, or by preventing sunlight reaching the Earth by using a solar shade, this warming effect can be cancelled out—although the cancellation is imperfect, with regional discrepancies remaining.

Therefore, solar radiation management, or albedo modification, is considered to be a potential option for addressing climate change. As the National Academy of Sciences states in its 2015 report: "The two main options for responding to the risks of climate change involve mitigation—reducing and eventually eliminating human-caused emissions of CO_2 and other greenhouse gases (GHGs)— and adaptation—reducing the vulnerability of human and natural systems to changes in climate. A third potentially viable option, currently under development but not yet widely deployed, is carbon dioxide removal (CDR) from the atmosphere accompanied by reliable sequestration. A fourth, more speculative family of approaches called albedo modification seeks to offset climate warming by greenhouse gases by increasing the amount of sunlight reflected back to space." In this context, solar radiation management is widely viewed as a complement, not a substitute, to climate change mitigation and adaptation efforts. As The Royal Society concluded in its 2009 report: "Geoengineering methods are not a substitute for climate change mitigation, and should only be considered as part of a wider package of options for addressing climate change." Or put another way: "The safest and most predictable method of moderating climate change is to take early and effective action to reduce emissions of greenhouse gases. No geoengineering method can provide an easy or readily acceptable alternative solution to the problem of climate change. Geoengineering methods could, however, potentially be useful in future to augment continuing efforts to mitigate climate change by reducing emissions, and so should be subject to more detailed research and analysis."

The phenomenon of global dimming is widely known, and is not necessarily a climate engineering technique. It already occurs under current conditions, due to aerosols caused by pollution, or caused naturally as a result of volcanoes and major forest fires. However, its deliberate manipulation is a tool of the geoengineer.

By intentionally changing the Earth's albedo, or reflectivity, scientists propose that we could reflect more heat back out into space, or intercept sunlight before it reaches the Earth through a literal shade built in space. A 2% albedo increase would roughly halve the effect of CO_2 doubling.

The National Academy of Sciences describes several of the potential benefits and risks of solar radiation management: "Modeling studies have shown that large amounts of cooling, equivalent in scale to the predicted warming due to doubling the CO_2 concentration in the atmosphere, can be produced by the introduction of tens of millions of tons of aerosols into the stratosphere. ... Preliminary modeling results suggest that albedo modification may be able to counter many of the damaging effects of high greenhouse gas concentrations on temperature and the hydrological cycle and reduce some impacts to sea ice. Models also strongly suggest that the benefits and risks will not be uniformly distributed around the globe."

The applicability of many techniques listed here has not been comprehensively tested. Even if the effects in computer simulation models or of small-scale interventions are known, there may be cumulative problems such as ozone depletion, which become apparent only from large scale experiments.

Various small-scale experiments have been carried out on techniques such as cloud seeding, increasing the volume of stratospheric sulfate aerosols and implementing cool roof technology.

As early as 1974, Russian expert Mikhail Budyko suggested that if global warming became a problem, we could cool down the planet by burning sulfur in the stratosphere, which would create a haze. The annual cost of delivering a sufficient amount of sulfur to counteract expected greenhouse warming is estimated at 2 to 8 billion USD.

A preliminary study by Edward Teller and others in 1997 presented the pros and cons of various relatively "low-tech" proposals to mitigate global warming through scattering/reflecting sunlight away from the Earth via insertion of various materials in the upper stratosphere, low earth orbit, and L_1 locations.

Advantages

Solar radiation management has certain advantages relative to emissions cuts, adaptation, and carbon dioxide removal. Its effect of counteracting climate change would be experienced very rapidly, on the order of months after implementation, whereas the effects of emissions cuts and carbon dioxide removal are delayed because the climate change that they prevent is itself delayed. Some proposed solar radiation management techniques are expected to have very low direct financial costs of implementation. This creates a different problem structure. Whereas the provision of emissions reduction and carbon dioxide removal present collective action problems (because ensuring a lower atmospheric carbon dioxide concentration is a public good), a single countries or a handful of countries could implement solar radiation management. Finally, the direct climatic effects of solar radiation management are reversible on short timescales.

Limitations and Risks

As well as the imperfect cancellation of the climatic effect of greenhouse gases, there are other significant problems with solar radiation management as a form of climate engineering.

Incomplete Solution to CO_2 Emissions

Solar radiation management does not remove greenhouse gases from the atmosphere and thus does not reduce other effects from these gases, such as ocean acidification. While not an argument against solar radiation management *per se*, this is an argument against reliance on climate engineering to the exclusion of greenhouse gas reduction.

Control and Predictability

Most of the information on solar radiation management is from models and computer simulations. The actual results may differ from the predicted effect. The full effects of various solar radiation management proposals are not yet well understood. It may be difficult to predict the ultimate effects of projects, with models presently giving varying results. In the cases of systems which involve tipping points, effects may be irreversible. Furthermore, most modeling to date consider the effects of using solar radiation management to fully counteract the increase in global average surface temperature arising from a doubling or a quadrupling of the preindustrial carbon dioxide concentration. Under these assumptions, it overcompensates for the changes in precipitation from climate change. Solar radiation management is more likely to be optimized in a way that balances counteracting changes to temperature and precipitation, to compensate for some portion of climate change, and/or to slow down the rate of climate change.

Side Effects

There may be unintended climatic consequences of solar radiation management, such as significant changes to the hydrological cycle that might not be predicted by the models used to plan them. Such effects may be cumulative or chaotic in nature. Ozone depletion is a risk of techniques involving sulfur delivery into the stratosphere. Not all side effects are negative, and an increase in agricultural productivity has been predicted by some studies due to the combination of more diffuse light and elevated carbon dioxide concentration.

Termination Shock

If solar radiation management were masking a significant amount of warming and then were to abruptly stop, the climate would rapidly warm. This would cause a sudden rise in global temperatures towards levels which would have existed without the use of the climate engineering technique. The rapid rise in temperature may lead to more severe consequences than a gradual rise of the same magnitude.

Disagreement

Leaders of countries and other actors may disagree as to whether, how, and to what degree solar radiation would be used. This could exacerbate international tensions.

Weaponization

In 1976, 85 countries signed the U.N. Convention on the Prohibition of Military or Any Other Hostile Use of Environmental Modification Techniques. The Environmental Modification Convention generally prohibits weaponising climate engineering techniques. However, this does not eliminate the risk. If perfected to a degree of controllability and accuracy that is not considered possible at the moment, climate engineering techniques could theoretically be used by militaries to cause droughts or famines. Theoretically they could also be used simply to make battlefield conditions more favourable to one side or the other in a war.

Carnegie's Ken Caldeira said, "It will make it harder to achieve broad consensus on developing and governing these technologies if there is suspicion that gaining military advantage is an underlying motivation for its development..."

Effect on Sunlight, Sky and Clouds

Managing solar radiation using aerosols or cloud cover would involve changing the ratio between direct and indirect solar radiation. This would affect plant life and solar energy. It is believed that there would be a significant effect on the appearance of the sky from stratospheric aerosol injection projects, notably a hazing of blue skies and a change in the appearance of sunsets. Aerosols affect the formation of clouds, especially cirrus clouds.

Roles

SRM has been suggested to control regional climate, but precise control over the geographical boundaries of the effect is not possible.

SRM is temporary in its effect, and thus and long-term restoration of the climate would rely on long-term SRM, unless carbon dioxide removal was subsequently used. However, short-term SRM programs are potentially beneficial.

Forms of Proposed Solar Radiation Management

Atmospheric Proposals

These projects seek to modify the atmosphere, either by enhancing naturally occurring stratospheric aerosols, or by using artificial techniques such as reflective balloons.

Stratospheric Aerosols

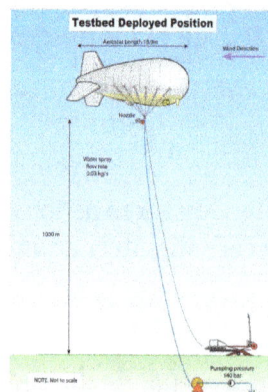

Stratospheric Particle Injection for Climate Engineering

Injecting reflective aerosols into the stratosphere is the proposed solar radiation management method that has received the most sustained attention. This technique could give much more than

3.7 W/m² of globally averaged negative forcing, which is sufficient to entirely offset the warming caused by a doubling of CO_2, which is a common benchmark for assessing future climate scenarios. Sulfates are the most commonly proposed aerosols for climate engineering, since there is a good natural analogue with (and evidence from) volcanic eruptions. Explosive volcanic eruptions inject large amounts of sulfur dioxide gas into the stratosphere, which form sulfate aerosol and cool the planet. Alternative materials such as using photophoretic particles, titaniun dioxide, and diamond have been proposed. Delivery could be achieved using artillery, aircraft (such as the high-flying F15-C) or balloons. Broadly speaking, stratospheric aerosol injection is seen as a relatively more credible climate engineering technique, although one with potential major risks and challenges for its implementation. Risks include changes in precipitation and, in the case of sulfur, possible ozone depletion.

Marine Cloud Brightening

Various cloud reflectivity methods have been suggested, such as that proposed by John Latham and Stephen Salter, which works by spraying seawater in the atmosphere to increase the reflectivity of clouds. The extra condensation nuclei created by the spray will change the size distribution of the drops in existing clouds to make them whiter. The sprayers would use fleets of unmanned rotor ships known as Flettner vessels to spray mist created from seawater into the air to thicken clouds and thus reflect more radiation from the Earth. The whitening effect is created by using very small cloud condensation nuclei, which whiten the clouds due to the Twomey effect.

This technique can give more than 3.7 W/m² of globally averaged negative forcing, which is sufficient to reverse the warming effect of a doubling of CO_2.

Ocean Sulfur Cycle Enhancement

Enhancing the natural marine sulfur cycle by fertilizing a small portion with iron—typically considered to be a greenhouse gas remediation method—may also increase the reflection of sunlight. Such fertilization, especially in the Southern Ocean, would enhance dimethyl sulfide production and consequently cloud reflectivity. This could potentially be used as regional solar radiation management, to slow Antarctic ice from melting. Such techniques also tend to sequester carbon, but the enhancement of cloud albedo also appears to be a likely effect.

Terrestrial Proposals

Cool Roof

Painting roof materials in white or pale colours to reflect solar radiation, known as 'cool roof' technology, is encouraged by legislation in some areas (notably California). This technique is limited in its ultimate effectiveness by the constrained surface area available for treatment. This technique can give between 0.01-0.19 W/m² of globally averaged negative forcing, depending on whether cities or all settlements are so treated. This is small relative to the 3.7 W/m² of positive forcing from a doubling of CO_2. Moreover, while in small cases it can be achieved at little or no cost by simply selecting different materials, it can be costly if implemented on a larger scale. A 2009 Royal Society report states that, "the overall cost of a 'white roof method' covering an area of 1% of the land surface (about 10^{12} m²) would be about $300 billion/yr, making this one of the least effective and most expensive methods considered." However, it can reduce the need for air conditioning, which emits CO_2 and contributes to global warming.

The albedo of several types of roofs

Reflective Sheeting

Adding reflective plastic sheets covering 67,000 square miles (170,000 km²) of desert every year between 2010 and 2070 to reflect the Sun's energy. may be able give globally averaged 1.74 W/m² of negative forcing. Although insufficient to fully offset the 3.7 W/m² of positive forcing from a doubling of CO_2, this would still be a significant contribution thereto, and would offset the current level of warming (approx. 1.7 W/m²). However, the effect would be strongly regional, and would not be ideal for controlling Arctic shrinkage, which is one of the most significant problems resulting from global warming. Furthermore, desert albedo modification would be expensive, would compete with other land uses, and would have strongly negative ecological consequences. Finally, the total area required during 2010-70 is larger than all non-polar deserts combined.

Ocean Changes

An early geoengineering idea was to use pale coloured floating litter within certain *stable* oceanic gyres. This litter would tend to group into large and stable areas, such as the Great Pacific Garbage Patch.

Oceanic foams have also been suggested, using microscopic bubbles suspended in the upper layers of the photic zone.

Farming, Forestry, and Land Proposals

Forestry

Reforestation in tropical areas has a cooling effect. Deforestation of high-latitude and high-altitude forests exposes snow and this increases albedo.

Grassland Management

Changes to grassland have been proposed to increase albedo. This technique can give 0.64 W/m² of globally averaged negative forcing, which is insufficient to offset the 3.7 W/m² of positive forcing from a doubling of CO_2, but could make a minor contribution.

High-albedo Crop Varieties

Selecting or genetically modifying commercial crops with high albedo has been suggested. This has the advantage of being relatively simple to implement, with farmers simply switching from one variety to another. Temperate areas may experience a 1 °C cooling as a result of this technique. This technique is an example of bio-geoengineering. This technique can give 0.44 W/m² of globally averaged negative forcing, which is insufficient to offset the 3.7 W/m² of positive forcing from a doubling of CO_2, but could make a minor contribution.

Space-based Proposals

Space-based climate engineering projects are seen by many commentators and scientists as being very expensive and technically difficult, with the Royal Society suggesting that "the costs of setting in place such a space-based armada for the relatively short period that SRM geoengineering may be considered applicable (decades rather than centuries) would likely make it uncompetitive with other SRM approaches."

Space Mirrors

Mirrors in space: proposed by Roger Angel with the purpose to deflect a percentage of solar sunlight into space, using mirrors orbiting around the Earth.

Moon Dust

Mining moon dust to create a shielding cloud was proposed by Curtis Struck at Iowa State University in Ames.

Dispersive Solutions

The basic function of a space lens to mitigate global warming. In reality, a 1000 kilometre diameter lens is enough, much smaller than what is shown in the simplified image. In addition, as a Fresnel lens it would only be a few millimeters thick.

Several authors have proposed dispersing light before it reaches the Earth by putting a very large diffraction grating (thin wire mesh) or lens in space, perhaps at the L1 point between the Earth and the Sun. Using a Fresnel lens in this manner was proposed in 1989 by J. T. Early. Using a diffraction grating was proposed in 1997 by Edward Teller, Lowell Wood, and Roderick Hyde. In 2004, physicist and science fiction author Gregory Benford calculated that a concave rotating Fresnel lens 1000 kilometres across, yet only a few millimeters thick, floating in space at the L_1 point, would reduce the solar energy reaching the Earth by approximately 0.5% to 1%. He estimated that this would cost around US$10 billion up front, and another $10 billion in supportive cost during its lifespan. One issue with implementing such a solution is the need to counteract the effects of the solar wind moving such megastructures out of position.

Governance

Climate engineering poses several challenges in the context of governance because of issues of power and jurisdiction. Climate engineering as a climate change solution differs from other mitigation and adaptation strategies. Unlike a carbon trading system that would be focused on participation from multiple parties along with transparency, monitoring measures and compliance procedures; this is not necessarily required by climate engineering. Bengtsson (2006) argues that "the artificial release of sulphate aerosols is a commitment of at least several hundred years". Yet this is true only if a long-term deployment strategy is adopted. Under a short-term, temporary strategy, implementation would instead be limited to decades. Both cases, however, highlight the importance for a political framework that is sustainable enough to contain a multilateral commitment over such a long period and yet is flexible as the techniques innovate through time. There are many controversies surrounding this topic and hence, climate engineering has been made into a very political issue. Most discussions and debates are not about which climate engineering technique is better than the other, or which one is more economically and socially feasible. Discussions are broadly on who will have control over the deployment of climate engineering and under what governance regime the deployment can be monitored and supervised. This is especially important due to the regional variability of the effects of many climate engineering techniques, benefiting some countries while damaging others. The challenge posed by climate engineering is not how to get countries to do it. It is to address the fundamental question of who should decide whether and how climate engineering should be attempted – a problem of governance.

Solar radiation management raises a number of governance challenges. David Keith argues that the cost is within the realm of small countries, large corporations, or even very wealthy individuals. David Victor suggests that climate engineering is within the reach of a lone "Greenfinger," a wealthy individual who takes it upon him or herself to be the "self-appointed protector of the planet". However, it has been argued that a rogue state threatening solar radiation management may strengthen action on mitigation.

Legal and regulatory systems may face a significant challenge in effectively regulating solar radiation management in a manner that allows for an acceptable result for society. There are, however, significant incentives for states to cooperate in choosing a specific climate engineering policy, which make unilateral deployment a rather unlikely event.

Some researchers have suggested that building a global agreement on climate engineering deployment will be very difficult, and instead power blocs are likely to emerge.

Public Attitudes

There have been a handful of studies into attitudes to and opinions of solar radiation management. These generally find low levels of awareness, uneasiness with the implementation of solar radiation management, cautious support of research, and a preference for greenhouse gas emissions reduction. As is often the case with public opinions regarding emerging issues, the responses are highly sensitive to the questions' particular wording and context.

One cited objection to implementing a short-term temperature fix is that there might then be less incentive to reduce carbon dioxide emissions until it caused some other environmental catastrophe, such as a chemical change in ocean water that could be disastrous to ocean life.

Carbon Offset

A carbon offset is a reduction in emissions of carbon dioxide or greenhouse gases made in order to compensate for or to offset an emission made elsewhere.

Carbon offsets are measured in metric tons of carbon dioxide-equivalent (CO_2e) and may represent six primary categories of greenhouse gases: carbon dioxide (CO_2), methane (CH_4), nitrous oxide (N_2O), perfluorocarbons (PFCs), hydrofluorocarbons (HFCs), and sulfur hexafluoride (SF_6). One carbon offset represents the reduction of one metric ton of carbon dioxide or its equivalent in other greenhouse gases.

There are two markets for carbon offsets. In the larger, compliance market, companies, governments, or other entities buy carbon offsets in order to comply with caps on the total amount of carbon dioxide they are allowed to emit. This market exists in order to achieve compliance with obligations of Annex 1 Parties under the Kyoto Protocol, and of liable entities under the EU Emission Trading Scheme. In 2006, about $5.5 billion of carbon offsets were purchased in the compliance market, representing about 1.6 billion metric tons of CO_2e reductions.

In the much smaller, voluntary market, individuals, companies, or governments purchase carbon offsets to mitigate their own greenhouse gas emissions from transportation, electricity use, and other sources. For example, an individual might purchase carbon offsets to compensate for the greenhouse gas emissions caused by personal air travel. Many companies offer carbon offsets as an up-sell during the sales process so that customers can mitigate the emissions related with their product or service purchase (such as offsetting emissions related to a vacation flight, car rental, hotel stay, consumer good, etc.). In 2008, about $705 million of carbon offsets were purchased in the voluntary market, representing about 123.4 million metric tons of CO_2e reductions. Some fuel suppliers in the UK offer fuel which has been carbon offset such as Fuel dyes.

Offsets are typically achieved through financial support of projects that reduce the emission of greenhouse gases in the short- or long-term. The most common project type is renewable energy, such as wind farms, biomass energy, or hydroelectric dams. Others include energy efficiency projects, the destruction of industrial pollutants or agricultural byproducts, destruction of landfill methane, and forestry projects. Some of the most popular carbon offset projects from a corporate perspective are energy efficiency and wind turbine projects.

Carbon offsetting has gained some appeal and momentum mainly among consumers in western countries who have become aware and concerned about the potentially negative environmental effects of energy-intensive lifestyles and economies. The Kyoto Protocol has sanctioned offsets as a way for governments and private companies to earn carbon credits that can be traded on a marketplace. The protocol established the Clean Development Mechanism (CDM), which validates and measures projects to ensure they produce authentic benefits and are genuinely "additional" activities that would not otherwise have been undertaken. Organizations that are unable to meet their emissions quota can offset their emissions by buying CDM-approved Certified Emissions Reductions. Emissions from burning fuel, such as red diesel, has pushed one UK fuel supplier to create a carbon offset fuel named Carbon Offset Red Diesel.

Offsets may be cheaper or more convenient alternatives to reducing one's own fossil-fuel consumption. However, some critics object to carbon offsets, and question the benefits of certain types of offsets. Due diligence is recommended to help businesses in the assessment and identification of "good quality" offsets to ensure offsetting provides the desired additional environmental benefits, and to avoid reputational risk associated with poor quality offsets.

Offsets are viewed as an important policy tool to maintain stable economies. One of the hidden dangers of climate change policy is unequal prices of carbon in the economy, which can cause economic collateral damage if production flows to regions or industries that have a lower price of carbon—unless carbon can be purchased from that area, which offsets effectively permit, equalizing the price.

Wind turbines near Aalborg, Denmark. Renewable energy projects are the most common source of carbon offsets.

Definitions

Features

1. Carbon offsets have several common features:

2. *Vintage.* The vintage is the year in which the carbon reduction takes place.

3. *Source.* The source refers to the project or technology used in offsetting the carbon emissions. Projects can include land-use, methane, biomass, renewable energy and industrial energy efficiency. Projects may also have secondary benefits (co-benefits). For example, projects that reduce agricultural greenhouse gas emissions may improve water quality by reducing fertilizer usage.

Certification regime. The certification regime describes the systems and procedures that are used to certify and register carbon offsets. Different methodologies are used for measuring and verifying emissions reductions, depending on project type, size and location. For example, the CDM uses another. In the voluntary market, a variety of industry standards exist. These include the Voluntary Carbon Standard and the CDM Gold Standard that are implemented to provide third-party verification of carbon offset projects. There are some additional standards for the validation of co-benefits, including the CCBS, issued by the Climate, Community & Biodiversity Alliance and the Social Carbon Standard, issued by Ecologica Institute.

Carbon Offset Markets

Global Market

In 2009, 8.2 billion metric tons of carbon dioxide equivalent changed hands worldwide, up 68 per cent from 2008, according to the study by carbon-market research firm Point Carbon, of Washington and Oslo. But at EUR94 billion, or about $135 billion, the market's value was nearly unchanged compared with 2008, with world carbon prices averaging EUR11.40 a ton, down about 40 per cent from the previous year, according to the study. The World Bank's "State and Trends of the Carbon Market 2010" put the overall value of the market at $144 billion, but found that a significant part of this figure resulted from manipulation of a VAT loophole.

· 90% of voluntary offset volumes were contracted by the private sector – where corporate social responsibility and industry leadership were primary motivations for offset purchases.

· Offset buyers' desire to positively impact the climate resilience of their supply chain or sphere of influence was evident in our data which identifies a strong relationship between buyers' business sectors and the project categories from which they contract offsets.

E.U. Market

The global carbon market is dominated by the European Union, where companies that emit greenhouse gases are required to cut their emissions or buy pollution allowances or carbon credits from the market, under the European Union Emission Trading Scheme (EU ETS). Europe, which has seen volatile carbon prices due to fluctuations in energy prices and supply and demand, will continue to dominate the global carbon market for another few years, as the U.S. and China—the world's top polluters—have yet to establish mandatory emission-reduction policies.

U.S. Market

On the whole, the U.S. market remains primarily a voluntary market, but multiple cap and trade regimes are either fully implemented or near-imminent at the regional level. The first mandatory, market-based cap-and-trade program to cut CO_2 in the U.S., called the Regional Greenhouse Gas Initiative (RGGI), kicked into gear in Northeastern states in 2009, growing nearly tenfold to $2.5 billion, according to Point Carbon. Western Climate Initiative (WCI)—a regional cap-and-trade program including seven western states (California notably among them) and four Canadian provinces—has established a regional target for reducing heat-trapping emissions of 15 percent below 2005 levels by 2020. A component of California's own Global Warming Solutions Act of 2006, kicked off in early 2013, requires high-emissions industries to purchase carbon credits to cover emissions in excess of 25,000 CO2 metric tons.

Voluntary Market

Participants

A wide range of participants are involved in the voluntary market, including providers of different types of offsets, developers of quality assurance mechanisms, third party verifiers, and consumers who purchase offsets from domestic or international providers. Suppliers include for-profit companies, governments, charitable non-governmental organizations, colleges and universities, and other groups.

Motivations

According to industry analyst Ecosystem Marketplace, the voluntary markets present the opportunity for citizen consumer action, as well as an alternative source of carbon finance and an incubator for carbon market innovation. In their survey of voluntary markets, data has shown that "Corporate Social Responsibility" and "Public Relations/Branding" are clearly in first place among motivations for voluntary offset purchases, with evidence indicating that companies seek to offset emissions "for goodwill, both of the general public and their investors".

In addition, regarding market composition, research indicates: "Though many analysts perceive pre-compliance buying as a dominant driving force in the voluntary market, the results of our survey have repeatedly indicated that precompliance motives (as indicated by 'investment/resale and 'anticipation of regulation') remain secondary to those of the pure voluntary market (companies/individuals offsetting their emissions)."

Pre-compliance & trading

The other main category of buyers on the voluntary markets are those engaged in pre-compliance and/or trading. Those purchasing offsets for pre-compliance purposes are doing so with the expectation, or as a hedge against the possibility, of future mandatory cap and trade regulations. As a mandatory cap would sharply increase the price of offsets, firms—especially those with large carbon footprints and the corresponding financial exposure to regulation—make the decision to acquire offsets in advance at what are expected to be lower prices.

The trading market in offsets in general resembles the trade in other commodities markets, with financial professionals including hedge funds and desks at major investment banks, taking positions in the hopes of buying cheap and selling dear, with their motivation typically short or medium term financial gain.

Retail

Multiple players in the retail market have offerings that enable consumers and businesses to calculate their carbon footprint, most commonly through a web-based interface including a calculator or questionnaire, and sell them offsets in the amount of that footprint. In addition many companies selling products and services, especially carbon-intensive ones such as airline travel, offer options to bundle a proportional offsetting amount of carbon credits with each transaction.

Suppliers of voluntary offsets operate under both nonprofit and social enterprise models, or a blended approach sometimes referred to as triple bottom line. Other suppliers include broader environmentally focused organizations with website subsections or initiatives that enable retail voluntary offset purchases by members, and government created projects.

Features of companies that voluntarily offset emissions

Companies that voluntarily offset their own emissions tend to be of relatively low carbon intensity, as they can offset a significant proportion of their emissions at relatively low cost. Voluntary offsetting is particularly common in the financial sector. 61 per cent of financial companies in the FTSE 100 offset at least a portion of their 2009 emissions. Twenty-two per cent of financial companies in the FTSE 100 considered their entire 2009 operations to be carbon neutral.

Sources of Carbon Offsets

The CDM identifies over 200 types of projects suitable for generating carbon offsets, which are grouped into broad categories. These project types include renewable energy, methane abatement, energy efficiency, reforestation and fuel switching.

Renewable Energy

Renewable energy offsets commonly include wind power, solar power, hydroelectric power and biofuel. Some of these offsets are used to reduce the cost differential between renewable and conventional energy production, increasing the commercial viability of a choice to use renewable energy sources.

Renewable Energy Credits (RECs) are also sometimes treated as carbon offsets, although the concepts are distinct. Whereas a carbon offset represents a reduction in greenhouse gas emissions, a REC represents a quantity of energy produced from renewable sources. To convert RECs into offsets, the clean energy must be translated into carbon reductions, typically by assuming that the clean energy is displacing an equivalent amount of conventionally produced electricity from the local grid. This is known as an indirect offset (because the reduction doesn't take place at the project site itself, but rather at an external site), and some controversy surrounds the question of whether they truly lead to "additional" emission reductions and who should get credit for any reductions that may occur. Intel corporation is the largest purchaser of renewable power in the US.

Methane Collection and Combustion

Some offset projects consist of the combustion or containment of methane generated by farm animals (by use of an anaerobic digester), landfills or other industrial waste. Methane has a global warming potential (GWP) 23 times that of CO_2; when combusted, each molecule of methane is converted to one molecule of CO_2, thus reducing the global warming effect by 96%.

An example of a project using an anaerobic digester can be found in Chile where in December 2000, the largest pork production company in Chile, initiated a voluntary process to implement advanced waste management systems (anaerobic and aerobic digestion of hog manure), in order to reduce greenhouse gas (GHG) emissions.

Energy Efficiency

While carbon offsets that fund renewable energy projects help lower the carbon intensity of energy *supply*, energy conservation projects seek to reduce the overall *demand* for energy. Carbon offsets in this category fund projects of several types:

Chicago Climate Justice activists protesting cap and trade legislation in front of Chicago Climate Exchange building in Chicago Loop

1. Cogeneration plants generate both electricity and heat from the same power source, thus improving upon the energy efficiency of most power plants, which waste the energy generated as heat.

2. Fuel efficiency projects replace a combustion device with one using less fuel per unit of energy provided. Assuming energy demand does not change, this reduces the carbon dioxide emitted.

3. Energy-efficient buildings reduce the amount of energy wasted in buildings through efficient heating, cooling or lighting systems. In particular, the replacement of incandescent light bulbs with compact fluorescent lamps can have a drastic effect on energy consumption. New buildings can also be constructed using less carbon-intensive input materials.

Destruction of Industrial Pollutants

Industrial pollutants such as hydrofluorocarbons (HFCs) and perfluorocarbons (PFCs) have a GWP many thousands of times greater than carbon dioxide by volume. Because these pollutants are easily captured and destroyed at their source, they present a large and low-cost source of carbon offsets. As a category, HFCs, PFCs, and N_2O reductions represent 71 per cent of offsets issued under the CDM.

Land use, Land-use Change And forestry

Land use, land-use change and forestry (LULUCF) projects focus on natural carbon sinks such as forests and soil. Deforestation, particularly in Brazil, Indonesia and parts of Africa, account for about 20 per cent of greenhouse gas emissions. Deforestation can be avoided either by paying directly for forest preservation, or by using offset funds to provide substitutes for forest-based products. There is a class of mechanisms referred to as REDD schemes (Reducing emissions from deforestation and forest degradation), which may be included in a post-Kyoto agreement. REDD credits provide carbon offsets for the protection of forests, and provide a possible mechanism to allow funding from developed nations to assist in the protection of native forests in developing nations.

Almost half of the world's people burn wood (or fiber or dung) for their cooking and heating needs. Fuel-efficient cook stoves can reduce fuel wood consumption by 30 to 50%, though the warming of the earth due to decreases in particulate matter (i.e. smoke) from such fuel-efficient stoves has not been addressed. There are a number of different types of LULUCF projects:

1. Avoided deforestation is the protection of existing forests.

2. Reforestation is the process of restoring forests on land that was once forested.

3. Afforestation is the process of creating forests on land that was previously unforested, typically for longer than a generation.

4. Soil management projects attempt to preserve or increase the amount of carbon sequestered in soil.

Purchase of Carbon Allowances from Emissions Trading Schemes

Voluntary purchasers can offset their carbon emissions by purchasing carbon allowances from legally mandated cap-and-trade programs such as the Regional Greenhouse Gas Initiative or the European Emissions Trading Scheme. By purchasing the allowances that power plants, oil refineries, and industrial facilities need to hold to comply with a cap, voluntary purchases tighten the cap and force additional emissions reductions.

Voluntary purchases can also be made through small-scale and sometimes uncertified schemes such as those offered at South African based Promoting Access to Carbon Equity Centre (PACE), which nevertheless offer clear services such as poverty alleviation in the form of renewable energy development. Also, as "easy carbon credits are coming to an end", these projects have the potential to develop projects that are either too small or too complicated to benefit from legally mandated cap-and-trade programs.

Links with Emission Trading Schemes

Once it has been accredited by the UNFCCC a carbon offset project can be used as carbon credit and linked with official emission trading schemes, such as the European Union Emission Trading Scheme or Kyoto Protocol, as Certified Emission Reductions. European emission allowances for the 2008–2012 second phase were selling for between 21 and 24 Euros per metric ton of CO_2 as of July 2007.

The voluntary Chicago Climate Exchange also includes a carbon offset scheme that allows offset project developers to sell emissions reductions to CCX members who have voluntarily agreed to meet emissions reduction targets.

The Western Climate Initiative, a regional greenhouse gas reduction initiative by states and provinces along the western rim of North America, includes an offset scheme. Likewise, the Regional Greenhouse Gas Initiative, a similar program in the northeastern U.S., includes an offset program. A credit mechanism that uses offsets may be incorporated in proposed schemes such as the Australian Carbon Exchange.

Other

A UK offset provider set up a carbon offsetting scheme that set up a secondary market for treadle pumps in developing countries. These pumps are used by farmers, using human power, in place of diesel pumps. However, given that treadle pumps are best suited to pumping shallow water, while diesel pumps are usually used to pump water from deep boreholes, it is not clear that the treadle pumps are actually achieving real emissions reductions. Other companies have explored and rejected treadle pumps as a viable carbon offsetting approach due to these concerns.

Carbon Retirement

Carbon retirement involves retiring allowances from emission trading schemes as a method for offsetting carbon emissions. Under schemes such as the European Union Emission Trading Scheme, EU Emission Allowances (EUAs), which represent the right to release carbon dioxide into the atmosphere, are issued to all the largest polluters. The theory is that by buying these allowances and permanently removing them, the price of EUAs increases and provides an incentive for industrial companies to reduce their emissions.

Accounting for and Verifying Reductions

Due to their indirect nature, many types of offset are difficult to verify. Some providers obtain independent certification that their offsets are accurately measured, to distance themselves from potentially fraudulent competitors. The credibility of the various certification providers is often questioned. Certified offsets may be purchased from commercial or non-profit organizations for US\$2.75–99.00 per tonne of CO_2, due to fluctuations of market price. Annual carbon dioxide emissions in developed countries range from 6 to 23 tons per capita.

Accounting systems differ on precisely what constitutes a valid offset for voluntary reduction systems and for mandatory reduction systems. However formal standards for quantification exist based on collaboration between emitters, regulators, environmentalists and project developers. These standards include the Voluntary Carbon Standard, Green-e Climate, Chicago Climate Exchange and the CDM Gold Standard, the latter of which expands upon the requirements for the Clean Development Mechanism of the Kyoto Protocol.

Accounting of offsets may address the following basic areas:

1. Baseline and Measurement—What emissions would occur in the absence of a proposed project? And how are the emissions that occur after the project is performed going to be measured?

2. Additionality—Would the project occur anyway without the investment raised by selling carbon offset credits? There are two common reasons why a project may lack additionality: (a) if it is intrinsically financially worthwhile due to energy cost savings, and (b) if it had to be performed due to environmental laws or regulations.

3. Permanence—Are some benefits of the reductions reversible? (for example, trees may be harvested to burn the wood, and does growing trees for fuel wood decrease the need for fossil fuel?) If wood-

lands are increasing in area or density, then carbon is being sequestered. After roughly 50 years, newly planted forests will reach maturity and remove carbon dioxide more slowly.

4. Leakage—Does implementing the project cause higher emissions outside the project boundary?

Co-benefits

Overall, carbon offsets improve the environment by reducing the amount of greenhouse gases in the earth's atmosphere. Offset projects often also lead to a number of co-benefits such as better air and water quality, and healthier communities.

While the primary goal of carbon offsets is to reduce global carbon emissions, many offset projects also claim to lead to improvements in the quality of life for a local population. These additional improvements are termed *co-benefits*, and may be considered when evaluating and comparing carbon offset projects. Some possible co-benefits from a project that replaces wood-burning stoves with ovens using a less carbon-intensive fuel include:

1. Lower non—greenhouse gas pollution (smoke, ash, and chemicals), which improves health in the home.

2. Better preservation of forests, which are an important habitat for wildlife.

In a recent survey conducted by EcoSecurities, Conservation International, CCBA and ClimateBiz, of the 120 corporates surveyed more than 77 per cent rated community and environmental benefits as the prime motivator for purchasing carbon offsets.

Carbon offset projects can also negatively affect quality of life. For example, people who earn their livelihoods from collecting firewood and selling it to households could become unemployed if firewood is no longer used. A paper from the Overseas Development Institute offers some indicators to be used in assessing the potential developmental impacts of voluntary carbon offset schemes:

1. What potential does the project have for income generation?

2. What effects might a project have on future changes in land use and could conflicts arise from this?

3. Can small-scale producers engage in the scheme?

4. What are the 'add on' benefits to the country—for example, will it assist capacity-building in local institutions?

Putting a price on carbon encourages innovation by providing funding for new ways to reduce greenhouse gases in many sectors. Carbon reduction goals drive the demand for offsets and carbon trading, encouraging the development of this new industry and offering opportunities for different sectors to develop and use innovative new technologies.

Carbon offset projects also provide savings – energy efficiency measures may reduce fuel or electricity consumption, leading to a potential reduction in maintenance and operating costs.

Quality Assurance Schemes

Quality Assurance Standard for Carbon Offsetting (Qas)

In an effort to inform and safeguard business and household consumers purchasing Carbon Offsets, in 2009, the UK Government has launched a scheme for regulating Carbon offset products. DEFRA have created the "Approved Carbon Offsetting" brand to use as an endorsement on offsets approved by the UK government. The Scheme sets standards for best practice in offsetting. Approved offsets have to demonstrate the following criteria:

1. Accurate calculation of emissions to be offset

2. Use of good quality carbon credits i.e. initially those that are Kyoto compliant

3. Cancellation of carbon credits within a year of the consumers purchase of the offset

4. Clear and transparent pricing of the offset

5. Provision of information about the role of offsetting in tackling climate change and advice on how a consumer can reduce his or her carbon footprint

The first company to qualify for the scheme was Clear, followed by Carbon Footprint, Carbon Passport, Pure, British Airways and Carbon Retirement Ltd.

On 20 May 2011 the Department of Energy and Climate Change announced that the Quality Assurance Scheme would close on 30 June 2011. The stated purpose of the Quality Assurance Scheme was 'to provide a straightforward route for those wishing to offset their emissions to identify quality offsets'. Critics of the closure therefore argued that without the scheme, businesses and individuals would struggle to identify quality carbon offsets.

In 2012 the scheme was relaunched as the Quality Assurance Standard (QAS). The QAS is now run independently by Quality Assurance Standard Ltd which is a company limited by guarantee based in the United Kingdom. The Quality Assurance Standard is a comprehensive independent audit system for carbon offsets. Approved offsets are checked against a 40-point checklist to ensure they meet the very highest standards in the world.

On 17 July 2012, the first organisations were approved as meeting the new QAS.

Australian Government National Carbon Offset Program

The Australian government is currently in a consultation period on the regulation of Carbon Offsets. On 20 December 2013, the Australian Government released the Emissions Reduction Fund Green Paper outlining its preferred design options for the Emissions Reduction Fund: a carbon buy-back model. The Government invites public comment and written submissions on the Green Paper by 5pm on Friday 21 February 2014.

Controversies

Project-offsetting

Less than 30 pence in every pound spent on some carbon offset schemes goes directly to projects designed to reduce emissions. The figures reported by the BBC and based on UN data reported that typically 28p goes to the set up and maintenance costs of an environmental project. 34p goes to the company that takes on the risk that the project may fail. The project's investors take 19p, with smaller amounts of money being distributed between organisations involved in brokering and auditing the carbon credits. In that respect carbon Offsets are similar to most consumer products, with only a fraction of sale prices going to the off-shore producers, the rest being shared between investors and distributors who bring it to the markets, who themselves need to pay their employees and service providers such as advertising agencies most of the time located in expensive areas.

Indulgence Controversy

Some activists disagree with the principle of carbon offsets, likening them to Roman Catholic indulgences, a way for the guilty to pay for absolution rather than changing their behavior. George Monbiot, an English environmentalist and writer, says that carbon offsets are an excuse for business as usual with regard to pollution. Proponents hold that the indulgence analogy is flawed because they claim carbon offsets actually reduce carbon emissions, changing the business as usual, and therefore address the root cause of climate change. Proponents of offsets claim that third-party certified carbon offsets are leading to increased investment in renewable energy, energy efficiency, methane biodigesters and reforestation and avoided deforestation projects, and claim that these alleged effects are the intended goal of carbon offsets. On October 16, 2009 responsibletravel.com, once a strong voice in favour of carbon offsetting, announced that it would stop offering carbon offsetting to its clients, stating that "too often offsets are being used by the tourism industry in developed countries to justify growth plans on the basis that money will be donated to projects in developing countries. Global reduction targets will not be met this way".

On 4 February 2010, travel networking site Vida Loca Travel announced that they would donate 5 per cent of profits to International Medical Corps, as they feel that international aid can be more effective at cutting global warming in the long term than carbon offsetting, citing the work of economist Jeffrey Sachs.

Effectiveness of Tree-planting Offsets

Some environmentalists have questioned the effectiveness of tree-planting projects for carbon offset purposes. Critics point to the following issues with tree planting projects:

Timing. Trees reach maturity over a course of many decades. Project developers and offset retailers typically pay for the project and sell the promised reductions up-front, a practice known as "forward selling".

1. Permanence. It is difficult to guarantee the permanence of the forests, which may be susceptible to clearing, burning, or mismanagement. The well-publicized instance of the "Coldplay forest", in which a forestry project supported by the British band Coldplay resulted in a grove of dead mango trees, illustrates the difficulties of guaranteeing the permanence of tree-planting offsets. When discussing "tree offsets, forest campaigner Jutta Kill of European environmental group FERN, clarified the physical reality that "Carbon in trees is temporary: Trees can easily

release carbon into the atmosphere through fire, disease, climatic changes, natural decay and timber harvesting."

2. Monocultures and invasive species. In an effort to cut costs, some tree-planting projects introduce fast-growing invasive species that end up damaging native forests and reducing biodiversity. For example, in Ecuador, the Dutch FACE Foundation has an offset project in the Andean Páramo involving 220 square kilometres of eucalyptus and pine planted. The NGO Acción Ecológica criticized the project for destroying a valuable Páramo ecosystem by introducing exotic tree species, causing the release of much soil carbon into the atmosphere, and harming local communities who had entered into contracts with the FACE Foundation to plant the trees. However, some certification standards, such as the Climate Community and Biodiversity Standard require multiple species plantings.

3. Methane. A recent study has claimed that plants are a significant source of methane, a potent greenhouse gas, raising the possibility that trees and other terrestrial plants may be significant contributors to global methane levels in the atmosphere. However, this claim has been disputed recently by findings in another study.

4. The albedo effect. Another study suggested that "high latitude forests probably have a net warming effect on the Earth's climate", because their absorption of sunlight creates a warming effect that balances out their absorption of carbon dioxide.

5. Necessity. Corporate tree-planting is not a new idea; farming operations have been used by companies making paper from trees for a long time. If farmed trees are replanted, and the products made from them are placed into landfills rather than recycled, a very safe, efficient, economical and time-proven method of geological sequestration of greenhouse carbon is the result of the paper product use cycle. This only holds if the paper in the land fill is not decomposed. In most landfills, this is the case and leads to the fact that more than half of the greenhouse gas emissions from the life cycle of paper products occur from landfill methane emissions.

Indigenous Land Rights Issues

Tree-planting projects can cause conflicts with indigenous people who are displaced or otherwise find their use of forest resources curtailed. For example, a World Rainforest Movement report documents land disputes and human rights abuses at Mount Elgon. In March 2002, a few days before receiving Forest Stewardship Council certification for a project near Mount Elgon, the Uganda Wildlife Authority evicted more than 300 families from the area and destroyed their homes and crops. That the project was taking place in an area of on-going land conflict and alleged human rights abuses did not make it into project report. A 2011 report by Oxfam International describes a case where over 20,000 farmers in Uganda were displaced for a FSC-certified plantation to offset carbon by London-based New Forests Company

Additionality and Lack of Regulation in the Voluntary Market

Several certification standards exist, offering variations for measuring emissions baseline, reductions, additionality, and other key criteria. However, no single standard governs the industry, and some offset providers have been criticized on the grounds that carbon reduction claims are exaggerated or misleading. Problems include:

1. Widespread instances of people and organizations buying worthless credits that do not yield any reductions in carbon emissions.

2. Industrial companies profiting from doing very little – or from gaining carbon credits on the basis of efficiency gains from which they have already benefited substantially.

3. Brokers providing services of questionable or no value.

4. A shortage of verification, making it difficult for buyers to assess the true value of carbon credits.

Perverse Incentives

Because offsets provide a revenue stream for the reduction of some types of emissions, they can in some cases provide incentives to emit more, so that emitting entities can later get credit for reducing emissions from an artificially high baseline. This is especially the case for offsets with a high profit margin. For example, one Chinese company generated $500 million in carbon offsets by installing a $5 million incinerator to burn the HFCs produced by the manufacture of refrigerants. The huge profits provided incentive to create new factories or expand existing factories solely for the purpose of increasing production of HFCs and then destroying the resultant pollutants to generate offsets. Not only is this outcome environmentally undesirable, it undermines other offset projects by causing offset prices to collapse. The practice had become so common that offset credits are now no longer awarded for new plants to destroy HFC-23.

In Nigeria oil companies *flare off* 40 per cent of the natural gas found. The Agip Oil Company plans to build plants to generate electricity from this gas and thus claim 1.5 million offset credits a year. United States company Pan Ocean Oil Corporation has also applied for credits in exchange for processing its own waste gas in Nigeria. Oilwatch.org's Michael Karikpo calls this "outrageous", as flaring is illegal in Nigeria, adding that "*It's like a criminal demanding money to stop committing crimes*".

Other Negative Impacts from Offset Projects

Although many carbon offset projects tout their environmental co-benefits, some are accused of having negative secondary effects. Point Carbon has reported on an inconsistent approach with regard to some hydro-electric projects as carbon offsets; some countries in the EU are not allowing large projects into the EU ETS, because of their environmental impacts, even though they have been individually approved by the UNFCCC and World Commission on Dams. It is difficult to assess the exact results of carbon offsets given the fact that they are a relatively new form of carbon reduction, and it is possible that some carbon offset purchases are made in an attempt to increase positive business public relations rather than to help solve the issue of greenhouse gas emissions.

Offset projects may also have negative social impacts, for example when local residents are evicted to enable a National Park to be marketed as a carbon offset.

Carbon Sequestration

Carbon sequestration is the process involved in carbon capture and the long-term storage of atmospheric carbon dioxide. Carbon sequestration describes long-term storage of carbon dioxide or other forms of carbon to either mitigate or defer global warming and avoid dangerous climate change. It has been proposed as a way to slow the atmospheric and marine accumulation of greenhouse gases, which are released by burning fossil fuels.

Carbon dioxide (CO2) is naturally captured from the atmosphere through biological, chemical, and physical processes. Artificial processes have been devised to produce similar effects, including large-scale, artificial capture and sequestration of industrially produced CO2 using subsurface saline aquifers, reservoirs, ocean water, aging oil fields, or other carbon sinks.

Description

Carbon sequestration is the process involved in carbon capture and the long-term storage of atmospheric carbon dioxide (CO2) and may refer specifically to:

1. "The process of removing carbon from the atmosphere and depositing it in a reservoir." When carried out deliberately, this may also be referred to as carbon dioxide removal, which is a form of geoengineering.

2. Carbon capture and storage, where carbon dioxide is removed from flue gases (e.g., at power stations) before being stored in underground reservoirs.

3. Natural biogeochemical cycling of carbon between the atmosphere and reservoirs, such as by chemical weathering of rocks.

Carbon sequestration describes long-term storage of carbon dioxide or other forms of carbon to either mitigate or defer global warming and avoid dangerous climate change. It has been proposed as a way to slow the atmospheric and marine accumulation of greenhouse gases, which are released by burning fossil fuels.

Carbon dioxide is naturally captured from the atmosphere through biological, chemical or physical processes. Some anthropogenic sequestration techniques exploit these natural processes, while some use entirely artificial processes.

Carbon dioxide may be captured as a pure by-product in processes related to petroleum refining or from flue gases from power generation. CO2 sequestration includes the storage part of carbon capture and storage, which refers to large-scale, artificial capture and sequestration of industrially produced CO2 using subsurface saline aquifers, reservoirs, ocean water, aging oil fields, or other carbon sinks.

Biological Processes

Biosequestration or carbon sequestration through biological processes affects the global carbon cycle. Examples include major climatic fluctuations, such as the Azolla event, which created the current Arctic climate. Such processes created fossil fuels, as well as clathrate and limestone. By manipulating such processes, geoengineers seek to enhance sequestration.

An oceanic phytoplankton bloom in the South Atlantic Ocean, off the coast of Argentina. Encouraging such blooms with iron fertilization could lock up carbon on the seabed.

Peat Production

Peat bogs are a very important carbon store. By creating new bogs, or enhancing existing ones, carbon can be sequestered.

Reforestation

Reforestation is the replanting of trees on marginal crop and pasture lands to incorporate carbon from atmospheric CO_2 into biomass. For this process to succeed the carbon must not return to the atmosphere from mass burning or rotting when the trees die. To this end, land allotted to the trees must not be converted to other uses and management of the frequency of disturbances might be necessary in order to avoid extreme events. Alternatively, the wood from them must itself be sequestered, e.g., via biochar, bio-energy with carbon storage (BECS), landfill or 'stored' by use in e.g. construction. Short of growth in perpetuity, however, reforestation with long-lived trees (>100 years) will sequester carbon for a more graduated release, minimizing impact during the expected carbon crisis of the 21st century.

Urban Forestry

Urban Forestry adds carbon with new tree sites and the sequestration of carbon over the lifetime of the tree. There are several methods of accounting for carbon storage, one of which, I-Tree is easy to use.

Wetland Restoration

Wetland soil is an important carbon sink; 14.5% of the world's soil carbon is found in wetlands, while only 6% of the world's land is composed of wetlands.

Agriculture

Globally, soils are estimated to contain approximately 1,500 gigatons of organic carbon to 1 m depth, more than the amount in vegetation and the atmosphere.

Modification of agricultural practices is a recognized method of carbon sequestration as soil can act as an effective carbon sink offsetting as much as 20% of 2010 carbon dioxide emissions annually.

Carbon emission reduction methods in agriculture can be grouped into two categories: reducing and/or displacing emissions and enhancing carbon removal. Some of these reductions involve increasing the efficiency of farm operations (e.g. more fuel-efficient equipment) while some involve interruptions in the natural carbon cycle. Also, some effective techniques (such as the elimination of stubble burning) can negatively impact other environmental concerns (increased herbicide use to control weeds not destroyed by burning).

Reducing Emissions

Increasing yields and efficiency generally reduces emissions as well, since more food results from the same or less effort. Techniques include more accurate use of fertilizers, less soil disturbance, better irrigation, and crop strains bred for locally beneficial traits and increased yields.

Replacing more energy intensive farming operations can also reduce emissions. Reduced or no-till farming requires less machine use and burns correspondingly less fuel per acre. However, no-till usually increases use of weed-control chemicals and the residue now left on the soil surface is more likely to release its CO_2 to the atmosphere as it decays, reducing the net carbon reduction.

In practice, most farming operations that incorporate post-harvest crop residues, wastes and by-products back into the soil provide a carbon storage benefit. This is particularly the case for practices such as field burning of stubble - rather than releasing almost all of the stored CO_2 to the atmosphere, tillage incorporates the biomass back into the soil.

Enhancing Carbon Removal

All crops absorb CO_2 during growth and release it after harvest. The goal of agricultural carbon removal is to use the crop and its relation to the carbon cycle to permanently sequester carbon within the soil. This is done by selecting farming methods that return biomass to the soil and enhance the conditions in which the carbon within the plants will be reduced to its elemental nature and stored in a stable state. Methods for accomplishing this include:

1. Use cover crops such as grasses and weeds as temporary cover between planting seasons

2. Concentrate livestock in small paddocks for days at a time so they graze lightly but evenly. This encourages roots to grow deeper into the soil. Stock also till the soil with their hooves, grinding old grass and manures into the soil.

3. Cover bare paddocks with hay or dead vegetation. This protects soil from the sun and allows the soil to hold more water and be more attractive to carbon-capturing microbes.

4. Restore degraded land, which slows carbon release while returning the land to agriculture or other use.

Agricultural sequestration practices may have positive effects on soil, air, and water quality, be beneficial to wildlife, and expand food production. On degraded croplands, an increase of 1 ton of soil carbon pool may increase crop yield by 20 to 40 kilograms per hectare of wheat, 10 to 20 kg/ha for maize, and 0.5 to 1 kg/ha for cowpeas.

The effects of soil sequestration can be reversed. If the soil is disrupted or tillage practices are abandoned, the soil becomes a net source of greenhouse gases. Typically after 15 to 30 years of sequestration, soil becomes saturated and ceases to absorb carbon. This implies that there is a global limit to the amount of carbon that soil can hold.

Many factors affect the costs of carbon sequestration including soil quality, transaction costs and various externalities such as leakage and unforeseen environmental damage. Because reduction of atmospheric CO_2 is a long-term concern, farmers can be reluctant to adopt more expensive agricultural techniques when there is not a clear crop, soil, or economic benefit. Governments such as Australia and New Zealand are considering allowing farmers to sell carbon credits once they document that they have sufficiently increased soil carbon content.

Ocean-related

Iron Fertilization

Ocean iron fertilization is an example of such a geoengineering technique. Iron fertilization attempts to encourage phytoplankton growth, which removes carbon from the atmosphere for at least a period of time. This technique is controversial due to limited understanding of its complete effects on the marine ecosystem, including side effects and possibly large deviations from expected behavior. Such effects potentially include release of nitrogen oxides, and disruption of the ocean's nutrient balance.

Natural iron fertilisation events (e.g., deposition of iron-rich dust into ocean waters) can enhance carbon sequestration. Sperm whales act as agents of iron fertilisation when they transport iron from the deep ocean to the surface during prey consumption and defecation. Sperm whales have been shown to increase the levels of primary production and carbon export to the deep ocean by depositing iron rich feces into surface waters of the Southern Ocean. The iron rich feces causes phytoplankton to grow and take up more carbon from the atmosphere. When the phytoplankton dies, some of it sinks to the deep ocean and takes the atmospheric carbon with it. By reducing the abundance of sperm whales in the Southern Ocean, whaling has resulted in an extra 200,000 tonnes of carbon remaining in the atmosphere each year.

Urea Fertilization

Ian Jones proposes fertilizing the ocean with urea, a nitrogen rich substance, to encourage phytoplankton growth.

Australian company Ocean Nourishment Corporation (ONC) plans to sink hundreds of tonnes of urea into the ocean to boost CO_2-absorbing phytoplankton growth as a way to combat climate change. In 2007, Sydney-based ONC completed an experiment involving 1 tonne of nitrogen in the Sulu Sea off the Philippines.

Mixing Layers

Encouraging various ocean layers to mix can move nutrients and dissolved gases around, offering avenues for geoengineering. Mixing may be achieved by placing large vertical pipes in the oceans to pump nutrient rich water to the surface, triggering blooms of algae, which store carbon when they grow and export carbon when they die. This produces results somewhat similar to iron fertilization. One side-effect is a short-term rise in CO2, which limits its attractiveness.

Seaweed

Seaweed grows very fast and can theoretically be harvested and processed to generate biomethane, via Anaerobic Digestion to generate electricity, via Cogeneration/CHP or as a replacement for natural gas. One study suggested that if seaweed farms covered 9% of the ocean they could produce enough biomethane to supply Earth's equivalent demand for fossil fuel energy, remove 53 gigatonnes of CO_2 per year from the atmosphere and sustainably produce 200 kg per year of fish, per person, for 10 billion people. Ideal species for such farming and conversion include Laminaria digitata, Fucus serratus and Saccharina latissima.

Physical Processes

Biochar can be landfilled, used as a soil improver or burned using carbon capture and storage

Biomass-related

Bio-energy with carbon capture and storage (BECCS)

BECCS refers to biomass in power stations and boilers that use carbon capture and storage. The carbon sequestered by the biomass would be captured and stored, thus removing carbon dioxide from the atmosphere.

This technology is sometimes referred to as bio-energy with carbon storage, BECS, though this term can also refer to the carbon sequestration potential in other technologies, such as biochar.

Burial

Burying biomass (such as trees) directly, mimics the natural processes that created fossil fuels. Landfills also represent a physical method of sequestration.

Biochar Burial

Biochar is charcoal created by pyrolysis of biomass waste. The resulting material is added to a landfill or used as a soil improver to create terra preta. Addition of pyrogenic organic carbon (biochar) is a novel strategy to increase the soil-C stock for the long-term and to mitigate global-warming by offsetting the atmospheric C (up to 9.5 Pg C annually).

In the soil, the carbon is unavailable for oxidation to CO2 and consequential atmospheric release. This is one technique advocated by scientist James Lovelock, creator of the Gaia hypothesis. According to Simon Shackley, "people are talking more about something in the range of one to two billion tonnes a year."

The mechanisms related to biochar are referred to as bio-energy with carbon storage, BECS.

Ocean Storage

If CO_2 were to be injected to the ocean bottom, the pressures would be great enough for CO_2 to be in its liquid phase. The idea behind ocean injection would be to have stable, stationary pools of CO_2 at the ocean floor. The ocean could potentially hold over a thousand billion tons of CO_2. However, this avenue of sequestration isn't being as actively pursued because of concerns about the impact on ocean life, and concerns about its stability.

River mouths bring large quantities of nutrients and dead material from upriver into the ocean as part of the process that eventually produces fossil fuels. Transporting material such as crop waste out to sea and allowing it to sink exploits this idea to increase carbon storage. International regulations on marine dumping may restrict or prevent use of this technique.

Geological Sequestration

Geological sequestration refers to the storage of CO_2 underground in depleted oil and gas reservoirs, saline formations, or deep, un-minable coal beds.

Once CO_2 is captured from a gas or coal-fired power plant, it would be compressed to ~100 bar so that it would be a supercritical fluid. In this fluid form, the CO_2 would be easy to transport via pipeline to the place of storage. The CO_2 would then be injected deep underground, typically around 1 km, where it would be stable for hundreds to millions of years. At these storage conditions, the density of supercritical CO_2 is 600 to 800 kg / m³. For consumers, the cost of electricity from a coal-fired power plant with carbon capture and storage (CCS) is estimated to be 0.01 - 0.05 $ / kWh higher than without CCS. For reference, the average cost of electricity in the US in 2004 was 0.0762 $ / kWh. In other terms, the cost of CCS would be 20 - 70 $/ton of CO_2 captured. The transportation and injection of CO_2 is relatively cheap, with the capture costs accounting for 70 - 80% of CCS costs.

The important parameters in determining a good site for carbon storage are: rock porosity, rock permeability, absence of faults, and geometry of rock layers. The medium in which the CO_2 is to be stored ideally has a high porosity and permeability, such as sandstone or limestone. Sandstone can have a permeability ranging from 1 to 10^{-5} Darcy, and can have a porosity as high as ~30%. The porous rock must be capped by a layer of low permeability which acts as a seal, or caprock, for the CO_2. Shale is an example of a very good caprock, with a permeability of 10^{-5} to 10^{-9} Darcy. Once injected, the CO_2 plume will rise via buoyant forces, since it is less dense than its surroundings. Once it encounters a caprock, it will spread laterally until it encounters a gap. If there are fault planes near the injection zone, there is a possibility the CO_2 could migrate along the fault to the surface, leaking into the atmosphere, which would be potentially dangerous to life in the surrounding area. Another danger related to carbon sequestration is induced seismicity. If the injection of CO_2 creates pressures that are too high underground, the formation will fracture, causing an earthquake.

While trapped in a rock formation, CO_2 can be in the supercritical fluid phase or dissolve in groundwater/brine. It can also react with minerals in the geologic formation to precipitate carbonates.

Worldwide storage capacity in oil and gas reservoirs is estimated to be 675 - 900 Gt CO_2, and in un-minable coal seams is estimated to be 15 - 200 Gt CO_2. Deep saline formations have the largest capacity, which is estimated to be 1,000 - 10,000 Gt CO_2. In the US, there is an estimated 160 Gt CO_2 storage capacity.

There are a number of large-scale carbon capture and sequestration projects that have demonstrated the viability and safety of this method of carbon storage, which are summarized here by the Global CCS Institute. The dominant monitoring technique is seismic imaging, where vibrations are generated that propagate through the subsurface. The geologic structure can be imaged from the refracted/reflected waves.

The first large-scale CO2 sequestration project which began in 1996 is called Sleipner, and is located in the North Sea where Norway's StatoilHydro strips carbon dioxide from natural gas with amine solvents and disposed of this carbon dioxide in a deep saline aquifer. In 2000, a coal-fueled synthetic natural gas plant in Beulah, North Dakota, became the world's first coal-using plant to capture and store carbon dioxide, at the Weyburn-Midale Carbon Dioxide Project.

CO2 has been used extensively in enhanced crude oil recovery operations in the United States beginning in 1972. There are in excess of 10,000 wells that inject CO2 in the state of Texas alone. The gas comes in part from anthropogenic sources, but is principally from large naturally occurring geologic formations of CO2. It is transported to the oil-producing fields through a large network of over 5,000 kilometres (3,100 mi) of CO2 pipelines. The use of CO2 for enhanced oil recovery (EOR) methods in heavy oil reservoirs in the Western Canadian Sedimentary Basin (WCSB) has also been proposed. However, transport cost remains an important hurdle. An extensive CO2 pipeline system does not yet exist in the WCSB. Athabasca oil sands mining that produces CO2 is hundreds of kilometers north of the subsurface Heavy crude oil reservoirs that could most benefit from CO2 injection.

Chemical Processes

Developed in the Netherlands, an electrocatalysis by a copper complex helps reduce carbon dioxide to oxalic acid; This conversion uses carbon dioxide as a feedstock to generate oxalic acid.

Mineral Carbonation

Carbon, in the form of CO2 can be removed from the atmosphere by chemical processes, and stored in stable carbonate mineral forms. This process is known as 'carbon sequestration by mineral carbonation' or mineral sequestration. The process involves reacting carbon dioxide with abundantly available metal oxides–either magnesium oxide (MgO) or calcium oxide (CaO)–to form stable carbonates. These reactions are exothermic and occur naturally (e.g., the weathering of rock over geologic time periods).

$$CaO + CO_2 \rightarrow CaCO_3$$

$$MgO + CO_2 \rightarrow MgCO_3$$

Calcium and magnesium are found in nature typically as calcium and magnesium silicates (such as forsterite and serpentinite) and not as binary oxides. For forsterite and serpentine the reactions are:

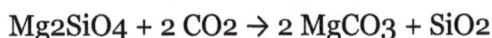

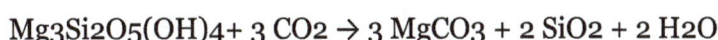

$$Mg_2SiO_4 + 2\ CO_2 \rightarrow 2\ MgCO_3 + SiO_2$$

$$Mg_3Si_2O_5(OH)_4 + 3\ CO_2 \rightarrow 3\ MgCO_3 + 2\ SiO_2 + 2\ H_2O$$

The following table lists principal metal oxides of Earth's crust. Theoretically up to 22% of this mineral mass is able to form carbonates.

Earthen Oxide	Percent of Crust	Carbonate	Enthalpy change (kJ/mol)
SiO_2	59.71		
Al_2O_3	15.41		
CaO	4.90	$CaCO_3$	-179
MgO	4.36	$MgCO_3$	-117
Na_2O	3.55	Na_2CO_3	
FeO	3.52	$FeCO_3$	
K_2O	2.80	K_2CO_3	
Fe_2O_3	2.63	$FeCO_3$	
	21.76	All Carbonates	

These reactions are slightly more favorable at low temperatures. This process occurs naturally over geologic time frames and is responsible for much of the Earth's surface limestone. The reaction rate can be made faster, for example by reacting at higher temperatures and/or pressures, or by pre-treatment, although this method requires additional energy. Experiments suggest this process is reasonably quick (one year) given porous basaltic rocks.

CO_2 naturally reacts with peridotite rock in surface exposures of ophiolites, notably in Oman. It has been suggested that this process can be enhanced to carry out natural mineralisation of CO_2.

Industrial Use

Traditional cement manufacture releases large amounts of carbon dioxide, but newly developed cement types from Novacem can absorb CO_2 from ambient air during hardening. A similar technique was pioneered by TecEco, which has been producing "EcoCement" since 2002.

In Estonia, oil shale ash, generated by power stations could be used as sorbents for CO_2 mineral sequestration. The amount of CO_2 captured averaged 60 to 65% of the carbonaceous CO_2 and 10 to 11% of the total CO_2 emissions.

Chemical Scrubbers

Various carbon dioxide scrubbing processes have been proposed to remove CO_2 from the air, usually using a variant of the Kraft process. Carbon dioxide scrubbing variants exist based on potassium carbonate, which can be used to create liquid fuels, or on sodium hydroxide. These notably include artificial trees proposed by Klaus Lackner to remove carbon dioxide from the atmosphere using chemical scrubbers.

Ocean-related

Basalt Storage

Carbon dioxide sequestration in basalt involves the injecting of CO_2 into deep-sea formations. The CO_2 first mixes with seawater and then reacts with the basalt, both of which are alkaline-rich elements. This reaction results in the release of Ca^{2+} and Mg^{2+} ions forming stable carbonate minerals.

Underwater basalt offers a good alternative to other forms of oceanic carbon storage because it has a number of trapping measures to ensure added protection against leakage. These measures include "geothermal, sediment, gravitational and hydrate formation." Because CO_2 hydrate is denser than CO_2 in seawater, the risk of leakage is minimal. Injecting the CO_2 at depths greater than 2,700 meters (8,900 ft) ensures that the CO_2 has a greater density than seawater, causing it to sink.

One possible injection site is Juan de Fuca plate. Researchers at the Lamont-Doherty Earth Observatory found that this plate at the western coast of the United States has a possible storage capacity of 208 gigatons. This could cover the entire current U.S. carbon emissions for over 100 years.

This process is undergoing tests as part of the CarbFix project, resulting in 95% of the injected 250 tonnes of CO_2 to solidify into calcite in 2 years, using 25 tonnes of water per tonne of CO_2.

Acid Neutralisation

Carbon dioxide forms carbonic acid when dissolved in water, so ocean acidification is a significant consequence of elevated carbon dioxide levels, and limits the rate at which it can be absorbed into the ocean (the solubility pump). A variety of different bases have been suggested that could neutralize the acid and thus increase CO_2 absorption. For example, adding crushed limestone to oceans enhances the absorption of carbon dioxide. Another approach is to add sodium hydroxide to oceans which is produced by electrolysis of salt water or brine, while eliminating the waste hydrochloric acid by reaction with a volcanic silicate rock such as enstatite, effectively increasing the rate of natural weathering of these rocks to restore ocean pH.

Obstruction

Danger of Leaks

Carbon dioxide may be stored deep underground. At depth, hydrostatic pressure acts to keep it in a liquid state. Reservoir design faults, rock fissures and tectonic processes may act to release the gas stored into the ocean or atmosphere.

Financial Costs

Some argue that the cost of carbon sequestration would actually increase over time. The use of the technology would add an additional 1-5 cents of cost per kilowatt hour, according to estimate made by the Intergovernmental Panel on Climate Change. The financial costs of modern coal technology would nearly double if use of CCS technology were to be required by regulation.

Energy Requirements

The energy requirements of sequestration processes may be significant. In one paper, sequestration consumed 25 percent of the plant's rated 600 megawatt output capacity.

After adding CO_2 capture and compression, the capacity of the coal-fired power plant is reduced to 457 MW.

Climate Engineering

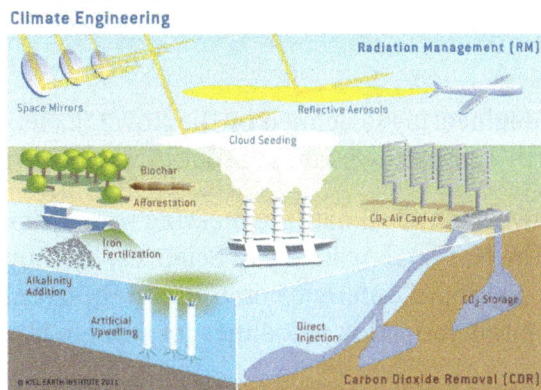

Diagram of several proposed climate engineering methods.

Climate engineering, commonly referred to as geoengineering, also known as climate intervention, is the deliberate and large-scale intervention in the Earth's climatic system with the aim of limiting adverse climate change. Climate engineering is an umbrella term for two types of measures: carbon dioxide removal and solar radiation management. Carbon dioxide removal addresses the cause of climate change by removing one of the greenhouse gases (carbon dioxide) from the atmosphere. Solar radiation management attempts to offset effects of greenhouse gases by causing the Earth to absorb less solar radiation.

Climate engineering approaches are sometimes viewed as additional potential options for limiting climate change, alongside mitigation and adaptation. There is substantial agreement among scientists that climate engineering cannot substitute for climate change mitigation. Some approaches might be used as accompanying measures to sharp cuts in greenhouse gas emissions. Given that all types of measures for addressing climate change have economic, political, or physical limitations as some climate engineering approaches might eventually be used as part of an ensemble of measures. Research on costs, benefits, and various types of risks of most climate engineering approaches is at an early stage and their understanding needs to improve to judge their adequacy and feasibility.

No outdoor solar radiation management projects have taken place to date. Almost all research into solar radiation management has consisted of computer modelling or laboratory tests, and an attempt to move to outdoor experimentation was controversial. Some carbon dioxide removal practices, such as planting of trees and bio-energy with carbon capture and storage projects, are underway. Their scalability to effectively affect global climate is however debated. Ocean iron fertilization has been given small-scale research trials, sparking substantial controversy.

Most experts and major reports advise against relying on climate engineering techniques as a simple solution to climate change, in part due to the large uncertainties over effectiveness and side effects. However, most experts also argue that the risks of such interventions must be seen in the context of risks of dangerous climate change. Interventions at large scale may run a greater risk of disrupting natural systems resulting in a dilemma that those approaches that could prove highly (cost-) effective in addressing extreme climate risk, might themselves cause substantial risk. Some have suggested that the concept of engineering the climate presents a so-called "moral hazard" because it could reduce political and public pressure for emissions reduction, which could exacerbate overall climate risks; others assert that the threat of climate engineering could spur emissions cuts. Groups such as ETC Group and some climate researchers (such as Raymond Pierrehumbert) are in favour of a moratorium on out-of-doors testing and deployment of solar radiation management (SRM.)

Background

With respect to climate, geoengineering is defined by the Royal Society as "... the deliberate large-scale intervention in the Earth's climate system, in order to moderate global warming."

Several organizations have investigated climate engineering with a view to evaluating its potential, including the US Congress, the National Academy of Sciences, the Royal Society, and the UK Parliament. The Asilomar International Conference on Climate Intervention Technologies was convened to identify and develop risk reduction guidelines for climate intervention experimentation.

Some environmental organisations (such as Friends of the Earth and Greenpeace) have been reluctant to endorse solar radiation management, but are often more supportive of some carbon dioxide removal projects, such as afforestation and peatland restoration. Some authors have argued that any public support for climate engineering may weaken the fragile political consensus to reduce greenhouse gas emissions.

Proposed Strategies

Several climate engineering strategies have been proposed. IPCC documents detail several notable proposals. These fall into two main categories: solar radiation management and carbon dioxide removal. Here is a list of specific proposals.

Solar Radiation Management

Solar radiation management (SRM) techniques would seek to reduce sunlight absorbed (ultra-violet, near infra-red and visible). This would be achieved by deflecting sunlight away from the Earth, or by increasing the reflectivity (albedo) of the atmosphere or the Earth's surface. These methods would not reduce greenhouse gas concentrations in the atmosphere, and thus would not seek to address problems such as the ocean acidification caused by CO_2. In general, solar radiation management projects presently appear to be able to take effect rapidly and to have very low direct implementation costs relative to greenhouse gas emissions cuts and carbon dioxide removal. Furthermore, many proposed SRM methods would be reversible in their direct climatic effects. While greenhouse gas remediation offers a more comprehensive possible solution to climate change, it does not give instantaneous results; for that, solar radiation management is required.

Solar radiation management methods may include:

1. Surface-based: for example, using pale-colored roofing materials, attempting to change the oceans' brightness, or growing high-albedo crops.

2. Troposphere-based: for example, marine cloud brightening, which would spray fine sea water to whiten clouds and thus increase cloud reflectivity.

3. Upper atmosphere-based: creating reflective aerosols, such as stratospheric sulfate aerosols, specifically designed self-levitating aerosols, or other substances.

4. Space-based: space sunshade—obstructing solar radiation with space-based mirrors, dust, etc.

Carbon Dioxide Removal

Carbon dioxide removal (sometimes known as negative emissions technologies or greenhouse gas removal) projects seek to remove carbon dioxide from the atmosphere. Proposed methods include those that directly remove such gases from the atmosphere, as well as indirect methods that seek to promote natural processes that draw down and sequester CO_2 (e.g. tree planting). Many projects overlap with carbon capture and storage projects, and may not be considered to be climate engineering by all commentators. Techniques in this category include:

1. Creating biochar, which can be mixed with soil to create terra preta

2. Bio-energy with carbon capture and storage to sequester carbon and simultaneously provide energy

3. Carbon air capture to remove carbon dioxide from ambient air

4. Afforestation, reforestation and forest restoration to absorb carbon dioxide

Ocean fertilization including iron fertilisation of the oceans

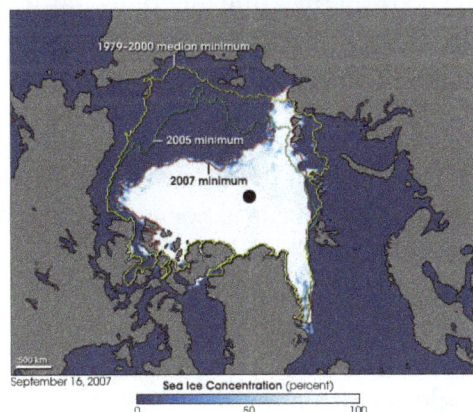

Significant reduction in ice volume in the Arctic Ocean in the range between 1979 and 2007 years

Justification

Tipping Points and Positive Feedback

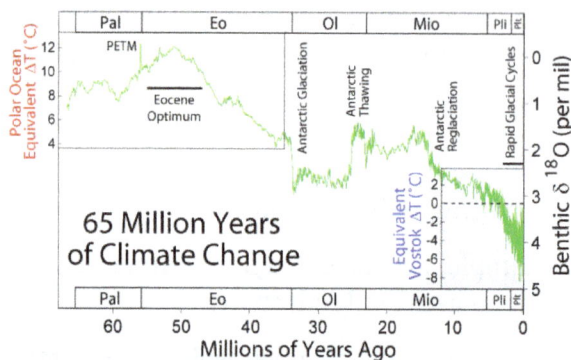

Climate change during the last 65 million years. The Paleocene–Eocene Thermal Maximum is labelled PETM.

It is argued that climate change may cross tipping points where elements of the climate system may 'tip' from one stable state to another stable state, much like a glass tipping over. When the new state is reached, further warming may be caused by positive feedback effects,. An example of a proposed causal chain leading to runaway global warming is the collapse of Arctic sea ice triggering subsequent release of methane. Such a scenario, however, is regarded as unlikely by many scientists.

The precise identity of such "tipping points" is not clear, with scientists taking differing views on whether specific systems are capable of "tipping" and the point at which this "tipping" will occur. An example of a previous tipping point is that which preceded the rapid warming leading up to the Paleocene–Eocene Thermal Maximum. Once a tipping point is crossed, cuts in anthropogenic greenhouse gas emissions will not be able to reverse the change. Conservation of resources and reduction of greenhouse emissions, used in conjunction with climate engineering, are therefore considered a viable option by some commentators.

Buying Time

Climate engineering offers the hope of temporarily reversing some aspects of climate change and allowing the natural climate to be substantially preserved whilst greenhouse gas emissions are brought under control and removed from the atmosphere by natural or artificial processes.

Costs

Estimates of direct costs for climate engineering implementation vary widely. In general, carbon dioxide removal methods are more expensive than the solar radiation management ones. In their 2009 report *Geoengineering the Climate* the Royal Society judged afforestation and stratospheric aerosol injection as the methods with the "highest affordability" (lowest costs). More recently, research into costs of solar radiation management have been published. This suggests that "well designed systems" might be available for costs in the order of a few hundred million to tens of billions of dollars per year. These are much lower than costs to achieve comprehensive reductions in CO_2 emissions. Such costs would be within the budget of most nations, and even some wealthy individuals.

Ethics and Responsibility

Climate engineering would represent a large-scale, intentional effort to modify the climate. It would differ from activities such as burning fossil fuels, as they change the climate inadvertently. Intentional climate change is often viewed differently from a moral standpoint. It raises questions of whether humans have the right to change the climate deliberately, and under what conditions. For example, there may be an ethical distinction between climate engineering to minimize anthropogenic climate change and doing so to optimize the climate. Furthermore, ethical arguments often confront larger considerations of worldview, including individual and social religious commitments. This may imply that discussions of climate engineering should reflect on how religious commitments might influence the discourse. For many people, religious beliefs are pivotal in defining the role of human beings in the wider world. Some religious communities might claim that humans have no responsibility in managing the climate, instead seeing such world systems as the exclusive domain of a Creator. In contrast, other religious communities might see the human role as one of "stewardship" or benevolent management of the world. The question of ethics also relates to issues of policy decision-making. For example, the selection of a globally agreed target temperature is a significant problem in any climate engineering governance regime, as different countries or interest groups may seek different global temperatures.

Politics

It has been argued that regardless of the economic, scientific and technical aspects, the difficulty of achieving concerted political action on climate change requires other approaches. Those arguing political expediency say the difficulty of achieving meaningful emissions cuts and the effective failure of the Kyoto Protocol demonstrate the practical difficulties of achieving carbon dioxide emissions reduction by the agreement of the international community. However, others point to support for climate engineering proposals among think tanks with a history of climate change skepticism and opposition to emissions reductions as evidence that the prospect of climate engineering is itself already politicized and being promoted as part of an argument against the need for (and viability of) emissions reductions; that, rather than climate engineering being a solution to the difficulties of emissions reductions, the prospect of climate engineering is being used as part of an argument to stall emissions reductions in the first place.

Risks and Criticisms

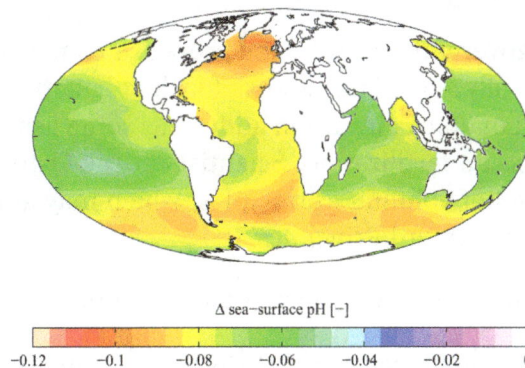

Change in sea surface pH caused by anthropogenic CO_2 between the 1700s and the 1990s. This ocean acidification will still be a major problem unless atmospheric CO_2 is reduced.

Various criticisms have been made of climate engineering, particularly solar radiation management (SRM) methods. Decision making suffers from intransitivity of policy choice. Some commentators appear fundamentally opposed. Groups such as ETC Group and individuals such as Raymond Pierrehumbert have called for a moratorium on climate engineering techniques.

Ineffectiveness

The effectiveness of the techniques proposed may fall short of predictions. In ocean iron fertilization, for example, the amount of carbon dioxide removed from the atmosphere may be much lower than predicted, as carbon taken up by plankton may be released back into the atmosphere from dead plankton, rather than being carried to the bottom of the sea and sequestered.

Moral Hazard or Risk Compensation

The existence of such techniques may reduce the political and social impetus to reduce carbon emissions. This has generally been called a potential moral hazard, although risk compensation may be a more accurate term. This concern causes many environmental groups and campaigners to be reluctant to advocate or discuss climate engineering for fear of reducing the imperative to cut greenhouse gas emissions. However, several public opinion surveys and focus groups have found evidence of either assertions of a desire to increase emission cuts in the face of climate engineering, or of no effect. Other modelling work suggests that the threat of climate engineering may in fact increase the likelihood of emissions reduction.

Governance

Climate engineering opens up various political and economic issues. The governance issues characterizing carbon dioxide removal compared to solar radiation management tend to be distinct. Carbon dioxide removal techniques are typically slow to act, expensive, and entail risks that are relatively familiar, such as the risk of carbon dioxide leakage from underground storage formations. In contrast, solar radiation management methods are fast-acting, comparatively cheap, and involve novel and more significant risks such as regional climate disruptions. As a result of these differing characteristics, the key governance problem for carbon dioxide removal (as with emissions reductions) is making sure actors do enough of it (the so-called "free rider problem"), whereas the key governance issue for solar radiation management is making sure actors do not do too much (the "free driver" problem).

Domestic and international governance vary by the proposed climate engineering method. There is presently a lack of a universally agreed framework for the regulation of either climate engineering activity or research. The London Convention addresses some aspects of the law in relation to biomass ocean storage and ocean fertilization. Scientists at the Oxford Martin School at Oxford University have proposed a set of voluntary principles, which may guide climate engineering research. The short version of the 'Oxford Principles' is:

1. Principle 1: Geoengineering to be regulated as a public good.

2. Principle 2: Public participation in geoengineering decision-making

3. Principle 3: Disclosure of geoengineering research and open publication of results

4. Principle 4: Independent assessment of impacts

5. Principle 5: Governance before deployment

These principles have been endorsed by the House of Commons of the United Kingdom Science and Technology Select Committee on "The Regulation of Geoengineering", and have been referred to by authors discussing the issue of governance.

The Asilomar conference was replicated to deal with the issue of climate engineering governance, and covered in a TV documentary, broadcast in Canada.

Implementation Issues

There is general consensus that no climate engineering technique is currently sufficiently safe or effective to greatly reduce climate change risks, for the reasons listed above. However, some may be able to contribute to reducing climate risks within relatively short times.

All proposed solar radiation management techniques require implementation on a relatively large scale, in order to impact the Earth's climate. The least costly proposals are budgeted at tens of billions of US dollars annually. Space sunshades would cost far more. Who was to bear the substantial costs of some climate engineering techniques may be hard to agree. However, the more effective solar radiation management proposals currently appear to have low enough direct implementation costs that it would be in the interests of several single countries to implement them unilaterally.

In contrast, carbon dioxide removal, like greenhouse gas emissions reductions, have impacts proportional to their scale. These techniques would not be "implemented" in the same sense as solar radiation management ones.The problem structure of carbon dioxide removal resembles that of emissions cuts, in that both are somewhat expensive public goods, whose provision presents a collective action problem.

Before they are ready to be used, most techniques would require technical development processes that are not yet in place. As a result, many promising proposed climate engineering do not yet have the engineering development or experimental evidence to determine their feasibility or efficacy.

Evaluation of Climate Engineering

Most of what is known about the suggested techniques is based on laboratory experiments, observations of natural phenomena, and on computer modelling techniques. Some proposed climate engineering methods employ methods that have analogues in natural phenomena such as stratospheric sulfur aerosols and cloud condensation nuclei. As such, studies about the efficacy of these methods can draw on information already available from other research, such as that following the 1991 eruption of Mount Pinatubo. However, comparative evaluation of the relative merits of each technology is complicated, especially given modelling uncertainties and the early stage of engineering development of many proposed climate engineering methods .

Reports into climate engineering have also been published in the United Kingdom by the Institution of Mechanical Engineers and the Royal Society. The IMechE report examined a small subset of proposed methods (air capture, urban albedo and algal-based CO_2 capture techniques), and its main conclusions were that climate engineering should be researched and trialled at the small scale alongside a wider decarbonisation of the economy.

The Royal Society review examined a wide range of proposed climate engineering methods and evaluated them in terms of effectiveness, affordability, timeliness and safety (assigning qualitative estimates in each assessment). The report divided proposed methods into "carbon dioxide removal" (CDR) and "solar radiation management" (SRM) approaches that respectively address longwave and shortwave radiation. The key recommendations of the report were that "Parties to the UNFCCC should make increased efforts towards mitigating and adapting to climate change, and in particular to agreeing to global emissions reductions", and that "[nothing] now known about climate engineering options gives any reason to diminish these efforts". Nonetheless, the report also recommended that "research and development of climate engineering options should be undertaken to investigate whether low risk methods can be made available if it becomes necessary to reduce the rate of warming this century".

In a 2009 review study, Lenton and Vaughan evaluated a range of proposed climate engineering techniques from those that sequester CO_2 from the atmosphere and decrease longwave radiation trapping, to those that decrease the Earth's receipt of shortwave radiation. In order to permit a comparison of disparate techniques, they used a common evaluation for each technique based on its effect on net radiative forcing. As such, the review examined the scientific plausibility of proposed methods rather than the practical considerations such as engineering feasibility or economic cost. Lenton and Vaughan found that "[air] capture and storage shows the greatest potential, combined with afforestation, reforestation and bio-char production", and noted that "other suggestions that have received considerable media attention, in particular "ocean pipes" appear to be ineffective". They concluded that "[climate] geoengineering is best considered as a potential complement to the mitigation of CO_2 emissions, rather than as an alternative to it".

In October 2011, a Bipartisan Policy Center panel issued a report urging immediate researching and testing in case "the climate system reaches a 'tipping point' and swift remedial action is required".

National Academy of Sciences

The National Academy of Sciences conducted a 21-month project to study the potential impacts, benefits, and costs of two different types of climate engineering: carbon dioxide removal and albedo modification (solar radiation management). The differences between these two classes of climate engineering "led the committee to evaluate the two types of approaches separately in companion reports, a distinction it hopes carries over to future scientific and policy discussions."

According to the two-volume study released in February 2015:

"Climate intervention is no substitute for reductions in carbon dioxide emissions and adaptation efforts aimed at reducing the negative consequences of climate change. However, as our planet enters a period of changing climate never before experienced in recorded human history, interest is growing in the potential for deliberate intervention in the climate system to counter climate change. ...Carbon dioxide removal strategies address a key driver of climate change, but research is needed to fully assess if any of these technologies could be appropriate for large-scale deployment. Albedo modification strategies could rapidly cool the planet's surface but pose environmental and other risks that are not well understood and therefore should not be deployed at climate-altering scales; more research is needed to determine if albedo modification approaches could be viable in the future."

The project was sponsored by the National Academy of Sciences, U.S. Intelligence Community, National Oceanic and Atmospheric Administration, NASA, and U.S. Department of Energy.

Intergovernmental Panel on Climate Change

The Intergovernmental Panel on Climate Change (IPCC) assessed the scientific literature on climate engineering (referred to as "geoengineering" in its reports), in which it considered carbon dioxide removal and solar radiation separately. Its Fifth Assessment Report states:

Models consistently suggest that SRM would generally reduce climate differences compared to a world with elevated GHG concentrations and no SRM; however, there would also be residual regional differences in climate (e.g., temperature and rainfall) when compared to a climate without elevated GHGs....

Models suggest that if SRM methods were realizable they would be effective in countering increasing temperatures, and would be less, but still, effective in countering some other climate changes. SRM would not counter all effects of climate change, and all proposed geoengineering methods also carry risks and side effects. Additional consequences cannot yet be anticipated as the level of scientific understanding about both SRM and CDR is low. There are also many (political, ethical, and practical) issues involving geoengineering that are beyond the scope of this report.

Kyoto Protocol

The Kyoto Protocol is an international treaty which extends the 1992 United Nations Framework Convention on Climate Change (UNFCCC) that commits State Parties to reduce greenhouse gas emissions, based on the premise that (a) global warming exists and (b) human-made CO_2 emissions have caused it. The Kyoto Protocol was adopted in Kyoto, Japan, on 11 December 1997 and entered into force on 16 February 2005. There are currently 192 parties (Canada withdrew effective December 2012) to the Protocol.

The Kyoto Protocol implemented the objective of the UNFCCC to fight global warming by reducing greenhouse gas concentrations in the atmosphere to "a level that would prevent dangerous anthropogenic interference with the climate system" (Art. 2). The Protocol is based on the principle of common but differentiated responsibilities: it puts the obligation to reduce current emissions on developed countries on the basis that they are historically responsible for the current levels of greenhouse gases in the atmosphere.

The Protocol's first commitment period started in 2008 and ended in 2012. A second commitment period was agreed on in 2012, known as the Doha Amendment to the protocol, in which 37 countries have binding targets: Australia, the European Union (and its 28 member states), Belarus, Iceland, Kazakhstan, Liechtenstein, Norway, Switzerland, and Ukraine. Belarus, Kazakhstan and Ukraine have stated that they may withdraw from the Protocol or not put into legal force the Amendment with second round targets. Japan, New Zealand and Russia have participated in Kyoto's first-round but have not taken on new targets in the second commitment period. Other developed countries without second-round targets are Canada (which withdrew from the Kyoto Protocol in 2012) and the United States (which has not ratified the Protocol). As of July 2016, 66 states have accepted the Doha Amendment, while entry into force requires the acceptances of 144 states. Of the 37 countries with binding commitments, 7 have ratified.

Negotiations were held in the framework of the yearly UNFCCC Climate Change Conferences on measures to be taken after the second commitment period ends in 2020. This resulted in the 2015 adoption of the Paris Agreement, which is a separate instrument under the UNFCCC rather than an amendment of the Kyoto protocol.

Background

The view that human activities are likely responsible for most of the observed increase in global mean temperature ("global warming") since the mid-20th century is an accurate reflection of current scientific thinking. Human-induced warming of the climate is expected to continue throughout the 21st century and beyond.

The Intergovernmental Panel on Climate Change (IPCC, 2007) have produced a range of projections of what the future increase in global mean temperature might be. The IPCC's projections are "baseline" projections, meaning that they assume no future efforts are made to reduce greenhouse gas emissions. The IPCC projections cover the time period from the beginning of the 21st century to the end of the 21st century. The "likely" range (as assessed to have a greater than 66% probability of being correct, based on the IPCC's expert judgement) is a projected increase in global mean temperature over the 21st century of between 1.1 and 6.4 °C.

The range in temperature projections partly reflects different projections of future greenhouse gas emissions. Different projections contain different assumptions of future social and economic development (e.g., economic growth, population level, energy policies), which in turn affects projections of future greenhouse gas (GHG) emissions. The range also reflects uncertainty in the response of the climate system to past and future GHG emissions (measured by the climate sensitivity).

Chronology

1992 The UN Conference on the Environment and Development is held in Rio de Janeiro. It results in the Framework Convention on Climate Change ("FCCC" or "UNFCCC") among other agreements.

1995 Parties to the UNFCCC meet in Berlin (the 1st Conference of Parties (COP) to the UNFCCC) to outline specific targets on emissions.

1997 In December the parties conclude the Kyoto Protocol in Kyoto, Japan, in which they agree to the broad outlines of emissions targets.

2002 Russia and Canada ratify the Kyoto Protocol to the UNFCCC bringing the treaty into effect on 16 February 2005.

2011 Canada became the first signatory to announce its withdrawal from the Kyoto Protocol.

2012 On 31 December 2012, the first commitment period under the Protocol expired.

Article 2 of the UNFCCC

Most countries are Parties to the United Nations Framework Convention on Climate Change (UN-FCCC). Article 2 of the Convention states its ultimate objective, which is to stabilize the concentration of greenhouse gases in the atmosphere "at a level that would prevent dangerous anthropogenic (i.e., human) interference with the climate system." The natural, technical, and social sciences can provide information on decisions relating to this objective, e.g., the possible magnitude and rate of future climate changes. However, the IPCC has also concluded that the decision of what constitutes "dangerous" interference requires value judgements, which will vary between different regions of the world. Factors that might affect this decision include the local consequences of climate change impacts, the ability of a particular region to adapt to climate change (adaptive capacity), and the ability of a region to reduce its GHG emissions (mitigative capacity).

Objectives

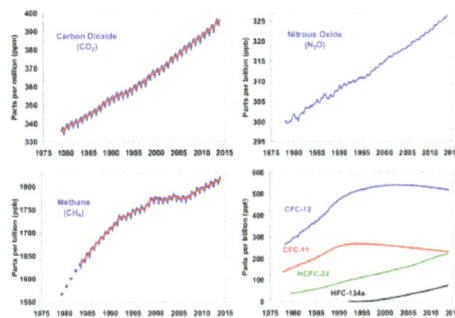

Kyoto is intended to cut global emissions of greenhouse gases.

In order to stabilize the atmospheric concentration of CO_2, emissions worldwide would need to be dramatically reduced from their present level.

The main goal of the Kyoto Protocol is to control emissions of the main anthropogenic (i.e., human-emitted) greenhouse gases (GHGs) in ways that reflect underlying national differences in GHG emissions, wealth, and capacity to make the reductions. The treaty follows the main principles agreed in the original 1992 UN Framework Convention. According to the treaty, in 2012, Annex I Parties who have ratified the treaty must have fulfilled their obligations of greenhouse gas emissions limitations established for the Kyoto Protocol's first commitment period (2008–2012). These emissions limitation commitments are listed in Annex B of the Protocol.

The Kyoto Protocol's first round commitments are the first detailed step taken within the UN Framework Convention on Climate Change (Gupta et al., 2007). The Protocol establishes a structure of rolling emission reduction commitment periods. It set a timetable starting in 2006 for negotiations to establish emission reduction commitments for a second commitment period. The first period emission reduction commitments expired on 31 December 2012.

The ultimate objective of the UNFCCC is the "stabilization of greenhouse gas concentrations in the atmosphere at a level that would stop dangerous anthropogenic interference with the climate system." Even if Annex I Parties succeed in meeting their first-round commitments, much greater emission reductions will be required in future to stabilize atmospheric GHG concentrations.

For each of the different anthropogenic GHGs, different levels of emissions reductions would be required to meet the objective of stabilizing atmospheric concentrations. Carbon dioxide (CO_2) is the most important anthropogenic GHG. Stabilizing the concentration of CO2 in the atmosphere would ultimately require the effective elimination of anthropogenic CO2 emissions.

Some of the principal concepts of the Kyoto Protocol are:

1. Binding commitments for the Annex I Parties. The main feature of the Protocol is that it established legally binding commitments to reduce emissions of greenhouse gases for Annex I Parties. The commitments were based on the Berlin Mandate, which was a part of UNFCCC negotiations leading up to the Protocol.

2. Implementation. In order to meet the objectives of the Protocol, Annex I Parties are required to prepare policies and measures for the reduction of greenhouse gases in their respective countries. In addition, they are required to increase the absorption of these gases and utilize all mechanisms available, such as joint implementation, the clean development mechanism and emissions trading, in order to be rewarded with credits that would allow more greenhouse gas emissions at home.

3. Minimizing Impacts on Developing Countries by establishing an adaptation fund for climate change.

4. Accounting, Reporting and Review in order to ensure the integrity of the Protocol.

5. Compliance. Establishing a Compliance Committee to enforce compliance with the commitments under the Protocol.

First Commitment Period: 2008–12

Under the Kyoto Protocol, 38 industrialized countries and the European Community (the European Union-15, made up of 15 states at the time of the Kyoto negotiations) commit themselves to binding targets for GHG emissions. The targets apply to the four greenhouse gases carbon dioxide (CO2), methane (CH4), nitrous oxide (N2O), sulphur hexafluoride (SF6), and two groups of gases, hydrofluorocarbons (HFCs) and perfluorocarbons (PFCs). The six GHG are translated into CO_2 equivalents in determining reductions in emissions. These reduction targets are in addition to the industrial gases, chlorofluorocarbons, or CFCs, which are dealt with under the 1987 Montreal Protocol on Substances that Deplete the Ozone Layer.

Under the Protocol, only the Annex I Parties have committed themselves to national or joint reduction targets (formally called "quantified emission limitation and reduction objectives" (QELRO) – Article 4.1). Parties to the Kyoto Protocol not listed in Annex I of the Convention (the non-Annex I Parties) are mostly low-income developing countries, and may participate in the Kyoto Protocol through the Clean Development Mechanism (explained below).

The emissions limitations of Annex I Parties varies between different Parties. Some Parties have emissions limitations reduce below the base year level, some have limitations at the base year level (i.e., no permitted increase above the base year level), while others have limitations above the base year level.

Emission limits do not include emissions by international aviation and shipping. Although Belarus and Turkey are listed in the Convention's Annex I, they do not have emissions targets as they were not Annex I Parties when the Protocol was adopted. Kazakhstan does not have a target, but has declared that it wishes to become an Annex I Party to the Convention.

Annex I countries under the Kyoto Protocol, their 2008-2012 commitments (% of base year) and 1990 emission levels (% of all Annex I countries)

Australia – 108% (2.1% of 1990 emissions)
Austria – 87%
Belarus – 95% (subject to acceptance by other parties)
Belgium – 92.5%
Bulgaria – 92% (0.6%)
Canada – 94% (3.33%)
Croatia – 95% ()
Czech Republic – 92% (1.24%)
Denmark – 79%
Estonia – 92% (0.28%)

Finland – 100%
France – 100%
Germany – 79%
Greece – 125%
Hungary – 94% (0.52%)
Iceland – 110% (0.02%)
Ireland – 113%
Italy – 93.5%
Japan – 94% (8.55%)
Latvia – 92% (0.17%)

Liechtenstein – 92% (0.0015%)
Lithuania – 92% ()
Luxembourg – 72%
Netherlands – 94%
New Zealand – 100% (0.19%)
Norway – 101% (0.26%)
Poland – 94% (3.02%)
Portugal – 92%
Romania – 92% (1.24%)

Russian Federation – 100% (17.4%)
Slovakia – 92% (0.42%)
Slovenia – 92% ()
Spain – 115%
Sweden – 104%
Switzerland – 92% (0.32%)
Ukraine – 100% ()
United Kingdom – 87.5%
United States of America – 93% (36.1%) (non-party)

For most Parties, 1990 is the base year for the national GHG inventory and the calculation of the assigned amount. However, five Parties have an alternative base year:

1. Bulgaria: 1988;

2. Hungary: the average of the years 1985–87;

3. Poland: 1988;

4. Romania: 1989;

5. Slovenia: 1986.

Annex I Parties can use a range of sophisticated "flexibility" mechanisms to meet their targets. Annex I Parties can achieve their targets by allocating reduced annual allowances to major operators within their borders, or by allowing these operators to exceed their allocations by offsetting any excess through a mechanism that is agreed by all the parties to the UNFCCC, such as by buying emission allowances from other operators which have excess emissions credits.

Flexibility Mechanisms

The Protocol defines three "flexibility mechanisms" that can be used by Annex I Parties in meeting their emission limitation commitments. The flexibility mechanisms are International Emissions Trading (IET), the Clean Development Mechanism (CDM), and Joint Implementation (JI). IET allows Annex I Parties to "trade" their emissions (Assigned Amount Units, AAUs, or "allowances" for short).

The economic basis for providing this flexibility is that the marginal cost of reducing (or abating) emissions differs among countries. "Marginal cost" is the cost of abating the last tonne of CO_2-eq for an Annex I/non-Annex I Party. At the time of the original Kyoto targets, studies suggested that the flexibility mechanisms could reduce the overall (aggregate) cost of meeting the targets. Studies also showed that national losses in Annex I gross domestic product (GDP) could be reduced by use of the flexibility mechanisms.

The CDM and JI are called "project-based mechanisms," in that they generate emission reductions from projects. The difference between IET and the project-based mechanisms is that IET is based on the setting of a quantitative restriction of emissions, while the CDM and JI are based on the idea of "production" of emission reductions. The CDM is designed to encourage production of emission reductions in non-Annex I Parties, while JI encourages production of emission reductions in Annex I Parties.

The production of emission reductions generated by the CDM and JI can be used by Annex I Parties in meeting their emission limitation commitments. The emission reductions produced by the CDM and JI are both measured against a hypothetical baseline of emissions that would have occurred in the absence of a particular emission reduction project. The emission reductions produced by the CDM are called Certified Emission Reductions (CERs); reductions produced by JI are called Emission Reduction Units (ERUs). The reductions are called "credits" because they are emission reductions credited against a hypothetical baseline of emissions.

Each Annex I country is required to submit an annual report of inventories of all anthropogenic greenhouse gas emissions from sources and removals from sinks under UNFCCC and the Kyoto Protocol. These countries nominate a person (called a "designated national authority") to create and manage its greenhouse gas inventory. Virtually all of the non-Annex I countries have also established a designated national authority to manage their Kyoto obligations, specifically the "CDM process". This determines which GHG projects they wish to propose for accreditation by the CDM Executive Board.

International Emissions Trading

A number of emissions trading schemes (ETS) have been, or are planned to be, implemented.

Asia

1. *Japan:* emissions trading in Tokyo started in 2010. This scheme is run by the Tokyo Metropolitan Government.

Europe

1. *European Union:* the European Union Emission Trading Scheme (EU ETS), which started in 2005. This is run by the European Commission.

2. *Norway*: domestic emissions trading in Norway started in 2005. This was run by the Norwegian Government, which is now a participant in the EU ETS.

3. *Switzerland*: the Swiss ETS, which runs from 2008 to 2012, to coincide with the Kyoto Protocol's first commitment period.

4. *United Kingdom:*

    1. the UK Emissions Trading Scheme, which ran from 2002–06. This was a scheme run by the UK Government, which is now a participant in the EU ETS.

    2. the UK CRC Energy Efficiency Scheme, which started in 2010, and is run by the UK Government.

North America

1. *Canada:* emissions trading in Alberta, Canada, which started in 2007. This is run by the Government of Alberta.

2. *United States:*

    1. the Regional Greenhouse Gas Initiative (RGGI), which started in 2009. This scheme caps emissions from power generation in ten north-eastern U.S. states (Connecticut, Delaware, Maine, Maryland, Massachusetts, New Hampshire, New Jersey, New York, Rhode Island and Vermont).

    2. emissions trading in California, which started in 2013.

3. the Western Climate Initiative (WCI), which is planned to start in 2012. This is a collective ETS agreed between 11 U.S. states and Canadian provinces.

Oceania

1. *Australia:* the New South Wales Greenhouse Gas Reduction Scheme (NSW), which started in 2003. This scheme is run by the Australian State of New South Wales, and has now joined the Alfa Climate Stabilization (ACS).

2. *New Zealand*: the New Zealand Emissions Trading Scheme, which started in 2008.

Intergovernmental Emissions Trading

The design of the European Union Emissions Trading Scheme (EU ETS) implicitly allows for trade of national Kyoto obligations to occur between participating countries (Carbon Trust, 2009, p. 24). Carbon Trust (2009, pp. 24–25) found that other than the trading that occurs as part of the EU ETS, no intergovernmental emissions trading had taken place.

One of the environmental problems with IET is the large surplus of allowances that are available. Russia, Ukraine, and the new EU-12 member states (the Kyoto Parties Annex I Economies-in-Transition, abbreviated "EIT": Belarus, Bulgaria, Croatia, Czech Republic, Estonia, Hungary, Latvia, Lithuania, Poland, Romania, Russia, Slovakia, Slovenia, and Ukraine) have a surplus of allowances, while many OECD countries have a deficit. Some of the EITs with a surplus regard it as potential compensation for the trauma of their economic restructuring. When the Kyoto treaty was negotiated, it was recognized that emissions targets for the EITs might lead to them having an excess number of allowances. This excess of allowances were viewed by the EITs as "headroom" to grow their economies. The surplus has, however, also been referred to by some as "hot air," a term which Russia (a country with an estimated surplus of 3.1 billion tonnes of carbon dioxide equivalent allowances) views as "quite offensive."

OECD countries with a deficit could meet their Kyoto commitments by buying allowances from transition countries with a surplus. Unless other commitments were made to reduce the total surplus in allowances, such trade would not actually result in emissions being reduced.

Green Investment Scheme

A Green Investment Scheme (GIS) refers to a plan for achieving environmental benefits from trading surplus allowances (AAUs) under the Kyoto Protocol. The Green Investment Scheme (GIS), a mechanism in the framework of International Emissions Trading (IET), is designed to achieve greater flexibility in reaching the targets of the Kyoto Protocol while preserving environmental integrity of IET. However, using the GIS is not required under the Kyoto Protocol, and there is no official definition of the term.

Under the GIS a Party to the Protocol expecting that the development of its economy will not exhaust its Kyoto quota, can sell the excess of its Kyoto quota units (AAUs) to another Party. The proceeds from the AAU sales should be "greened", i.e. channeled to the development and implementation of the projects either acquiring the greenhouse gases emission reductions (hard greening) or building up the necessary framework for this process (soft greening).

Trade in AAUs

Latvia was one of the front-runners of GISs. World Bank (2011) reported that Latvia has stopped offering AAU sales because of low AAU prices. In 2010, Estonia was the preferred source for AAU buyers, followed by the Czech Republic and Poland.

Japan's national policy to meet their Kyoto target includes the purchase of AAUs sold under GISs. In 2010, Japan and Japanese firms were the main buyers of AAUs. In terms of the international carbon market, trade in AAUs are a small proportion of overall market value. In 2010, 97% of trade in the international carbon market was driven by the European Union Emission Trading Scheme (EU ETS). However, firms regulated under the EU ETS are unable to use AAUs in meeting their emissions caps.

Clean Development Mechanism

Between 2001, which was the first year Clean Development Mechanism (CDM) projects could be registered, and 2012, the end of the first Kyoto commitment period, the CDM is expected to produce some 1.5 billion tons of carbon dioxide equivalent (CO_2e) in emission reductions. Most of these reductions are through renewable energy commercialisation, energy efficiency, and fuel switching (World Bank, 2010, p. 262). By 2012, the largest potential for production of CERs are estimated in China (52% of total CERs) and India (16%). CERs produced in Latin America and the Caribbean make up 15% of the potential total, with Brazil as the largest producer in the region (7%).

Joint Implementation

The formal crediting period for Joint Implementation (JI) was aligned with the first commitment period of the Kyoto Protocol, and did not start until January 2008 (Carbon Trust, 2009, p. 20). In November 2008, only 22 JI projects had been officially approved and registered. The total projected emission savings from JI by 2012 are about one tenth that of the CDM. Russia accounts for about two-thirds of these savings, with the remainder divided up roughly equally between the Ukraine and the EU's New Member States. Emission savings include cuts in methane, HFC, and N_2O emissions.

Stabilization of GHG Concentrations

As noted earlier on, the first-round Kyoto emissions limitation commitments are not sufficient to stabilize the atmospheric concentration of GHGs. Stabilization of atmospheric GHG concentrations will require further emissions reductions after the end of the first-round Kyoto commitment period in 2012.

Background

Indicative probabilities of exceeding various increases in global mean temperature for different stabilization levels of atmospheric GHG concentrations.

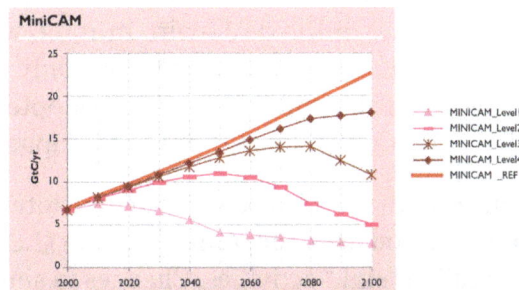

Different targets for stabilization require different levels of cuts in emissions over time. Lower stabilization targets require global emissions to be reduced more sharply in the near-term.

Analysts have developed scenarios of future changes in GHG emissions that lead to a stabilization in the atmospheric concentrations of GHGs. Climate models suggest that lower stabilization levels are associated with lower magnitudes of future global warming, while higher stabilization levels are associated with higher magnitudes of future global warming.

To achieve stabilization, global GHG emissions must peak, then decline. The lower the desired stabilization level, the sooner this peak and decline must occur. For a given stabilization level, larger emissions reductions in the near term allow for less stringent emissions reductions later on. On the other hand, less stringent near term emissions reductions would, for a given stabilization level, require more stringent emissions reductions later on.

The first period Kyoto emissions limitations can be viewed as a first-step towards achieving atmospheric stabilization of GHGs. In this sense, the first period Kyoto commitments may affect what future atmospheric stabilization level can be achieved.

Relation to Temperature Targets

At the 16th Conference of the Parties held in 2010, Parties to the UNFCCC agreed that future global warming should be limited below 2°C relative to the pre-industrial temperature level. One of the stabilization levels discussed in relation to this temperature target is to hold atmospheric concentrations of GHGs at 450 parts per million (ppm) CO2- eq. Stabilization at 450 ppm could be associated with a 26 to 78% risk of exceeding the 2 °C target.

Scenarios assessed by Gupta *et al.* (2007) suggest that Annex I emissions would need to be 25% to 40% below 1990 levels by 2020, and 80% to 95% below 1990 levels by 2050. The only Annex I Parties to have made voluntary pledges in line with this are Japan (25% below 1990 levels by 2020) and Norway (30-40% below 1990 levels by 2020).

Gupta *et al.* (2007) also looked at what 450 ppm scenarios projected for non-Annex I Parties. Projections indicated that by 2020, non-Annex I emissions in several regions (Latin America, the Middle East, East Asia, and centrally planned Asia) would need to be substantially reduced below "business-as-usual". "Business-as-usual" are projected non-Annex I emissions in the absence of any new policies to control emissions. Projections indicated that by 2050, emissions in all non-Annex I regions would need to be substantially reduced below "business-as-usual".

Details of the Agreement

The agreement is a protocol to the United Nations Framework Convention on Climate Change (UNFCCC) adopted at the Earth Summit in Rio de Janeiro in 1992, which did not set any legally binding limitations on emissions or enforcement mechanisms. Only Parties to the UNFCCC can become Parties to the Kyoto Protocol. The Kyoto Protocol was adopted at the third session of the Conference of Parties to the UNFCCC (COP 3) in 1997 in Kyoto, Japan.

National emission targets specified in the Kyoto Protocol exclude international aviation and shipping. Kyoto Parties can use land use, land use change, and forestry (LULUCF) in meeting their targets. LULUCF activities are also called "sink" activities. Changes in sinks and land use can have an effect on the climate, and indeed the Intergovernmental Panel on Climate Change's Special Report on Land Use, Land-Use Change and Forestry estimates that since 1750 a third of global warming has been caused by land use change. Particular criteria apply to the definition of forestry under the Kyoto Protocol.

Forest management, cropland management, grazing land management, and revegetation are all eligible LULUCF activities under the Protocol. Annex I Parties use of forest management in meeting their targets is capped.

Negotiations

Article 4.2 of the UNFCCC commits industrialized countries to "[take] the lead" in reducing emissions. The initial aim was for industrialized countries to stabilize their emissions at 1990 levels by the year 2000. The failure of key industrialized countries to move in this direction was a principal reason why Kyoto moved to binding commitments.

At the first UNFCCC Conference of the Parties in Berlin, the G77 was able to push for a mandate (the "Berlin mandate") where it was recognized that:

1. developed nations had contributed most to the then-current concentrations of GHGs in the atmosphere.

2. developing country emissions per-capita (i.e., average emissions per head of population) were still relatively low.

3. and that the share of global emissions from developing countries would grow to meet their development needs.

During negotiations, the G-77 represented 133 developing countries. China was not a member of the group but an associate. It has since become a member.

The Berlin mandate was recognized in the Kyoto Protocol in that developing countries were not subject to emission reduction commitments in the first Kyoto commitment period. However, the large potential for growth in developing country emissions made negotiations on this issue tense. In the final agreement, the Clean Development Mechanism was designed to limit emissions in developing countries, but in such a way that developing countries do not bear the costs for limiting emissions. The general assumption was that developing countries would face quantitative commitments in later commitment periods, and at the same time, developed countries would meet their first round commitments.

Emissions Cuts

Views on the Kyoto Protocol#Commentaries on negotiations contains a list of the emissions cuts that were proposed by UNFCCC Parties during negotiations. The G77 and China were in favour of strong uniform emission cuts across the developed world. The U.S. originally proposed for the second round of negotiations on Kyoto commitments to follow the negotiations of the first. In the end, negotiations on the second period were set to open no later than 2005. Countries over-achieving in their first period commitments can "bank" their unused allowances for use in the subsequent period.

The EU initially argued for only three GHGs to be included – CO_2, CH_4, and N_2O – with other gases such as HFCs regulated separately. The EU also wanted to have a "bubble" commitment, whereby it could make a collective commitment that allowed some EU members to increase their emissions, while others cut theirs.

The most vulnerable nations – the Alliance of Small Island States (AOSIS) – pushed for deep uniform cuts by developed nations, with the goal of having emissions reduced to the greatest possible extent. Countries that had supported differentiation of targets had different ideas as to how it should be calculated, and many different indicators were proposed. Two examples include differentiation of targets based on gross domestic product (GDP), and differentiation based on energy intensity (energy use per unit of economic output).

The final targets negotiated in the Protocol are the result of last minute political compromises. The targets closely match those decided by Argentinian Raul Estrada, the diplomat who chaired the negotiations. The numbers given to each Party by Chairman Estrada were based on targets already pledged by Parties, information received on latest negotiating positions, and the goal of achieving the strongest possible environmental outcome. The final targets are weaker than those proposed by some Parties, e.g., the Alliance of Small Island States and the G-77 and China, but stronger than the targets proposed by others, e.g., Canada and the United States.

Financial Commitments

The Protocol also reaffirms the principle that developed countries have to pay billions of dollars, and supply technology to other countries for climate-related studies and projects. The principle was originally agreed in UNFCCC. One such project is The Adaptation Fund"", that has been established by the Parties to the Kyoto Protocol of the UN Framework Convention on Climate Change to finance concrete adaptation projects and programmes in developing countries that are Parties to the Kyoto Protocol.

Implementational Provisions

The protocol left several issues open to be decided later by the sixth Conference of Parties COP6 of the UNFCCC, which attempted to resolve these issues at its meeting in the Hague in late 2000, but it was unable to reach an agreement due to disputes between the European Union (who favoured a tougher implementation) and the United States, Canada, Japan and Australia (who wanted the agreement to be less demanding and more flexible).

In 2001, a continuation of the previous meeting (COP6bis) was held in Bonn where the required decisions were adopted. After some concessions, the supporters of the protocol (led by the European Union) managed to get the agreement of Japan and Russia by allowing more use of carbon dioxide sinks.

COP7 was held from 29 October 2001 through 9 November 2001 in Marrakech to establish the final details of the protocol.

The first Meeting of the Parties to the Kyoto Protocol (MOP1) was held in Montreal from 28 November to 9 December 2005, along with the 11th conference of the Parties to the UNFCCC (COP11).

During COP13 in Bali 36 developed C.G. countries (plus the EU as a party in the European Union) agreed to a 10% emissions increase for Iceland; but, since the EU's member states each have individual obligations, much larger increases (up to 27%) are allowed for some of the less developed EU countries. Reduction limitations expire in 2013.

Mechanism of Compliance

The protocol defines a mechanism of "compliance" as a "monitoring compliance with the commitments and penalties for non-compliance." According to Grubb (2003), the explicit consequences of non-compliance of the treaty are weak compared to domestic law. Yet, the compliance section of the treaty was highly contested in the Marrakesh Accords.

Enforcement

If the enforcement branch determines that an Annex I country is not in compliance with its emissions limitation, then that country is required to make up the difference during the second commitment period plus an additional 30%. In addition, that country will be suspended from making transfers under an emissions trading program.

Ratification Process

The Protocol was adopted by COP 3 of UNFCCC on 11 December 1997 in Kyoto, Japan. It was opened on 16 March 1998 for signature during one year by parties to UNFCCC, when it was signed Antigua and Barbuda, Argentina, the Maldives, Samoa, St. Lucia and Switzerland. At the end of the signature period, 82 countries and the European Community had signed. Ratification (which is required to become a party to the Protocol) started on 17 September with ratification of Fiji. Countries that did not sign acceded to the convention, which has the same legal effect.

Article 25 of the Protocol specifies that the Protocol enters into force "on the ninetieth day after the date on which not less than 55 Parties to the Convention, incorporating Parties included in Annex I which accounted in total for at least 55% of the total carbon dioxide emissions for 1990 of the Annex I countries, have deposited their instruments of ratification, acceptance, approval or accession."

The EU and its Member States ratified the Protocol in May 2002. Of the two conditions, the "55 parties" clause was reached on 23 May 2002 when Iceland ratified the Protocol. The ratification by Russia on 18 November 2004 satisfied the "55%" clause and brought the treaty into force, effective 16 February 2005, after the required lapse of 90 days.

As of May 2013, 191 countries and one regional economic organization (the EC) have ratified the agreement, representing over 61.6% of the 1990 emissions from Annex I countries. One of the 191 ratifying states—Canada—has denounced the protocol.

Convention Parties

Afghanistan	Dominican Republic	Liechtenstein	São Tomé and Príncipe
Albania	Ecuador	Lithuania	Saudi Arabia
Algeria	Egypt	Luxembourg	Senegal
Angola	El Salvador	Republic of Macedonia	Serbia
Antigua and Barbuda	Equatorial Guinea	Madagascar	Seychelles
Argentina	Eritrea	Malawi	Sierra Leone
Armenia	Estonia	Malaysia	Singapore
Australia	Ethiopia	Maldives	Slovakia
Austria	European Union	Mali	Slovenia
Azerbaijan	Fiji	Malta	Solomon Islands
Bahamas	Finland	Marshall Islands	Somalia (non-party to
Bahrain	France	Mauritania	Kyoto)
Bangladesh	Gabon	Mauritius	South Africa
Barbados	Gambia	Mexico	Spain
Belarus	Georgia	Federated States of	Sri Lanka
Belgium		Micronesia	Sudan

Belize	Germany	Moldova	Suriname
Benin	Ghana	Monaco	Swaziland
Bhutan	Greece	Mongolia	Sweden
Bolivia	Grenada	Montenegro	Switzerland
Bosnia and Herzegovina	Guatemala	Morocco	Syria
Botswana	Guinea	Mozambique	Tajikistan
Brazil	Guinea-Bissau	Namibia	Tanzania
Brunei	Guyana	Nauru	Thailand
Bulgaria	Haiti	Nepal	Timor-Leste
Burkina Faso	Honduras	Netherlands	Togo
Myanmar	Hungary	New Zealand	Tonga
Burundi	Iceland	Nicaragua	Trinidad and Tobago
Cambodia	India	Niger	Tunisia
Cameroon	Indonesia	Nigeria	Turkey
Canada	Iran	Niue	Turkmenistan
Cape Verde	Iraq	Norway	Tuvalu
Central African Republic	Ireland	Oman	Uganda
Chad	Israel	Pakistan	Ukraine
Chile	Italy	Palau	United Arab Emirates
China	Jamaica	Panama	United Kingdom
Colombia	Japan	Papua New Guinea	United States (non-party
Comoros	Jordan	Paraguay	to Kyoto)
Democratic Republic of	Kazakhstan	Peru	Uruguay
the Congo	Kenya	Philippines	Uzbekistan
Republic of the Congo	Kiribati	Poland	Vanuatu
Cook Islands	North Korea	Portugal	Venezuela
Costa Rica	South Korea	Qatar	Vietnam
Ivory Coast	Kuwait	Romania	Yemen
Croatia	Kyrgyzstan	Russia	Zambia
Cuba	Laos	Rwanda	Zimbabwe
Cyprus	Latvia	Saint Kitts and Nevis	Observers:
Czech Republic	Lebanon	Saint Lucia	
Denmark	Lesotho	Saint Vincent and the	
Djibouti	Liberia	Grenadines	Andorra (non-party to
Dominica	Libya	Samoa	Kyoto)
		San Marino	Holy See (non-party to Kyoto)

Non-ratification by the Usa

The US signed the Protocol on 12 November 1998, during the Clinton presidency. To become binding in the US, however, the treaty had to be ratified by the Senate, which had already passed the 1997 non-binding Byrd-Hagel Resolution, expressing disapproval of any international agreement that did not require developing countries to make emission reductions and "would seriously harm the economy of the United States". The resolution passed 95-0. Therefore, even though the Clinton administration signed the treaty, it was never submitted to the Senate for ratification.

When George W. Bush was elected US president in 2000, he was asked by US Senator Hagel what his administration's position was on climate change. Bush replied that he took climate change "very seriously," but that he opposed the Kyoto treaty, because "it exempts 80% of the world, including major population centers such as China and India, from compliance, and would cause serious harm to the US economy". The Tyndall Centre for Climate Change Research reported in 2001 that, "This policy reversal received a massive wave of criticism that was quickly picked up by the international media. Environmental groups blasted the White House, while Europeans and Japanese alike expressed deep concern and regret. [...] Almost all world leaders (e.g. China, Japan, South Africa, Pacific Islands, etc.) expressed their disappointment at Bush's decision." Bush's response that, "I was responding to reality, and reality is the nation has got a real problem when it comes to energy" was, it said, "an overstatement used to cover up the big benefactors of this policy reversal, i.e., the US oil and coal industry, which has a powerful lobby with the administration and conservative Republican congressmen."

As of 2016, the USA is the only signatory that has not ratified the Protocol.

The US accounted for 36% of emissions in 1990, U.S. ratification, only an EU+Russia+Japan+small party coalition could place the treaty into legal effect. A deal, without at that moment the US Administration, was reached in the Bonn climate talks (COP-6.5), held in 2001.

Withdrawal of Canada

In 2011, Canada, Japan and Russia stated that they would not take on further Kyoto targets. The Canadian government announced its withdrawal—possible at any time three years after ratification—from the Kyoto Protocol on 12 December 2011, effective 15 December 2012. Canada was committed to cutting its greenhouse emissions to 6% below 1990 levels by 2012, but in 2009 emissions were 17% higher than in 1990. The Harper government prioritized oil sands development in Alberta, and deprioritized improving the environment. Environment minister Peter Kent cited Canada's liability to "enormous financial penalties" under the treaty unless it withdrew. He also suggested that the recently signed Durban agreement may provide an alternative way forward. The Harper government claimed it would find a "Made in Canada" solution, but never found any such solution. Canada's decision received a generally negative response from representatives of other ratifying countries.

Government Action and Emissions

Annex I Countries

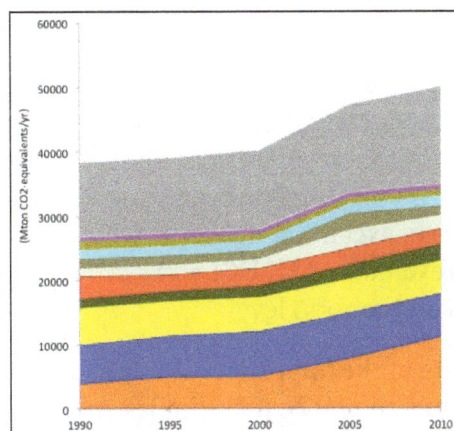

Anthropogenic emissions of CO_2-equivalents per year by the 10 largest emitters (the European Union is lumped as a single area, because of their integrated carbon trading scheme). Data sorted based on 2010 contributions.

▌ China (party, no binding targets)

▌ United States (non-party)

▌ European Union (party, binding targets)

▌ India (party, no binding targets)

▌ Russia (party, binding targets 2008-2012)

▌ Indonesia (party, no binding targets)

▌ Brazil (party, no binding targets)

▌ Japan (party, no binding targets)

▌ Congo (DR) (party, no binding targets)

▌ Canada (former party, binding targets 2008-2012)

▌ Other countries

Total aggregate GHG emissions excluding emissions/removals from land use, land use change and forestry (LULUCF, i.e., carbon storage in forests and soils) for all Annex I Parties including the United States taken together decreased from 19.0 to 17.8 thousand teragrams (Tg, which is equal to 10^9 kg) CO2 equivalent, a decline of 6.0% during the 1990-2008 period. Several factors have contributed to this decline. The first is due to the economic restructuring in the Annex I Economies in Transition. Over the period 1990-1999, emissions fell by 40% in the EITs following the collapse of central planning in the former Soviet Union and east European countries. This led to a massive contraction of their heavy industry-based economies, with associated reductions in their fossil fuel consumption and emissions.

Emissions growth in Annex I Parties have also been limited due to policies and measures (PaMs). In particular, PaMs were strengthened after 2000, helping to enhance energy efficiency and develop renewable energy sources. Energy use also decreased during the economic crisis in 2007-2008.

Projections

UNFCCC (2011) made projections of changes in emissions of the Annex I Parties and the effectiveness of their PaMs. It was noted that their projections should be interpreted with caution. For the 39 Annex I Parties, UNFCCC (2011) projected that existing PaMs would lead to annual emissions in 2010 of 17.5 thousand Tg CO2 eq, excluding LULUCF, which is a decrease of 6.7% from the 1990 level. Annual emissions in 2020 excluding LULUCF were projected to reach 18.9 thousand Tg CO2 eq, which is an increase of 0.6% on the 1990 level.

UNFCCC (2011) made an estimate of the total effect of implemented and adopted PaMs. Projected savings were estimated relative to a reference (baseline) scenario where PaMs are not implemented. PaMs were projected to deliver emissions savings relative to baseline of about 1.5 thousand Tg CO2 eq by 2010, and 2.8 thousand Tg CO2 eq by 2020. In percentage terms, and using annual emissions in the year 1990 as a reference point, PaMs were projected to deliver at least a 5.0% reduction relative to baseline by 2010, and a 10.0% reduction relative to baseline in 2020. Scenarios reviewed by UNFCCC (2011) still suggested that total Annex I annual emissions would increase out to 2020.

Annex I Parties with Targets

Percentage changes in emissions for Annex I Parties with Kyoto targets					
Country/ region	Kyoto target 2008-2012	Kyoto target 2013-2020	GHG emissions 1990-2008 including LULUCF	GHG emissions 1990-2008 excluding LULUCF	CO$_2$ emissions from fuel combustion only 1990-2009
North America	-	-	-	-	+20.4
Canada	-6	-	+33.6	+24.1	+20.4
Europe					-4.9
European Union	-8	-20			
Austria	-8 (-13)	-20	+6.6	+10.8	+12.2
Belgium	-8 (-7.5)	-20	-6.2	-7.1	-6.7
Denmark	-8 (-21)	-20	-6.8	-6.8	-7.2
Finland	-8 (0)	-20	-35.9	-0.2	+1.1
France	-8 (0)	-20	-12.7	-5.9	+0.6
Germany	-8 (-21)	-20	-17.6	-21.4	-21.1
Greece	-8 (+25)	-20	+22.9	+23.1	+28.6
Iceland	+10	-20	+19.2	+42.9	+6.2
Ireland	-8 (+13)	-20	+19.9	+23.2	+32.4
Italy	-8 (-6.5)	-20	+0.4	+4.7	-2.0
Luxembourg	-8 (-28)	-20	-9.2	-4.8	-4.4
Netherlands	-8 (-6)	-20	-2.4	-2.4	+13.0
Norway	+1	-16	-32.8	+9.4	+31.9
Portugal	-8 (+27)	-20	+18.3	+32.2	+35.3
Spain	-8 (+15)	-20	+44.0	+42.5	+37.7
Sweden	-8 (+4)	-20	+19.8	-11.3	-20.9
Switzerland	-8	-15.8	+6.8	+0.4	+2.5
United Kingdom	-8 (-12.5)	-19.0	-18.5	-15.2	

Asia and Oceania	-	-	-	+12.7	
Australia	+8	-0.5	+33.1	+31.4	+51.8
Japan	-6	-	-0.2	+1.0	+2.7
New Zealand	0	-	+62.4	+22.7	+34.3
Economies in Transition	-	-	-	-36.2	
Bulgaria	-8	-20	-45.5	-42.8	-43.7
Croatia	-5	-20	-13.7	-0.9	-8.4
Czech Republic	-8	-20	-28.7	-27.5	-29.2
Estonia	-8	-20	-69.9	-50.9	-59.4
Hungary	-6	-20	-38.1	-36.2	-27.8
Latvia	-8	-20	-307.9	-55.6	-63.8
Lithuania	-8	-20	-69.1	-51.8	-62.6
Poland	-6	-20	-34.4	-29.6	-16.2
Romania	-8	-20	-53.5	-45.9	-53.1
Russian Federation	0	-	-52.8	-32.8	-29.7
Slovak Republic	-8	-20	-34.4	-33.7	-41.5
Slovenia	-8	-20	+5.2	+5.2	+21.2
Ukraine	0	-24	-52.2	-53.9	-62.7

Data given in the table above may not be fully reflective of a country's progress towards meeting its first-round Kyoto target. The summary below contains more up-to-date information on how close countries are to meeting their first-round targets.

Carbon dioxide emissions from fuel combustion
of Kyoto Protocol Parties 1990-2009

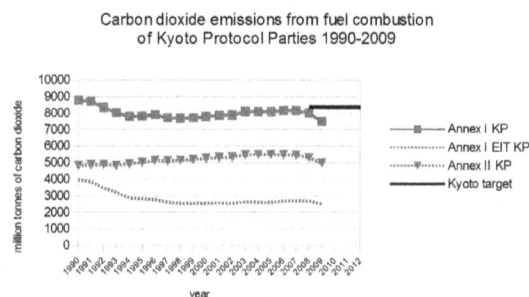

CO$_2$ emissions from fuel combustion of Annex I Kyoto Protocol (KP) Parties, 1990-2009. Total Annex I KP emissions are shown, along with emissions of Annex II KP and Annex I EITs.

Collectively the group of industrialized countries committed to a Kyoto target, i.e., the Annex I countries excluding the USA, have a target of reducing their GHG emissions by 4.2% on average for the period 2008-2012 relative to the base year, which in most cases is 1990. According to Olivier *et al.* (2011), the Kyoto Parties will comfortably exceed their collective target, with a projected average reduction of 16% for 2008-2012. This projection excludes both LULUCF and credits generated by the Clean Development Mechanism (CDM).

As noted in the preceding section, between 1990–1999, there was a large reduction in the emissions of the EITs. The reduction in the EITs is largely responsible for the total (aggregate) reduction (excluding LULUCF) in emissions of the Annex I countries, excluding the USA. Emissions of the Annex II countries (Annex I minus the EIT countries) have experienced a limited increase in emissions from 1990–2006, followed by stabilization and a more marked decrease from 2007 onwards. The emissions reductions in the early nineties by the 12 EIT countries who have since joined the EU, assist the present EU-27 in meeting its collective Kyoto target.

Almost all European countries are on track to achieve their first-round Kyoto targets. Spain plans to meet its target by purchasing a large quantity of Kyoto units through the flexibility mechanisms. Australia, Canada (Canada withdrew from the Kyoto treaty in 2012), and Italy are not on course to meet their first-round Kyoto targets. In order to meet their targets, these countries would need to purchase emissions credits from other Kyoto countries. As noted in the section on Intergovernmental Emissions Trading, purchasing surplus credits from the EIT countries would not actually result in total emissions being reduced. An alternative would be the purchase of CDM credits or the use of the voluntary Green Investment Scheme.

In December 2011, Canada's environment minister, Peter Kent, formally announced that Canada would withdraw from the Kyoto accord a day after the end of the 2011 United Nations Climate Change Conference.

Annex I Parties without Kyoto Targets

Belarus, Malta, and Turkey are Annex I Parties but do not have first-round Kyoto targets. The US has a Kyoto target of a 6% reduction relative to the 1990 level, but has not ratified the treaty. Emissions in the US have increased 11% since 1990, and according to Olivier *et al.* (2011), it will be unable to meet its original Kyoto target.

If the US had ratified the Kyoto Protocol, the average percentage reduction in total GHG emissions for the Annex I group would have been a 5.2% reduction relative to the base year. Including the US in their calculation, Olivier *et al.* (2011) projected that the Annex I countries would collectively achieve a 7% reduction relative to the base year, which is lower than the original target of a 5.2% reduction. This projection excludes expected purchases of emissions credits.

Non-Annex I

Annual per capita carbon dioxide emissions (i.e., average emissions per person) from fuel combustion between 1990-2009 for the Kyoto Annex I and non-Annex I Parties.

Annual carbon dioxide emissions from fuel combustion between 1990-2009 for the Kyoto Annex I and non-Annex I Parties.

UNFCCC (2005) compiled and synthesized information reported to it by non-Annex I Parties.

Most non-Annex I Parties belonged in the low-income group, with very few classified as middle-income. Most Parties included information on policies relating to sustainable development. Sustainable development priorities mentioned by non-Annex I Parties included poverty alleviation and access to basic education and health care. Many non-Annex I Parties are making efforts to amend and update their environmental legislation to include global concerns such as climate change.

A few Parties, e.g., South Africa and Iran, stated their concern over how efforts to reduce emissions by Annex I Parties could adversely affect their economies. The economies of these countries are highly dependent on income generated from the production, processing, and export of fossil fuels.

Emissions

GHG emissions, excluding land use change and forestry (LUCF), reported by 122 non-Annex I Parties for the year 1994 or the closest year reported, totalled 11.7 billion tonnes (billion = 1,000,000,000) of CO_2-eq. CO_2 was the largest proportion of emissions (63%), followed by methane (26%) and nitrous oxide (N_2O) (11%).

The energy sector was the largest source of emissions for 70 Parties, whereas for 45 Parties the agriculture sector was the largest. Per capita emissions (in tonnes of CO_2-eq, excluding LUCF) averaged 2.8 tonnes for the 122 non-Annex I Parties.

1. The Africa region's aggregate emissions were 1.6 billion tonnes, with per capita emissions of 2.4 tonnes.

2. The Asia and Pacific region's aggregate emissions were 7.9 billion tonnes, with per capita emissions of 2.6 tonnes.

3. The Latin America and Caribbean region's aggregate emissions were 2 billion tonnes, with per capita emissions of 4.6 tonnes.

4. The "other" region includes Albania, Armenia, Azerbaijan, Georgia, Malta, Moldova, and Macedonia. Their aggregate emissions were 0.1 billion tonnes, with per capita emissions of 5.1 tonnes.

Parties reported a high level of uncertainty in LUCF emissions, but in aggregate, there appeared to only be a small difference of 1.7% with and without LUCF. With LUCF, emissions were 11.9 billion tonnes, without LUCF, total aggregate emissions were 11.7 billion tonnes.

Trends

In several large developing countries and fast growing economies (China, India, Thailand, Indonesia, Egypt, and Iran) GHG emissions have increased rapidly (PBL, 2009). For example, emissions in China have risen strongly over the 1990–2005 period, often by more than 10% year. Emissions per-capita in non-Annex I countries are still, for the most part, much lower than in industrialized countries. Non-Annex I countries do not have quantitative emission reduction commitments, but they are committed to mitigation actions. China, for example, has had a national policy programme to reduce emissions growth, which included the closure of old, less efficient coal-fired power plants.

Cost Estimates

Barker *et al.* (2007, p. 79) assessed the literature on cost estimates for the Kyoto Protocol. Due to non-US participation in the Kyoto treaty, costs estimates were found to be much lower than those estimated in the previous IPCC Third Assessment Report. Without US participation, and with full use of the Kyoto flexible mechanisms, costs were estimated at less than 0.05% of Annex B GDP. This compared to earlier estimates of 0.1–1.1%. Without use of the flexible mechanisms, costs without US participation were estimated at less than 0.1%. This compared to earlier estimates of 0.2–2%. These cost estimates were viewed as being based on much evidence and high agreement in the literature.

Views on the Protocol

Gupta *et al.* (2007) assessed the literature on climate change policy. They found that no authoritative assessments of the UNFCCC or its Protocol asserted that these agreements had, or will, succeed in solving the climate problem. In these assessments, it was assumed that the UNFCCC or its Protocol would not be changed. The Framework Convention and its Protocol include provisions for future policy actions to be taken.

Gupta *et al.* (2007) described the Kyoto first-round commitments as "modest," stating that they acted as a constraint on the treaty's effectiveness. It was suggested that subsequent Kyoto commitments could be made more effective with measures aimed at achieving deeper cuts in emissions, as well as having policies applied to a larger share of global emissions. In 2008, countries with a Kyoto cap made up less than one-third of annual global carbon dioxide emissions from fuel combustion.

World Bank (2010) commented on how the Kyoto Protocol had only had a slight effect on curbing global emissions growth. The treaty was negotiated in 1997, but in 2006, energy-related carbon dioxide emissions had grown by 24%. World Bank (2010) also stated that the treaty had provided only limited financial support to developing countries to assist them in reducing their emissions and adapting to climate change.

Some of the criticism of the Protocol has been based on the idea of climate justice (Liverman, 2008, p. 14). This has particularly centred on the balance between the low emissions and high vulnerability of the developing world to climate change, compared to high emissions in the developed world.

Some environmentalists have supported the Kyoto Protocol because it is "the only game in town," and possibly because they expect that future emission reduction commitments may demand more stringent emission reductions (Aldy *et al.*., 2003, p. 9). In 2001, seventeen national science academies stated that ratification of the Protocol represented a "small but essential first step towards stabilising atmospheric concentrations of greenhouse gases." Some environmentalists and scientists have criticized the existing commitments for being too weak (Grubb, 2000, p. 5).

The United States (under former President George W. Bush) and Australia (initially, under former Prime Minister John Howard) did not ratify the Kyoto treaty. According to Stern (2006), their decision was based on the lack of quantitative emission commitments for emerging economies. Australia, under former Prime Minister Kevin Rudd, has since ratified the treaty, which took effect in March 2008.

Views on the Flexibility Mechanisms

Another area which has been commented on is the role of the Kyoto flexibility mechanisms – emissions trading, Joint Implementation, and the Clean Development Mechanism (CDM). The flexibility mechanisms have attracted both positive and negative comments.

As mentioned earlier, a number of Annex I Parties have implemented emissions trading schemes (ETSs) as part of efforts to meet their Kyoto commitments. General commentaries on emissions trading are contained in emissions trading and carbon emission trading. Individual articles on the ETSs contain commentaries on these schemes.

One of the arguments made in favour of the flexibility mechanisms is that they can reduce the costs incurred by Annex I Parties in meeting their Kyoto commitments. Criticisms of flexibility have, for example, included the ineffectiveness of emissions trading in promoting investment in non-fossil energy sources, and adverse impacts of CDM projects on local communities in developing countries.

Conference of the Parties

The official meeting of all states party to the Kyoto Protocol is the *Conference of the Parties*. It is held every year as part of the United Nations Climate Change conference, which also serves as the formal meeting of UNFCCC. The first Meetings of Parties of the Kyoto Protocol (CMP) was held in 2005 in conjunction with the eleventh Conferences of parties to UNFCCC (COP). Also parties to the Convention that are not parties to the Protocol can participate in Protocol-related meetings as observers. The first conference was held in 1995 in Berlin, while the 2013 conference was held in Warsaw. Later COPs were held in Lima, Peru in 2014 and in Paris, France in 2015.

Amendment and Possible Successors

In the non-binding "Washington Declaration" agreed on 16 February 2007, heads of governments from Canada, France, Germany, Italy, Japan, Russia, the United Kingdom, the United States, Brazil, China, India, Mexico and South Africa agreed in principle on the outline of a successor to the Kyoto Protocol. They envisaged a global cap-and-trade system that would apply to both industrialized nations and developing countries, and initially hoped that it would be in place by 2009.

The United Nations Climate Change Conference in Copenhagen in December 2009 was one of the annual series of UN meetings that followed the 1992 Earth Summit in Rio. In 1997 the talks led to the Kyoto Protocol, and the conference in Copenhagen was considered to be the opportunity to agree a successor to Kyoto that would bring about meaningful carbon cuts.

The 2010 Cancún agreements include voluntary pledges made by 76 developed and developing countries to control their emissions of greenhouse gases. In 2010, these 76 countries were collectively responsible for 85% of annual global emissions.

By May 2012, the USA, Japan, Russia, and Canada had indicated they would not sign up to a second Kyoto commitment period. In November 2012, Australia confirmed it would participate in a second commitment period under the Kyoto Protocol and New Zealand confirmed that it would not.

New Zealand's climate minister Tim Groser said the 15-year-old Kyoto Protocol was outdated, and that New Zealand was "ahead of the curve" in looking for a replacement that would include developing nations. Non-profit environmental organisations such as the World Wildlife Fund criticised New Zealand's decision to pull out.

On 8 December 2012, at the end of the 2012 United Nations Climate Change Conference, an agreement was reached to extend the Protocol to 2020 and to set a date of 2015 for the development of a successor document, to be implemented from 2020. The outcome of the Doha talks has received a mixed response, with small island states critical of the overall package.The Kyoto second commitment period applies to about 11% of annual global emissions of greenhouse gases. Other results of the conference include a timetable for a global agreement to be adopted by 2015 which includes all countries. At the Doha meeting of the parties to the UNFCCC on 8 December 2012, the European Union chief climate negotiator, Artur Runge-Metzger, pledged to extend the treaty, binding on the 27 European Member States, up to the year 2020 pending an internal ratification procedure.

Ban Ki Moon, Secretary General of the United Nations, called on world leaders to come to an agreement on halting global warming during the 69th Session of the UN General Assembly on 23 September 2014 in New York. UN member states have been negotiating a future climate deal over the last five years. A preliminary calendar was adopted to confirm "national contributions" to the reduction of CO2 emissions by 2015 before the UN climate summit which will be held in Paris 2015 United Nations Climate Change Conference.

Bali Road Map

After the 2007 United Nations Climate Change Conference on the island Bali in Indonesia in December, 2007 the participating nations adopted the Bali Road Map as a two-year process to finalizing a binding agreement in 2009 in Copenhagen. The conference encompassed meetings of several bodies, including the 13th Conference of the Parties to the United Nations Framework Convention on Climate Change (COP 13) and the 3rd Meeting of the Parties to the Kyoto Protocol (MOP 3 or CMP 3).

The Bali Road Map includes the Bali Action Plan (BAP) that was adopted by Decision 1/CP.13 of the COP-13. It also includes the Ad Hoc Working Group on Further Commitments for Annex I Parties under the Kyoto Protocol (AWG-KP) negotiations and their 2009 deadline, the launch of the Adaptation Fund, the scope and content of the Article 9 review of the Kyoto Protocol, as well as decisions on technology transfer and on reducing emissions from deforestation.

Bali Action Plan

Pillars

The Conference of Parties decided to launch a comprehensive process to enable the implementation of the Convention through long-term cooperative action, now, up to and beyond 2012, by addressing: (the called pillars or building blocks)

1. A shared vision for long-term cooperative action, including a long-term global goal for emission reductions.

2. Enhanced national/international action on mitigation of climate change.

3. Enhanced action on adaptation.

4. Enhanced action on technology development and transfer to support action on mitigation and adaptation.

5. Enhanced action on the provision of financial resources and investment to support action on mitigation and adaptation and technology cooperation.

Cutting Emissions

The nations acknowledge that evidence for global warming is *unequivocal*, and that humans must reduce emissions to reduce the risks of "severe climate change impacts" and emphasized the urgency to address climate change. There was a strong consensus for updated changes for both developed and developing countries. Although there were not specific numbers agreed upon in order to cut emissions,the decision recognized that there was a need for "deep cuts in global emissions" (plural countries proposed 100% reduction in 2050) and that "developed country emissions must fall 10-40% by 2020".

Mitigation

1. Enhanced action on mitigation of climate change includes, inter alia:

2. Nationally appropriate mitigation commitments or actions by all developed countries.

3. Nationally appropriate mitigation actions (NAMAs) by developing countries.

4. Cooperative sectorial approaches and sector-specific actions (CSAs).

5. Ways to strengthen the catalytic role of the convention.

Forests

The nations pledge "policy approaches and positive incentives" on issues relating to reducing emissions from deforestation and forest degradation (REDD) in developing countries; and enhancement of forest carbon stock in developing countries This paragraph is referred to as "REDD-plus".

Adaptation

The nations opt for enhanced co-operation to "support urgent implementation" of measures to protect poorer countries against climate change, including NAPAs. impacts.

Technology

In technology development and transfer, the nations will consider how to facilitate the transfer of clean and renewable energy technologies from industrialised nations to the developing countries. This includes, inter alia:

1. Removal of obstacles to, and provision of financial and other incentives for, scaling up the de-

velopment and transfer of technology to developing country Parties in order to promote access to affordable environmentally sound technologies (renewable energies, electric vehicles).

2. Ways to accelerate the deployment, diffussion and transfer of such technologies.

3. Cooperation on research and development of current, new and innovative technology, including win-win solutions.

4. The effectiveness of mechanism and tools for technology cooperation in specific sectors.

Finance

1. Provision of financial resources and investment includes:

2. Improved access to predictable and sustainable financial resources and the provision of new and additional resources, including official and concessional funding for developing country Parties (dcP).

3. Positive incentives for dcP for national mitigation strategies and adaptation action.

4. Innovative means of funding for dcP that are particularly vulnerable to the adverse impacts of climate change in meeting the costs of adaptation.

5. Incentivisation of adaptation actions on the basis of sustainable development policies.

6. Mobilization of funding and investment, including facilitation of climate-friendly investment choices.

7. Financial and technical support for capacity-building in the assessment of costs of adaptation in developing countries, to aid in determining their financial needs.

Ad Hoc Working Groups

The Conference decided to establish two subsidiary bodies under the Convention to conduct the process, the Ad Hoc Working Group on Long-term Cooperative Action (AWG-LCA) and the Ad Hoc Working Group on Further Commitments for Annex I Parties under the Kyoto Protocol (AWG-KP), that were to complete their work in 2009 and present the outcome to the COP15/MOP 5.

The AWG-LCA and AWG-KP presented draft conclusions to COP15 and CMP5, which contained many unresolved issues. The working groups were subsequently asked to report to COP16 and CMP6 in Cancun, Mexico.

Timescales

Four major UNFCCC meetings to implement the Bali Road Map were planned for 2008, with the first to be held in either March or April and the second in June, with the third in either August or September followed by a major meeting in Poznań, Poland in December 2008. The negotiations process was scheduled to conclude at the United Nations Climate Change Conference 2009 in Copenhagen, Denmark.

Carbon Credit

A carbon credit is a generic term for any tradable certificate or permit representing the right to emit one tonne of carbon dioxide or the mass of another greenhouse gas with a carbon dioxide equivalent (tCO_2e) equivalent to one tonne of carbon dioxide.

Carbon credits and carbon markets are a component of national and international attempts to mitigate the growth in concentrations of greenhouse gases (GHGs). One carbon credit is equal to one tonne of carbon dioxide, or in some markets, carbon dioxide equivalent gases. Carbon trading is an application of an emissions trading approach. Greenhouse gas emissions are capped and then markets are used to allocate the emissions among the group of regulated sources.

The goal is to allow market mechanisms to drive industrial and commercial processes in the direction of low emissions or less carbon intensive approaches than those used when there is no cost to emitting carbon dioxide and other GHGs into the atmosphere. Since GHG mitigation projects generate credits, this approach can be used to finance carbon reduction schemes between trading partners and around the world.

There are also many companies that sell carbon credits to commercial and individual customers who are interested in lowering their carbon footprint on a voluntary basis. These carbon offsetters purchase the credits from an investment fund or a carbon development company that has aggregated the credits from individual projects. Buyers and sellers can also use an exchange platform to trade, which is like a stock exchange for carbon credits. The quality of the credits is based in part on the validation process and sophistication of the fund or development company that acted as the sponsor to the carbon project. This is reflected in their price; voluntary units typically have less value than the units sold through the rigorously validated Clean Development Mechanism.

Definitions

The Collins English Dictionary defines a carbon credit as "*a certificate showing that a government or company has paid to have a certain amount of carbon dioxide removed from the environment*". The Environment Protection Authority of Victoria defines a carbon credit as a "*generic term to assign a value to a reduction or offset of greenhouse gas emissions.. usually equivalent to one tonne of carbon dioxide equivalent (CO2-e).*"

The Investopedia Inc investment dictionary defines a carbon credit as a "*permit that allows the holder to emit one ton of carbon dioxide*"..which "*can be traded in the international market at their current market price*".

Types

There are two main markets for carbon credits; Compliance Market credits Secondary / Verified Market credits (VERs)

Background

The burning of fossil fuels is a major source of greenhouse gas emissions, especially for power, cement, steel, textile, fertilizer and many other industries which rely on fossil fuels (coal, electricity derived from coal, natural gas and oil). The major greenhouse gases emitted by these industries are carbon dioxide, methane, nitrous oxide, hydrofluorocarbons (HFCs), etc., all of which increase the atmosphere's ability to trap infrared energy and thus affect the climate.

The concept of carbon credits came into existence as a result of increasing awareness of the need for controlling emissions. The IPCC (Intergovernmental Panel on Climate Change) has observed that:

Policies that provide a real or implicit price of carbon could create incentives for producers and consumers to significantly invest in low-GHG products, technologies and processes. Such policies could include economic instruments, government funding and regulation,

while noting that a tradable permit system is one of the policy instruments that has been shown to be environmentally effective in the industrial sector, as long as there are reasonable levels of predictability over the initial allocation mechanism and long-term price.

The mechanism was formalized in the Kyoto Protocol, an international agreement between more than 170 countries, and the market mechanisms were agreed through the subsequent Marrakesh Accords. The mechanism adopted was similar to the successful US Acid Rain Program to reduce some industrial pollutants.

Emission Allowances

Under the Kyoto Protocol, the 'caps' or quotas for Greenhouse gases for the developed Annex 1 countries are known as Assigned Amounts and are listed in Annex B. The quantity of the initial assigned amount is denominated in individual units, called Assigned amount units (AAUs), each of which represents an allowance to emit one metric tonne of carbon dioxide equivalent, and these are entered into the country's national registry.

In turn, these countries set quotas on the emissions of installations run by local business and other organizations, generically termed 'operators'. Countries manage this through their national registries, which are required to be validated and monitored for compliance by the UNFCCC. Each operator has an allowance of credits, where each unit gives the owner the right to emit one metric tonne of carbon dioxide or other equivalent greenhouse gas. Operators that have not used up their quotas can sell their unused allowances as carbon credits, while businesses that are about to exceed their quotas can buy the extra allowances as credits, privately or on the open market. As demand for energy grows over time, the total emissions must still stay within the cap, but it allows industry some flexibility and predictability in its planning to accommodate this.

By permitting allowances to be bought and sold, an operator can seek out the most cost-effective way of reducing its emissions, either by investing in 'cleaner' machinery and practices or by purchasing emissions from another operator who already has excess 'capacity'.

Since 2005, the Kyoto mechanism has been adopted for CO_2 trading by all the countries within the European Union under its European Trading Scheme (EU ETS) with the European Commission as its validating authority. From 2008, EU participants must link with the other developed countries who ratified Annex I of the protocol, and trade the six most significant anthropogenic greenhouse gases. In the United States, which has not ratified Kyoto, and Australia, whose ratification came into force in March 2008, similar schemes are being considered.

Kyoto's 'Flexible Mechanisms'

A tradable credit can be an emissions allowance or an assigned amount unit which was originally allocated or auctioned by the national administrators of a Kyoto-compliant cap-and-trade scheme, or it can be an offset of emissions. Such offsetting and mitigating activities can occur in any developing country which has ratified the Kyoto Protocol, and has a national agreement in place to validate its carbon project through one of the UNFCCC's approved mechanisms. Once approved, these units are termed Certified Emission Reductions, or CERs. The Protocol allows these projects to be constructed and credited in advance of the Kyoto trading period.

The Kyoto Protocol provides for three mechanisms that enable countries or operators in developed countries to acquire greenhouse gas reduction credits

1. Under Joint Implementation (JI) a developed country with relatively high costs of domestic greenhouse reduction would set up a project in another developed country.

2. Under the Clean Development Mechanism (CDM) a developed country can 'sponsor' a greenhouse gas reduction project in a developing country where the cost of greenhouse gas reduction project activities is usually much lower, but the atmospheric effect is globally equivalent. The developed country would be given credits for meeting its emission reduction targets, while the developing country would receive the capital investment and clean technology or beneficial change in land use.

3. Under International Emissions Trading (IET) countries can trade in the international carbon credit market to cover their shortfall in Assigned amount units. Countries with surplus units can sell them to countries that are exceeding their emission targets under Annex B of the Kyoto Protocol.

These carbon projects can be created by a national government or by an operator within the country. In reality, most of the transactions are not performed by national governments directly, but by operators who have been set quotas by their country.

Emission Markets

For trading purposes, one allowance or CER is considered equivalent to one metric ton of CO_2 emissions. These allowances can be sold privately or in the international market at the prevailing market price. These trade and settle internationally and hence allow allowances to be transferred between countries. Each international transfer is validated by the UNFCCC. Each transfer of ownership within the European Union is additionally validated by the European Commission.

Climate exchanges have been established to provide a spot market in allowances, as well as futures and options market to help discover a market price and maintain liquidity. Carbon prices are normally quoted in Euros per tonne of carbon dioxide or its equivalent (CO_2e). Other greenhouse gasses can also be traded, but are quoted as standard multiples of carbon dioxide with respect to their global warming potential. These features reduce the quota's financial impact on business, while ensuring that the quotas are met at a national and international level.

Currently there are five exchanges trading in carbon allowances: the European Climate Exchange, NASDAQ OMX Commodities Europe, PowerNext, Commodity Exchange Bratislava and the European Energy Exchange. NASDAQ OMX Commodities Europe listed a contract to trade offsets generated by a CDM carbon project called Certified Emission Reductions (CERs). Many companies now engage in emissions abatement, offsetting, and sequestration programs to generate credits that can be sold on one of the exchanges. At least one private electronic market has been established in 2008: CantorCO2e. Carbon credits at Commodity Exchange Bratislava are traded at special platform - Carbon place.

Managing emissions is one of the fastest-growing segments in financial services in the City of London with a market estimated to be worth about €30 billion in 2007. Louis Redshaw, head of environmental markets at Barclays Capital predicts that "Carbon will be the world's biggest commodity market, and it could become the world's biggest market overall."

Setting a Market Price for Carbon

Unchecked, energy use and hence emission levels are predicted to keep rising over time. Thus the number of companies needing to buy credits will increase, and the rules of supply and demand will push up the market price, encouraging more groups to undertake environmentally friendly activities that create carbon credits to sell.

An individual allowance, such as an Assigned amount unit (AAU) or its near-equivalent European Union Allowance (EUA), may have a different market value to an offset such as a CER. This is due to the lack of a developed secondary market for CERs, a lack of homogeneity between projects which causes difficulty in pricing, as well as questions due to the principle of supplementarity and its lifetime. Additionally, offsets generated by a carbon project under the Clean Development Mechanism are potentially limited in value because operators in the EU ETS are restricted as to what percentage of their allowance can be met through these flexible mechanisms.

Yale University economics professor William Nordhaus argues that the price of carbon needs to be high enough to motivate the changes in behavior and changes in economic production systems necessary to effectively limit emissions of greenhouse gases.

Raising the price of carbon will achieve four goals. First, it will provide signals to consumers about what goods and services are high-carbon ones and should therefore be used more sparingly. Second, it will provide signals to producers about which inputs use more carbon (such as coal and oil) and which use less or none (such as natural gas or nuclear power), thereby inducing firms to substitute low-carbon inputs. Third, it will give market incentives for inventors and innovators to develop and introduce low-carbon products and processes that can replace the current generation of technologies. Fourth, and most important, a high carbon price will economize on the information that is required to do all three of these tasks. Through the market mechanism, a high carbon price will raise the price of products according to their carbon content. Ethical consumers today, hoping to minimize their "carbon footprint," have little chance of making an accurate calculation of the relative carbon use in, say, driving 250 miles as compared with flying 250 miles. A harmonized carbon tax would raise the price of a good proportionately to exactly the amount of CO_2 that is emitted in all the stages of production that are involved in producing that good. If 0.01 of a ton of carbon emissions results from the wheat growing and the milling and the trucking and the baking of a loaf of bread, then a tax of $30 per ton carbon will raise the price of bread by $0.30. The "carbon footprint" is automatically calculated by the price system. Consumers would still not know how much of the price is due to carbon emissions, but they could make their decisions confident that they are paying for the social cost of their carbon footprint.

Nordhaus has suggested, based on the social cost of carbon emissions, that an optimal price of carbon is around $30(US) per ton and will need to increase with inflation.

The social cost of carbon is the additional damage caused by an additional ton of carbon emissions. ... The optimal carbon price, or optimal carbon tax, is the market price (or carbon tax) on carbon emissions that balances the incremental costs of reducing carbon emissions with the incremental benefits of reducing climate damages. ... [I]f a country wished to impose a carbon tax of $30 per ton of carbon, this would involve a tax on gasoline of about 9 cents per gallon. Similarly, the tax on coal-generated electricity would be about 1 cent per kWh, or 10 percent of the current retail price. At current levels of carbon emissions in the United States, a tax of $30 per ton of carbon would generate $50 billion of revenue per year.

How Buying Carbon Credits Can Reduce Emissions

Carbon credits create a market for reducing greenhouse emissions by giving a monetary value to the cost of polluting the air. Emissions become an internal cost of doing business and are visible on the balance sheet alongside raw materials and other liabilities or assets.

For example, consider a business that owns a factory putting out 100,000 tonnes of greenhouse gas emissions in a year. Its government is an Annex I country that enacts a law to limit the emissions that the business can produce. So the factory is given a quota of say 80,000 tonnes per year. The factory either reduces its emissions to 80,000 tonnes or is required to purchase carbon credits to offset the excess. After costing up alternatives the business may decide that it is uneconomical or infeasible to invest in new machinery for that year. Instead it may choose to buy carbon credits on the open market from organizations that have been approved as being able to sell legitimate carbon credits.

We should consider the impact of manufacturing alternative energy sources. For example, the energy consumed and the Carbon emitted in the manufacture and transportation of a large wind turbine would prohibit a credit being issued for a predetermined period of time.

1. One seller might be a company that will offer to offset emissions through a project in the developing world, such as recovering methane from a swine farm to feed a power station that previously would use fossil fuel. So although the factory continues to emit gases, it would pay another group to reduce the equivalent of 20,000 tonnes of carbon dioxide emissions from the atmosphere for that year.

2. Another seller may have already invested in new low-emission machinery and have a surplus of allowances as a result. The factory could make up for its emissions by buying 20,000 tonnes of allowances from them. The cost of the seller's new machinery would be subsidized by the

sale of allowances. Both the buyer and the seller would submit accounts for their emissions to prove that their allowances were met correctly.

Credits Versus Taxes

Carbon credits and carbon taxes each have their advantages and disadvantages. Credits were chosen by the signatories to the Kyoto Protocol as an alternative to Carbon taxes. A criticism of tax-raising schemes is that they are frequently not hypothecated, and so some or all of the taxation raised by a government would be applied based on what the particular nation's government deems most fitting. However, some would argue that carbon trading is based around creating a lucrative artificial market, and, handled by free market enterprises as it is, carbon trading is not necessarily a focused or easily regulated solution.

By treating emissions as a market commodity some proponents insist it becomes easier for businesses to understand and manage their activities, while economists and traders can attempt to predict future pricing using market theories. Thus the main advantages of a tradeable carbon credit over a carbon tax are argued to be:

1. the price may be more likely to be perceived as fair by those paying it. Investors in credits may have more control over their own costs.

2. the flexible mechanisms of the Kyoto Protocol help to ensure that all investment goes into genuine sustainable carbon reduction schemes through an internationally agreed validation process.

3. some proponents state that if correctly implemented a target level of emission reductions may somehow be achieved with more certainty, while under a tax the actual emissions might vary over time.

4. it may provide a framework for rewarding people or companies who plant trees or otherwise meet standards exclusively recognized as "green."

The advantages of a carbon tax are argued to be:

1. possibly less complex, expensive, and time-consuming to implement. This advantage is especially great when applied to markets like gasoline or home heating oil.

2. perhaps some reduced risk of certain types of cheating, though under both credits and taxes, emissions must be verified.

3. reduced incentives for companies to delay efficiency improvements prior to the establishment of the baseline if credits are distributed in proportion to past emissions.

4. when credits are grandfathered, this puts new or growing companies at a disadvantage relative to more established companies.

5. allows for more centralized handling of acquired gains

6. worth of carbon is stabilized by government regulation rather than market fluctuations. Poor market conditions and weak investor interest have a lessened impact on taxation as opposed to carbon trading.

Creating Carbon Credits

The principle of Supplementarity within the Kyoto Protocol means that internal abatement of emissions should take precedence before a country buys in carbon credits. However it also established the Clean Development Mechanism as a Flexible Mechanism by which capped entities could develop measurable and permanent emissions reductions voluntarily in sectors outside the cap. Many criticisms of carbon credits stem from the fact that establishing that an emission of CO_2-equivalent greenhouse gas has truly been reduced involves a complex process. This process has evolved as the concept of a carbon project has been refined over the past 10 years.

The first step in determining whether or not a carbon project has legitimately led to the reduction of measurable and permanent emissions is understanding the CDM methodology process. This is the process by which project sponsors submit, through a Designated Operational Entity (DOE), their concepts for emissions reduction creation. The CDM Executive Board, with the CDM Methodology Panel and their expert advisors, review each project and decide how and if they do indeed result in reductions that are additional

Additionality and Its Importance

It is also important for any carbon credit (offset) to prove a concept called additionality. The concept of additionality addresses the question of whether the project would have happened in the absence of an intervention in the form of the price signal of carbon credits. Only projects with emissions below their baseline level, defined as emissions under a scenario without this price signal (holding all other factors constant), represent a net environmental benefit. Carbon projects that yield strong financial returns even in the absence of revenue from carbon credits; or that are compelled by regulations; or that represent common practice in an industry; are usually not considered additional. A full determination of additionality requires a careful investigation of proposed carbon offset projects.

It is generally agreed that voluntary carbon offset projects must demonstrate additionality to ensure the legitimacy of the environmental stewardship claims resulting from the retirement of carbon credits (offsets).

Criticisms

The Kyoto mechanism is the only internationally agreed mechanism for regulating carbon credit activities, and, crucially, includes checks for additionality and overall effectiveness. Its supporting organisation, the UNFCCC, is the only organisation with a global mandate on the overall effectiveness of emission control systems, although enforcement of decisions relies on national co-operation. The Kyoto trading period only applies for five years between 2008 and 2012. The first phase of the EU ETS system started before then, and is expected to continue in a third phase afterwards, and may co-ordinate with whatever is internationally agreed at but there is general uncertainty as to what will be agreed in Post–Kyoto Protocol negotiations on greenhouse gas emissions. As business investment often operates over decades, this adds risk and uncertainty to their plans. As several countries responsible for a large proportion of global emissions (notably USA, India, China) have avoided mandatory caps, this also means that businesses in capped countries may perceive themselves to be working at a competitive disadvantage against those in uncapped countries as they are now paying for their carbon costs directly.

A key concept behind the cap and trade system is that national quotas should be chosen to represent genuine and meaningful reductions in national output of emissions. Not only does this ensure that overall emissions are reduced but also that the costs of emissions trading are carried fairly across all parties to the trading system. However, governments of capped countries may seek to unilaterally weaken their commitments, as evidenced by the 2006 and 2007 National Allocation Plans for several countries in the EU ETS, which were submitted late and then were initially rejected by the European Commission for being too lax.

A question has been raised over the grandfathering of allowances. Countries within the EU ETS have granted their incumbent businesses most or all of their allowances for free. This can sometimes be perceived as a protectionist obstacle to new entrants into their markets. There have also been accusations of power generators getting a 'windfall' profit by passing on these emissions 'charges' to their customers. As the EU ETS moves into its second phase and joins up with Kyoto, it seems likely that these problems will be reduced as more allowances will be auctioned.

Economics of Climate Change Mitigation

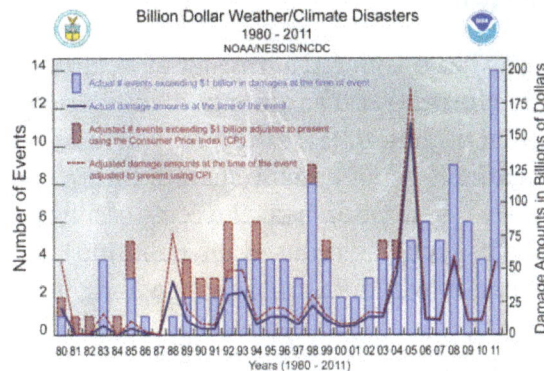

Total extreme weather cost and number of events costing more than $1 billion in the United States from 1980 to 2011

This article is about the economics of climate change mitigation. Mitigation of climate change involves actions that are designed to limit the amount of long-term climate change (Fisher *et al.*, 2007:225). Mitigation may be achieved through the reduction of greenhouse gas (GHG) emissions or through the enhancement of sinks that absorb GHGs, for example forests.

Definitions

In this article, the phrase "climate change" is used to describe a change in the climate, measured in terms of its statistical properties, e.g., the global mean surface temperature. In this context, "climate" is taken to mean the average weather. Climate can change over period of time ranging from months to thousands or millions of years. The classical time period is 30 years, as defined by the World Meteorological Organization. The climate change referred to may be due to natural causes, e.g., changes in the sun's output, or due human activities, e.g., changing the composition of the atmosphere. Any human-induced changes in climate will occur against the "background" of natural climatic variations.

Public Good Issues

The atmosphere is an international public good and GHG emissions are an international externality (Goldemberg *et al.*, 1996:,21, 28, 43). Each individual's or country's welfare, U^j, is a function of its own consumption, C^j, and the quality of the atmosphere, A, such that $U^j(C^j,A)$. A change in the quality of the atmosphere, A, does not affect the welfare of all individuals and countries equally. In other words, some individuals and countries may benefit from climate change, but others may lose out.

Heterogeneity

GHG emissions are unevenly distributed around the world, as are the potential impacts of climate change (Toth *et al.*, 2001:607). Nations with higher than average emissions that face potentially small negative/positive climate change impacts have little incentive to reduce their emissions. Nations with relatively low levels of emissions that face potentially large negative climate change impacts have a large incentive to reduce emissions. Nations that avoid mitigation can benefit from free-riding on the actions of others, and may even enjoy gains in trade and/or investment (Halsnæs *et al.*, 2007:127). The unequal distribution of benefits from mitigation, and the potential advantages of free-riding, make it difficult to secure an international agreement to reduce emissions.

Intergenerational Transfers

Mitigation of climate change can be considered a transfer of wealth from the present generation to future generations (Toth *et al..*, 2001:607). The amount of mitigation determines the composition of resources (e.g., environmental or material) that future generations receive. Across generations, the costs and benefits of mitigation are not equally shared: future generations potentially benefit from mitigation, while the present generation bear the costs of mitigation but do not directly benefit (ignoring possible co-benefits, such as reduced air pollution). If the current generation also benefitted from mitigation, it might lead them to be more willing to bear the costs of mitigation.

Irreversible Impacts and Policy

Emissions of carbon dioxide (CO_2) might be irreversible on the time scale of millennia (Halsnæs *et al.*, 2007). There are risks of irreversible climate changes, and the possibility of sudden changes in climate. On the other hand, these effects are also true of mitigation efforts. Investments made in long-lived, large-scale low-emission technologies are essentially irreversible. If the scientific basis for these investments turns out to be wrong, they would become "stranded" assets. Additionally, the costs of reducing emissions may change over time in a non-linear fashion.

From an economic perspective, as the scale of private sector investment in low-carbon technologies increases, so do the risks. Uncertainty over future climate policy decisions makes investors reluctant to undertake large-scale investment without upfront government support. The later section on finance discusses how risk affects investment in developing and emerging economies.

Sustainable Development

Solow (1992) (referred to by Arrow, 1996b, pp. 140–141) defined sustainable development as allowing for reductions in exhaustible resources so long as these reductions are adequately offset by increases in other resources. This definition implicitly assumes that resources can be substituted, a view which is supported by economic history. Another view is that reductions in some exhaustible resources can only be partially made up for by substitutes. If true, this might mean then some assets need to be preserved at all costs.

In many developing countries, Solow's definition might not be viewed as being acceptable, since it could place a constraint on their ambitions for development. A remedy for this would be for developed countries to pay all the costs of mitigation, including costs in developing countries. This solution is suggested by both Rawlsian and utilitarian constructs of the social welfare function. These functions are used to assess the welfare impacts on all individuals of climate change and related policies (Markandya *et al.*, 2001, p. 460). The Rawlsian approach concentrates on the welfare of the worst-off in society, whereas the utilitarian approach is a sum of utilities (Arrow *et al.*, 1996b, p. 138).

It might be argued that since such redistributions of resources are not observed now, why would either Rawlsian or utilitarian constructs be appropriate for climate change (Arrow *et al.*, 1996b, p. 140)? A possible response to this would point to the fact that in the absence of government intervention, market rates of redistribution will not equal social rates.

Emissions and Economic Growth

Economic growth is a key driver of CO_2 emissions (Sathaye *et al.*, 2007:707). As the economy expands, demand for energy and energy-intensive goods increases, pushing up CO_2 emissions. On the other hand, economic growth may drive technological change and increase energy efficiency. Economic growth may be associated specialization in certain economic sectors. If specialization is in energy-intensive sectors, then there might be a strong link between economic growth and emissions growth. If specialization is in less energy-intensive sectors, e.g., the services sector, then there might be a weak link between economic growth and emissions growth. Unlike technological change or energy efficiency improvements, specialization in high or low energy intensity sectors does not affect global emissions. Rather, it changes the distribution of global emissions.

Much of the literature focuses on the "environmental Kuznets curve" (EKC) hypothesis, which posits that at early stages of development, pollution per capita and GDP per capita move in the same direction. Beyond a certain income level, emissions per capita will decrease as GDP per capita increase, thus generating an inverted-U shaped relationship between GDP per capita and pollution. Sathaye *et al..* (2007) concluded that the econometrics literature did not support either an optimistic interpretation of the EKC hypothesis - i.e., that the problem of emissions growth will solve itself - or a pessimistic interpretation - i.e., that economic growth is irrevocably linked to emissions growth. Instead, it was suggested that there was some degree of flexibility between economic growth and emissions growth.

Policies that Impact Emissions

Price Signals and Subsidies

In developed countries, energy costs are low and heavily subsidized, whereas in developing countries, the poor pay high costs for low-quality services. Bashmakov *et al..* (2001:410) commented on the difficulty of measuring energy subsidies, but found some evidence that coal production subsidies had declined in several developing and OECD countries.

Structural Market Reforms

Market-orientated reforms, as undertaken by several countries in the 1990s, can have important effects on energy use, energy efficiency, and therefore GHG emissions. In a literature assessment, Bashmakov *et al..* (2001:409) gave the example of China, which has made structural reforms with the aim of increasing GDP. They found that since 1978, energy use in China had increased by an average of 4% per year, but at the same time, energy use had been reduced per unit of GDP.

Liberalization of Energy Markets

Liberalization and restructuring of energy markets has occurred in several countries and regions, including Africa, the EU, Latin America, and the US. These policies have mainly been designed to increase competition in the market, but they can have a significant impact on emissions. Bashmakov *et al..* (2001:410) concluded that structural reform of the energy sector could not guarantee a shift towards less carbon-intensive power generation. Reform could, however, allow the market to be more responsive to price signals placed on emissions.

Climate and other Environmental Policies

National

1. Regulatory standards: These set technology or performance standards, and can be effective in addressing the market failure of informational barriers (Bashmakov *et al.*, 2001:412). If the costs of regulation are less than the benefits of addressing the market failure, standards can result in net benefits.

2. Emission taxes and charges: an emissions tax requires domestic emitters to pay a fixed fee or tax for every tonne of CO_2-eq GHG emissions released into the atmosphere (Bashmakov *et al.*, 2001:413). If every emitter were to face the same level of tax, the lowest cost way of achieving emission reductions in the economy would be undertaken first. In the real world, however, markets are not perfect, meaning that an emissions tax may deviate from this ideal. Distributional and equity considerations usually result in differential tax rates for different sources.

3. Tradable permits: Emissions can be limited with a permit system (Bashmakov *et al.*, 2001:415). A number of permits are distributed equal to the emission limit, with each liable entity required to hold the number of permits equal to its actual emissions. A tradable permit system can be cost-effective so long as transaction costs are not excessive, and there are no significant imperfections in the permit market and markets relating to emitting activities.

4. Voluntary agreements: These are agreements between government and industry (Bashmakov *et al.*, 2001:417). Agreements may relate to general issues, such as research and development, but in other cases, quantitative targets may be agreed upon. An advantage of voluntary agreements are their low transaction costs. There is, however, the risk that participants in the agreement will free ride, either by not complying with the agreement or by benefitting from the agreement while bearing no cost.

5. Informational instruments: According to Bashmakov *et al.*. (2001:419), poor information is recognized as a barrier to improved energy efficiency or reduced emissions. Examples of policies in this area include increasing public awareness of climate change, e.g., through advertising, and the funding of climate change research.

6. Environmental subsidies: A subsidy for GHG emissions reductions pays entities a specific amount per tonne of CO_2-eq for every tonne of GHG reduced or sequestered (Bashmakov *et al.*, 2001:421). Although subsidies are generally less efficient than taxes, distributional and competitiveness issues sometimes result in energy/emission taxes being coupled with subsidies or tax exceptions.

7. Research and development policies: Government funding of research and development (R&D) on energy has historically favoured nuclear and coal technologies. Bashmakov *et al.*. (2001:421) found that although research into renewable energy and energy-efficient technologies had increased, it was still a relatively small proportion of R&D budgets in the OECD.

8. Green power: The policy ensures that part of the electricity supply comes from designated renewable sources (Bashmakov *et al.*, 2001:422). The cost of compliance is borne by all consumers.

9. Demand-side management: This aims to reduce energy demand, e.g., through energy audits, labelling, and regulation (Bashmakov *et al.*, 2001:422).

According to Bashmakov *et al.*. (2001:422), the most effective and economically efficient approach of achieving lower emissions in the energy sector is to apply a combination of market-based instruments (taxes, permits), standards, and information policies.

International

Kyoto Protocol

The Kyoto Protocol is an international treaty designed to reduce emissions of GHGs. The Kyoto treaty was agreed in 1997, and is a protocol to the United Nations Framework Convention on Climate Change (UNFCCC), which had previously been agreed in 1992. The Kyoto Protocol sets legally-blinding emissions limitations for developed countries ("Annex I Parties") out to 2008-2012. The US has not ratified the Kyoto Protocol, and its target is therefore non-binding. Canada has ratified the treaty, but withdrew in 2011.

The Kyoto treaty is a "cap-and-trade" system of emissions trading, which includes emissions reductions in developing countries ("non-Annex I Parties") through the Clean Development Mechanism (CDM). The economics of the Kyoto Protocol is discussed in Views on the Kyoto Protocol and Flexible mechanisms#Views on the flexibility mechanisms. Cost estimates for the treaty are summarized at Kyoto Protocol#Cost estimates. Economic analysis of the CDM is available at Clean Development Mechanism.

To summarize, the caps agreed to in Kyoto's first commitment period (2008-2012) have turned out to be too weak. There are a large surplus of emissions allowances in the former-Soviet economies ("Economies-in-Transition" - EITs), while several other OECD countries have a deficit, and are not on course to meet their Kyoto targets. Because of the large surplus of allowances, full trading of Kyoto allowances would likely depress the price of the permits near to zero. Some of the surplus allowances have been bought from the EITs, but overall little trading has taken place. Countries have mainly concentrated on meeting their targets domestically, and through the use of the CDM.

Some countries have implemented domestic energy/carbon taxes and emissions trading schemes (ETSs). The individual articles on the various ETSs contain commentaries on these schemes.

A number of analysts have focussed on the need to establish a global price on carbon in order to reduce emissions cost-effectively. The Kyoto treaty does not set a global price for carbon. As stated earlier, the US is not part of the Kyoto treaty, and is a major contributor to global annual emissions of carbon dioxide. Additionally, the treaty does not place caps on emissions in developing countries. The lack of caps for developing countries was based on equity (fairness) considerations. Developing countries, however, have undertaken a range of policies to reduce their emissions domestically. The later Cancún agreement, agreed under the UNFCCC, is based on voluntary pledges rather than binding commitments.

The UNFCCC has agreed that future global warming should be limited to below 2 °C relative to the pre-industrial temperature. Analyses by the United Nations Environment Programme and International Energy Agency suggest that current policies (as of 2011) are not strong enough to meet this target.

Other Policies

1. Regulatory instruments: This could involve the setting of regulatory standards for various products and processes for countries to adopt. The other option is to set national emission limits. The second option leads to inefficiency because the marginal costs of abatement differs between countries (Bashmakov *et al..*, 2001:430).

2. Carbon taxes: This would offer a potentially cost-effective means of reducing CO_2 emissions. Compared with emissions trading, international or harmonized (where each country keeps the revenue it collects) taxes provide greater certainty about the likely costs of emission reductions. This is also true of a hybrid policy (Bashmakov *et al..*, 2001:430).

Efficiency of International Agreements

For the purposes of analysis, it is possible to separate efficiency from equity (Goldemberg *et al.*, 1996, p. 30). It has been suggested that because of the low energy efficiency in many developing countries, efforts should first be made in those countries to reduce emissions. Goldemberg *et al.* (1996, p. 34) suggested a number of policies to improve efficiency, including:

1. Property rights reform. For example, deforestation could be reduced through reform of property rights.

2. Administrative reforms. For example, in many countries, electricity is priced at the cost of

production. Economists, however, recommend that electricity, like any other good, should be priced at the competitive price.

3. Regulating non-greenhouse externalities. There are externalities other than the emission of GHGs, for example, road congestion leading to air pollution. Addressing these externalities, e.g., through congestion pricing and energy taxes, could help to lower both air pollution and GHG emissions.

General Equilibrium Theory

One of the aspects of efficiency for an international agreement on reducing emissions is participation. In order to be efficient, mechanisms to reduce emissions still require all emitters to face the same costs of emission (Goldemberg *et al.*, 1996, p. 30). Partial participation significantly reduces the effectiveness of policies to reduce emissions. This is because of how the global economy is connected through trade.

General equilibrium theory points to a number of difficulties with partial participation (p. 31). Examples are of "leakage" (carbon leakage) of emissions from countries with regulations on GHG emissions to countries with less regulation. For example, stringent regulation in developed countries could result in polluting industries such as aluminium production moving production to developing countries. Leakage is a type of "spillover" effect of mitigation policies.

Estimates of spillover effects are uncertain (Barker *et al.*, 2007). If mitigation policies are only implemented in Kyoto Annex I countries, some researchers have concluded that spillover effects might render these policies ineffective, or possibly even cause global emissions to increase (Barker *et al.*, 2007). Others have suggested that spillover might be beneficial and result in reduced emission intensities in developing countries.

Comprehensiveness

Efficiency also requires that the costs of emission reductions be minimized (Goldemberg *et al.*, 1996, p. 31). This implies that all GHGs (CO_2, methane, etc.) are considered as part of a policy to reduce emissions, and also that carbon sinks are included. Perhaps most controversially, the requirement for efficiency implies that all parts of the Kaya identity are included as part of a mitigation policy. The components of the Kaya identity are:

1. CO_2 emissions per unit of energy, (carbon intensity)

2. energy per unit of output, (energy efficiency)

3. economic output per capita,

4. and human population.

Efficiency requires that the marginal costs of mitigation for each of these components is equal. In other words, from the perspective of improving the overall efficiency of a long-term mitigation strategy, population control has as much "validity" as efforts made to improve energy efficiency.

Equity in International Agreements

Unlike efficiency, there is no consensus view of how to assess the fairness of a particular climate policy. This does not prevent the study of how a particular policy impacts welfare. Edmonds *et al.* (1995) estimated that a policy of stabilizing national emissions without trading would, by 2020, shift more than 80% of the aggregate policy costs to non-OECD regions (Bashmakov *et al.*, 2001:439). A common global carbon tax would result in an uneven burden of abatement costs across the world and would change with time. With a global tradable quota system, welfare impacts would vary according to quota allocation.

Regional Aspects

In a literature assessment, Sathaye *et al.*. (2001:387-389) described regional barriers to mitigation:

1. Developing countries:

    2. In many developing countries, importing mitigation technologies might lead to an increase in their external debt and balance-of-payments deficit.

    3. Technology transfer to these countries can be hindered by the possibility of non-enforcement of intellectual property rights. This leaves little incentive for private firms to participate. On the other hand, enforcement of property rights can lead to developing countries facing high costs associated with patents and licensing fees.

    4. A lack of available capital and finance is common in developing countries.. Together with the absence of regulatory standards, this barrier supports the proliferation of inefficient equipment.

2. Economies in transition: In the New Independent States, Sathaye *et al.* (2007) concluded that a lack of liquidity and a weak environmental policy framework were barriers to investment in mitigation.

Finance

Article 4.2 of the United Nations Framework Convention on Climate Change commits industrialized countries to "[take] the lead" in reducing emissions. The Kyoto Protocol to the UNFCCC has provided only limited financial support to developing countries to assist them in climate change mitigation and adaptation. Additionally, private sector investment in mitigation and adaptation could be discouraged in the short and medium term because of the 2008 global financial crisis.

The International Energy Agency estimates that US$197 billion is required by states in the developing world above and beyond the underlying investments needed by various sectors regardless of climate considerations, this is twice the amount promised by the developed world at the UN Framework Convention on Climate Change (UNFCCC) Cancún Agreements. Thus, a new method is being developed to help ensure that funding is available for climate change mitigation. This involves financial leveraging, whereby public financing is used to encourage private investment.

The private sector is often unwilling to finance low carbon technologies in developing and emerging economies as the market incentives are often lacking. There are many perceived risks involved, in particular:

1. General political risk associated politically instability, uncertain property rights and an unfamiliar legal framework.

2. Currency risks are involved is financing is sought internationally and not provided in the nationally currency.

3. Regulatory and policy risk - if the public incentives provided by a state may not be actually provided, or if provided, then not for the full length of the investment.

4. Execution risk – reflecting concern that the local project developer/firm may lack the capacity and/or experience to execute the project efficiently.

5. Technology risk as new technologies involved in low carbon technology may not work as well as expected.

6. Unfamiliarity risks occur when investors have never undertaken such projects before.

Funds from the developed world can help mitigate these risks and thus leverage much larger private funds, the current aim to create $3 of private investment for every $1 of public funds.[4] Public funds can be used to minimise the risks in the following way.

1. Loan guarantees provided by international public financial institutions can be useful to reduce the risk to private lenders.

2. Policy insurance can insurance the investor against changes or disruption to government policies designed to encourage low carbon technology, such as a feed-in tariff.

3. Foreign exchange liquidity facilities can help reduce the risks associated with borrowing money in a different currency by creating a line of credit that can be drawn on when the project needs money as a result of local currency devaluation but then repaid when the project has a financial surplus.

4. Pledge fund can help projects are too small for equity investors to consider or unable to access sufficient equity. In this model, public finance sponsors provide a small amount of equity to anchor and encourage much larger pledges from private investors, such as sovereign wealth funds, large private equity firms and pension funds. Private equity investors will tend to be risk-adverse and focused primarily on long-term profitability, thus all projects would need to meet the fiduciary requirements of the investors.

5. Subordinated equity fund - an alternative use of public finance is through the provision of subordinated equity, meaning that the repayment on the equity is of lower priority than the repayment of other equity investors. The subordinated equity would aim to leverage other equity investors by ensuring that the latter have first claim on the distribution of profit, thereby increasing their risk-adjusted returns. The fund would have claim on profits only after rewards to other equity investors were distributed.

Assessing Costs and Benefits

GDP

The costs of mitigation and adaptation policies can be measured as a change in GDP. A problem with this method of assessing costs is that GDP is an imperfect measure of welfare (Markandya *et al..*, 2001:478):

1.   Not all welfare is included in GDP, e.g., housework and leisure activities.

2.   There are externalities in the economy which mean that some prices might not be truly reflective of their social costs.

Corrections can be made to GDP estimates to allow for these problems, but they are difficult to calculate. In response to this problem, some have suggested using other methods to assess policy. For example, the United Nations Commission for Sustainable Development has developed a system for "Green" GDP accounting and a list of sustainable development indicators.

Baselines

The emissions baseline is, by definition, the emissions that would occur in the absence of policy intervention. Definition of the baseline scenario is critical in the assessment of mitigation costs (Markandya *et al..*, 2001:469-470). This because the baseline determines the potential for emissions reductions, and the costs of implementing emission reduction policies.

There are several concepts used in the literature over baselines, including the "efficient" and "business-as-usual" (BAU) baseline cases. In the efficient baseline, it is assumed that all resources are being employed efficiently. In the BAU case, it is assumed that future development trends follow those of the past, and no changes in policies will take place. The BAU baseline is often associated with high GHG emissions, and may reflect the continuation of current energy-subsidy policies, or other market failures.

Some high emission BAU baselines imply relatively low net mitigation costs per unit of emissions. If the BAU scenario projects a large growth in emissions, total mitigation costs can be relatively high. Conversely, in an efficient baseline, mitigation costs per unit of emissions can be relatively high, but total mitigation costs low.

Ancillary Impacts

These are the secondary or side effects of mitigation policies, and including them in studies can result in higher or lower mitigation cost estimates (Markandya *et al..*, 2001:455). Reduced mortality and morbidity costs are potentially a major ancillary benefit of mitigation. This benefit is associated with reduced use of fossil fuels, thereby resulting in less air pollution (Barker *et al..*, 2001:564). There may also be ancillary costs. In developing countries, for example, if policy changes resulted in a relative increase in electricity prices, this could result in more pollution (Markandya *et al..*, 2001:462).

Flexibility

Flexibility is the ability to reduce emissions at the lowest cost. The greater the flexibility that governments allow in their regulatory framework to reduce emissions, the lower the potential costs are for achieving emissions reductions (Markandya *et al..*, 2001:455).

1. "Where" flexibility allows costs to be reduced by allowing emissions to be cut at locations where it is most efficient to do so. For example, the Flexibility Mechanisms of the Kyoto Protocol allow "where" flexibility (Toth *et al.*, 2001:660).

2. "When" flexibility potentially lowers costs by allowing reductions to be made at a time when it is most efficient to do so.

Including carbon sinks in a policy framework is another source of flexibility. Tree planting and forestry management actions can increase the capacity of sinks. Soils and other types of vegetation are also potential sinks. There is, however, uncertainty over how net emissions are affected by activities in this area (Markandya *et al.*, 2001:476).

No Regrets Options

These are, by definition, emission reduction options that have net negative costs (Markandya *et al.*, 2001:474-475). The presumption of no regret options affects emission reduction cost estimates (p. 455).

By convention, estimates of emission reduction costs do not include the benefits of avoided climate change damages. It can be argued that the existence of no regret options implies that there are market and non-market failures, e.g., lack of information, and that these failures can be corrected without incurring costs larger than the benefits gained. In most cases, studies of the no regret concept have not included all the external and implementation costs of a given policy.

Different studies make different assumptions about how far the economy is from the production frontier (defined as the maximum outputs attainable with the optimal use of available inputs – natural resources, labour, etc. (IPCC, 2007c:819)). "Bottom-up" studies (which consider specific technological and engineering details of the economy) often assume that in the baseline case, the economy is operating below the production frontier. Where the costs of implementing policies are less than the benefits, a no regret option (negative cost) is identified. "Top-down" approaches, based on macroeconomics, assume that the economy is efficient in the baseline case, with the result that mitigation policies always have a positive cost.

Technology

Assumptions about technological development and efficiency in the baseline and mitigation scenarios have a major impact on mitigation costs, in particular in bottom-up studies (Markandya *et al.*, 2001:473). The magnitude of potential technological efficiency improvements depends on assumptions about future technological innovation and market penetration rates for these technologies.

Discount Rates

Assessing climate change impacts and mitigation policies involves a comparison of economic flows that occur in different points in time. The discount rate is used by economists to compare economic effects occurring at different times. Discounting converts future economic impacts into their present-day value. The discount rate is generally positive because resources invested today can, on average, be transformed into more resources later. If climate change mitigation is viewed as an investment, then the return on investment can be used to decide how much should be spent on mitigation.

The choice of discount rate has a large effect on the result of any climate change cost analysis (Halsnæs *et al..*, 2007:136). Using too high a discount rate will result in too little investment in mitigation, but using too low a rate will result in too much investment in mitigation.

Discounting can either be prescriptive or descriptive. The descriptive approach is based on what discount rates are observed in the behaviour of people making every day decisions (the private discount rate) (IPCC, 2007c:813). In the prescriptive approach, a discount rate is chosen based on what is thought to be in the best interests of future generations (the social discount rate).

The descriptive approach can be interpreted as an effort to maximize the economic resources available to future generations, allowing them to decide how to use those resources (Arrow *et al.*, 1996b:133-134). The prescriptive approach can be interpreted as an effort to do as much as is economically justified to reduce the risk of climate change.

According to Markandya *et al..* (2001:466), discount rates used in assessing mitigation programmes need to at least partly reflect the opportunity costs of capital. In developed countries, Markandya *et al..* (2001:466) thought that a discount rate of around 4%-6% was probably justified, while in developing countries, a rate of 10%-12% was cited. The discount rates used in assessing private projects were found to be higher – with potential rates of between 10% and 25%.

When deciding how to discount future climate change impacts, value judgements are necessary (Arrow *et al.*, 1996b:130). IPCC (2001a:9) found that there was no consensus on the use of long-term discount rates in this area. The prescriptive approach to discounting leads to long-term discount rates of 2-3% in real terms, while the descriptive approach leads to rates of at least 4% after tax - sometimes much higher (Halsnæs *et al..*, 2007:136).

Decision Analysis

This is a quantitative type of analysis that is used to assess different potential decisions. Examples are cost-benefit and cost-effectiveness analysis (Toth *et al..*, 2001:609). In cost-benefit analysis, both costs and benefits are assessed economically. In cost-effectiveness analysis, the benefit-side of the analysis, e.g., a specified ceiling for the atmospheric concentration of GHGs, is not based on economic assessment.

One of the benefits of decision analysis is that the analysis is reproducible. Weaknesses, however, have been citied (Arrow *et al..*, 1996a:57):

1. The decision maker:

    2. In decision analysis, it is assumed that a single decision maker, with well-order preferences, is present throughout the analysis. In a cost-benefit analysis, the preferences of the decision maker are determined by applying the concepts of "willingness to pay" (WTP) and "willingness to accept" (WTA). These concepts are applied in an attempt to determine the aggregate value that society places on different resources (Markandya *et al..*, 2001:459).

    3. In reality, there is no single decision maker. Different decision makers have different sets

of values and preferences, and for this reason, decision analysis cannot yield a universally preferred solution.

2. Utility valuation: Many of the outcomes of climate policy decisions are difficult to value.

Arrow *et al.*. (1996a) concluded that while decision analysis had value, it could not identify a globally optimal policy for mitigation. In determining nationally optimal mitigation policies, the problems of decision analysis were viewed as being less important.

Cost-benefit Analysis

In an economically efficient mitigation response, the marginal (or incremental) costs of mitigation would be balanced against the marginal benefits of emission reduction. "Marginal" means that the costs and benefits of preventing (abating) the emission of the last unit of CO_2-eq are being compared. Units are measured in tonnes of CO_2-eq. The marginal benefits are the avoided damages from an additional tonne of carbon (emitted as carbon dioxide) being abated in a given emissions pathway (the social cost of carbon).

A problem with this approach is that the marginal costs and benefits of mitigation are uncertain, particularly with regards to the benefits of mitigation (Munasinghe *et al.*, 1996, p. 159). In the absence of risk aversion, and certainty over the costs and benefits, the optimum level of mitigation would be the point where marginal costs equal marginal benefits. IPCC (2007b:18) concluded that integrated analyses of the costs and benefits of mitigation did not unambiguously suggest an emissions pathway where benefits exceed costs.

Damage function

In cost-benefit analysis, the optimal timing of mitigation depends more on the shape of the aggregate damage function than the overall damages of climate change (Fisher *et al.*., 2007:235). If a damage function is used that shows smooth and regular damages, e.g., a cubic function, the results suggest that emission abatement should be postponed. This is because the benefits of early abatement are outweighed by the benefits of investing in other areas that accelerate economic growth. This result can change if the damage function is changed to include the possibility of catastrophic climate change impacts.

The Mitigation Portfolio

In deciding what role emissions abatement should play in a mitigation portfolio, different arguments have been made in favour of modest and stringent near-term abatement (Toth *et al.*., 2001:658):

1. Modest abatement:

    2. Modest deployment of improving technologies prevents lock-in to existing, low-productivity technology.

    3. Beginning with modest emission abatement avoids the premature retirement of existing capital stocks.

4. Gradual emission reduction reduces induced sectoral unemployment.

5. Reduces the costs of emissions abatement.

6. There is little evidence of damages from relatively rapid climate change in the past.

2. Stringent abatement:

    1. Endogenous (market-induced) change could accelerate development of low-cost technologies.

    2. Reduces the risk of being forced to make future rapid emission reductions that would require premature capital retirement.

    3. Welfare losses might be associated with faster rates of emission reduction. If, in the future, a low GHG stabilization target is found to be necessary, early abatement reduces the need for a rapid reduction in emissions.

    4. Reduces future climate change damages.

    5. Cutting emissions more quickly reduces the possibility of higher damages caused by faster rates of future climate change.

Energy Sector Subsidies

Large energy subsidies are present in many countries (Barker *et al.*, 2001:567-568). Currently governments subsidize fossil fuels by $557 billion per year. Economic theory indicates that the optimal policy would be to remove coal mining and burning subsidies and replace them with optimal taxes. Global studies indicate that even without introducing taxes, subsidy and trade barrier removal at a sectoral level would improve efficiency and reduce environmental damage (Barker *et al.*, 2001:568). Removal of these subsidies would substantially reduce GHG emissions and stimulate economic growth.

The actual effects of removing fossil fuel subsidies would depend heavily on the type of subsidy removed and the availability and economics of other energy sources. There is also the issue of carbon leakage, where removal of a subsidy to an energy-intensive industry could lead to a shift in production to another country with less regulation, and thus to a net increase in global emissions.

Policy Suggestions

Jacobson and Delucchi (2009) have advanced a plan to power 100% of the world's energy with wind, hydroelectric, and solar power by the year 2030, recommending transfer of energy subsidies from fossil fuel to renewable, and a price on carbon reflecting its cost for flood, cyclone, hurricane, drought, and related extreme weather expenses.

Cost Estimates

Global Costs

According to a literature assessment by Barker *et al.* (2007:622), mitigation cost estimates depend critically on the baseline (in this case, a reference scenario that the alternative scenario is compared with), the way costs are modelled, and assumptions about future government policy. Fisher *et al.* (2007) estimated macroeconomic costs in 2030 for multi-gas mitigation (reducing emissions of carbon dioxide and other GHGs, such as methane) as between a 3% decrease in global GDP to a small increase, relative to baseline. This was for an emissions pathway consistent with atmospheric stabilization of GHGs between 445 and 710 ppm CO_2-eq. In 2050, the estimated costs for stabilization between 710 and 445 ppm CO_2-eq ranged between a 1% gain to a 5.5% decrease in global GDP, relative to baseline. These cost estimates were supported by a moderate amount of evidence and much agreement in the literature (IPCC, 2007b:11,18).

Macroeconomic cost estimates made by Fisher *et al.* (2007:204) were mostly based on models that assumed transparent markets, no transaction costs, and perfect implementation of cost-effective policy measures across all regions throughout the 21st century. According to Fisher *et al.* (2007), relaxation of some or all these assumptions would lead to an appreciable increase in cost estimates. On the other hand, IPCC (2007b:8) noted that cost estimates could be reduced by allowing for accelerated technological learning, or the possible use of carbon tax/emission permit revenues to reform national tax systems.

In most of the assessed studies, costs rose for increasingly stringent stabilization targets. In scenarios that had high baseline emissions, mitigation costs were generally higher for comparable stabilization targets. In scenarios with low emissions baselines, mitigation costs were generally lower for comparable stabilization targets.

Distributional Effects

Regional Costs

Gupta *et al.* (2007:776-777) assessed studies where estimates are given for regional mitigation costs. The conclusions of these studies are as follows:

1. Regional abatement costs are largely dependent on the assumed stabilization level and baseline scenario. The allocation of emission allowances/permits is also an important factor, but for most countries, is less important than the stabilization level (Gupta *et al.*, 2007, pp. 776–777).

2. Other costs arise from changes in international trade. Fossil fuel-exporting regions are likely to be affected by losses in coal and oil exports compared to baseline, while some regions might experience increased bio-energy (energy derived from biomass) exports (Gupta *et al.*, 2007, pp. 776–777).

3. Allocation schemes based on current emissions (i.e., where the most allowances/permits are given to the largest current polluters, and the fewest allowances are given to smallest current polluters) lead to welfare losses for developing countries, while allocation schemes based on a per capita convergence of emissions (i.e., where per capita emissions are equalized) lead to welfare gains for developing countries.

Sectoral Costs

In a literature assessment, Barker *et al.* (2001:563-564), predicted that the renewables sector could potentially benefit from mitigation. The coal (and possibly the oil) industry was predicted to potentially lose substantial proportions of output relative to a baseline scenario (Barker *et al.*, 2001, pp. 563–564).

References

- UK Royal Society (September 2009), Geoengineering the climate: science, governance and uncertainty (PDF), London: UK Royal Society, ISBN 978-0-85403-773-5.

- Smit, Berend; Reimer, Jeffrey A.; Oldenburg, Curtis M.; Bourg, Ian C. (2014). Introduction to Carbon Capture and Sequestration. London: Imperial College Press. ISBN 978-1-78326-328-8.

- Royal Society (September 2009). Geoengineering the Climate: Science, Governance and Uncertainty (PDF) (Report). p. 1. ISBN 978-0-85403-773-5. Retrieved 2011-12-01.

- "Biochar reducing and removing CO2 while improving soils: A significant sustainable response to climate change" (PDF). UKBRC. UK Biochar research Center. Retrieved 25 April 2016.

- "Geoengineering could buy the time needed to develop a sustainable energy economy". Bulletin of the Atomic Scientists. Retrieved 2016-02-03.

- Oxford Geoengineering Programme. "Oxford Geoengineering Programme // History of the Oxford Principles". www.geoengineering.ox.ac.uk. Retrieved 2016-02-03.

5

Various Observations on Climate change

This chapter concentrates on the various observations on climate change. Some of these are scientific opinion on climate change, attribution of recent climate change, climate change denial and Milankovitch cycles. Scientific opinion on climate is the judgment among scientists in regard to global warming. The scientific opinions are explained in synthesis reports. All the various observations of climate change have been carefully analyzed in this text.

Scientific Opinion on Climate Change

The scientific opinion on climate change is the overall judgment among scientists regarding the extent to which global warming is occurring, its causes, and its probable consequences. This scientific opinion is expressed in synthesis reports, by scientific bodies of national or international standing, and by surveys of opinion among climate scientists. Individual scientists, universities, and laboratories contribute to the overall scientific opinion via their peer-reviewed publications, and the areas of collective agreement and relative certainty are summarised in these respected reports and surveys.

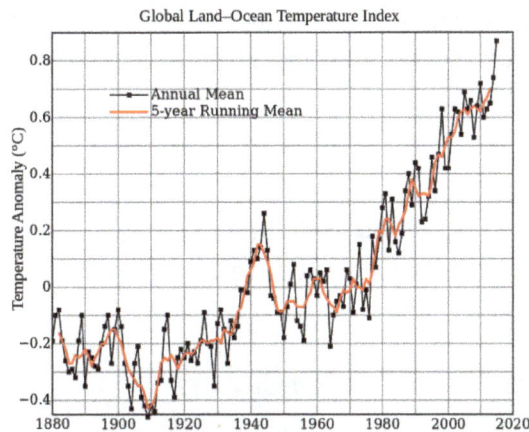

Global mean land-ocean temperature change since 1880, relative to the 1951–1980 mean. The black line is the annual mean and the red line is the 5-year running mean. Source: NASA GISS

The scientific consensus is that the Earth's climate system is unequivocally warming, and that it is *extremely likely* (meaning 95% probability or higher) that this warming is predominantly caused by humans. It is likely that this mainly arises from increased concentrations of greenhouse gases in the atmosphere, such as from deforestation and the burning of fossil fuels, partially offset by human caused increases in aerosols; natural changes had little effect.

National and international science academies and scientific societies have assessed current scientific opinion on global warming. These assessments are generally consistent with the conclusions of the Intergovernmental Panel on Climate Change. The IPCC Fourth Assessment Report stated that:

1. Warming of the climate system is unequivocal, as evidenced by increases in global average air and ocean temperatures, the widespread melting of snow and ice, and rising global average sea level.

2. Most of the global warming since the mid-20th century is very likely due to human activities.

3. Benefits and costs of climate change for [human] society will vary widely by location and scale. Some of the effects in temperate and polar regions will be positive and others elsewhere will be negative. Overall, net effects are more likely to be strongly negative with larger or more rapid warming.

4. The range of published evidence indicates that the net damage costs of climate change are likely to be significant and to increase over time.

5. The resilience of many ecosystems is likely to be exceeded this century by an unprecedented combination of climate change, associated disturbances (e.g. flooding, drought, wildfire, insects, ocean acidification) and other global change drivers (e.g. land-use change, pollution, fragmentation of natural systems, over-exploitation of resources).

Some scientific bodies have recommended specific policies to governments and science can play a role in informing an effective response to climate change. Policy decisions, however, may require value judgements and so are not included in the scientific opinion.

No scientific body of national or international standing maintains a formal opinion dissenting from any of these main points. The last national or international scientific body to drop dissent was the American Association of Petroleum Geologists, which in 2007 updated its statement to its current non-committal position. Some other organizations, primarily those focusing on geology, also hold non-committal positions.

Synthesis Reports

Synthesis reports are assessments of scientific literature that compile the results of a range of stand-alone studies in order to achieve a broad level of understanding, or to describe the state of knowledge of a given subject.

Intergovernmental Panel on Climate Change (IPCC) 2014

The IPCC Fifth Assessment Report *Summary for Policymakers* stated that warming of the climate system is 'unequivocal' with changes unprecedented over decades to millennia, including warming of the atmosphere and oceans, loss of snow and ice, and sea level rise. Greenhouse gas emissions, driven largely by economic and population growth, have led to greenhouse gas concentrations that are unprecedented in at least the last 800,000 years. These, together with other anthropogenic drivers, are "extremely likely" to have been the dominant cause of the observed global warming since the mid-20th century.

It said that

Continued emission of greenhouse gases will cause further warming and long-lasting changes in all components of the climate system, increasing the likelihood of severe, pervasive and irreversible impacts for people and ecosystems. Limiting climate change would require substantial and sustained reductions in greenhouse gas emissions which, together with adaptation, can limit climate change risks.

Reporting on the publication of the report, *The Guardian* said that

In the end it all boils down to risk management. The stronger our efforts to reduce greenhouse gas emissions, the lower the risk of extreme climate impacts. The higher our emissions, the larger climate changes we'll face, which also means more expensive adaptation, more species extinctions, more food and water insecurities, more income losses, more conflicts, and so forth.

The New York Times reported that

In Washington, President Obama's science adviser, John P. Holdren, cited increased scientific confidence "that the kinds of harm already being experienced from climate change will continue to worsen unless and until comprehensive and vigorous action to reduce emissions is undertaken worldwide."

It went on to say that Ban Ki-moon, the United Nations secretary general, had declared his intention to call a meeting of heads of state in 2014 to develop such a treaty. The last such meeting, in Copenhagen in 2009, the NY Times reported, had ended in disarray.

Intergovernmental Panel on Climate Change (IPCC) 2007

In February 2007, the IPCC released a summary of the forthcoming Fourth Assessment Report. According to this summary, the Fourth Assessment Report found that human actions are "very likely" the cause of global warming, meaning a 90% or greater probability. Global warming in this case was indicated by an increase of 0.75 degrees in average global temperatures over the last 100 years.

The New York Times reported that "the leading international network of climate scientists has concluded for the first time that global warming is 'unequivocal' and that human activity is the main driver, 'very likely' causing most of the rise in temperatures since 1950".

A retired journalist for *The New York Times*, William K. Stevens wrote: "The Intergovernmental Panel on Climate Change said the likelihood was 90 percent to 99 percent that emissions of heat-trapping greenhouse gases like carbon dioxide, spewed from tailpipes and smokestacks, were the dominant cause of the observed warming of the last 50 years. In the panel's parlance, this level of certainty is labeled 'very likely'. Only rarely does scientific odds-making provide a more definite answer than that, at least in this branch of science, and it describes the endpoint, so far, of a progression.".

The Associated Press summarized the position on sea level rise:

On sea levels, the report projects rises of 7 to 23 inches by the end of the century. An additional 3.9 to 7.8 inches are possible if recent, surprising melting of polar ice sheets continues.

U.S. Global Change Research Program

The U.S. Global Change Research Program reported in June 2009 that:

Observations show that warming of the climate is unequivocal. The global warming observed over the past 50 years is due primarily to human-induced emissions of heat-trapping gases. These emissions come mainly from the burning of fossil fuels (coal, oil, and gas), with important contributions from the clearing of forests, agricultural practices, and other activities.

The report, which is about the effects that climate change is having in the United States, also says:

Climate-related changes have already been observed globally and in the United States. These include increases in air and water temperatures, reduced frost days, increased frequency and intensity of heavy downpours, a rise in sea level, and reduced snow cover, glaciers, permafrost, and sea ice. A longer ice-free period on lakes and rivers, lengthening of the growing season, and increased water vapor in the atmosphere have also been observed. Over the past 30 years, temperatures have risen faster in winter than in any other season, with average winter temperatures in the Midwest and northern Great Plains increasing more than 7 °F (3.9 °C). Some of the changes have been faster than previous assessments had suggested.

Arctic Climate Impact Assessment

In 2004, the intergovernmental Arctic Council and the non-governmental International Arctic Science Committee released the synthesis report of the Arctic Climate Impact Assessment:

Climate conditions in the past provide evidence that rising atmospheric carbon dioxide levels are associated with rising global temperatures. Human activities, primarily the burning of fossil fuels (coal, oil, and natural gas), and secondarily the clearing of land, have increased the concentration of carbon dioxide, methane, and other heat-trapping ("greenhouse") gases in the atmosphere... There is international scientific consensus that most of the warming observed over the last 50 years is attributable to human activities.

Policy

There is an extensive discussion in the scientific literature on what policies might be effective in responding to climate change. Some scientific bodies have recommended specific policies to governments (refer to the later sections of the article). The natural and social sciences can play a role in informing an effective response to climate change. However, policy decisions may require value judgements. For example, the US National Research Council has commented:

The question of whether there exists a "safe" level of concentration of greenhouse gases cannot be answered directly because it would require a value judgment of what constitutes an acceptable risk to human welfare and ecosystems in various parts of the world, as well as a more quantitative assessment of the risks and costs associated with the various impacts of global warming. In general, however, risk increases with increases in both the rate and the magnitude of climate change.

This article mostly focuses on the views of natural scientists. However, social scientists, medical experts, engineers and philosophers have also commented on climate change science and policies. Climate change policy is discussed in several articles: climate change mitigation, climate change adaptation, climate engineering, politics of global warming, climate ethics, and economics of global warming.

Statements by Scientific Organizations of National or International Standing

This is a list of scientific bodies of national or international standing, that have issued formal statements of opinion, classifies those organizations according to whether they concur with the IPCC view, are non-committal, or dissent from it.

Concurring

Academies of Science (General Science)

Since 2001, 34 national science academies, three regional academies, and both the international InterAcademy Council and International Council of Academies of Engineering and Technological Sciences have made formal declarations confirming human induced global warming and urging nations to reduce emissions of greenhouse gases. The 34 national science academy statements include 33 who have signed joint science academy statements and one individual declaration by the Polish Academy of Sciences in 2007.

Joint National Science Academy Statements

1. 2001 Following the publication of the IPCC Third Assessment Report, seventeen national science academies issued a joint statement, entitled "The Science of Climate Change", explicitly acknowledging the IPCC position as representing the scientific consensus on climate change science. The statement, printed in an editorial in the journal *Science* on May 18, 2001, was signed by the science academies of Australia, Belgium, Brazil, Canada, the Caribbean, China, France, Germany, India, Indonesia, Ireland, Italy, Malaysia, New Zealand, Sweden, Turkey, and the United Kingdom.

2. 2005 The national science academies of the G8 nations, plus Brazil, China and India, three of the largest emitters of greenhouse gases in the developing world, signed a statement on the global response to climate change. The statement stresses that the scientific understanding of climate change is now sufficiently clear to justify nations taking prompt action, and explicitly endorsed the IPCC consensus. The eleven signatories were the science academies of Brazil, Canada, China, France, Germany, India, Italy, Japan, Russia, the United Kingdom, and the United States.

3. 2007 In preparation for the 33rd G8 summit, the national science academies of the G8+5 nations issued a declaration referencing the position of the 2005 joint science academies' statement, and acknowledging the confirmation of their previous conclusion by recent research. Following the IPCC Fourth Assessment Report, the declaration states, "It is unequivocal that the climate is changing, and it is very likely that this is predominantly caused by the increasing human interference with the atmosphere. These changes will transform the environmental conditions on Earth unless counter-measures are taken." The thirteen signatories were the national science academies of Brazil, Canada, China, France, Germany, Italy, India, Japan, Mexico, Russia, South Africa, the United Kingdom, and the United States.

4. 2007 In preparation for the 33rd G8 summit, the Network of African Science Academies submitted a joint "statement on sustainability, energy efficiency, and climate change" :

A consensus, based on current evidence, now exists within the global scientific community that human activities are the main source of climate change and that the burning of fossil fuels is largely responsible for driving this change. The IPCC should be congratulated for the contribution it has made to public understanding of the nexus that exists between energy, climate and sustainability.

— The thirteen signatories were the science academies of *Cameroon, Ghana, Kenya, Madagascar, Nigeria, Senegal, South Africa, Sudan, Tanzania, Uganda, Zambia, Zimbabwe,* as well as the *African Academy of Sciences,*

1. 2008 In preparation for the 34th G8 summit, the national science academies of the G8+5 nations issued a declaration reiterating the position of the 2005 joint science academies' statement, and reaffirming "that climate change is happening and that anthropogenic warming is influencing many physical and biological systems." Among other actions, the declaration urges all nations to "(t)ake appropriate economic and policy measures to accelerate transition to a low carbon society and to encourage and effect changes in individual and national behaviour." The thirteen signatories were the same national science academies that issued the 2007 joint statement.

2. 2009 In advance of the UNFCCC negotiations to be held in Copenhagen in December 2009, the national science academies of the G8+5 nations issued a joint statement declaring, "Climate change and sustainable energy supply are crucial challenges for the future of humanity. It is essential that world leaders agree on the emission reductions needed to combat negative consequences of anthropogenic climate change". The statement references the IPCC's Fourth Assessment of 2007, and asserts that "climate change is happening even faster than previously estimated; global CO_2 emissions since 2000 have been higher than even the highest predictions, Arctic sea ice has been melting at rates much faster than predicted, and the rise in the sea level has become more rapid." The thirteen signatories were the same national science academies that issued the 2007 and 2008 joint statements.

Polish Academy of Sciences

In December 2007, the General Assembly of the Polish Academy of Sciences (Polska Akademia Nauk), which has not been a signatory to joint national science academy statements issued a declaration endorsing the IPCC conclusions, and stating:

it is the duty of Polish science and the national government to, in a thoughtful, organized and active manner, become involved in realisation of these ideas.

Problems of global warming, climate change, and their various negative impacts on human life and on the functioning of entire societies are one of the most dramatic challenges of modern times.

PAS General Assembly calls on the national scientific communities and the national government to actively support Polish participation in this important endeavor.

Additional National Science Academy and Society Statements

1. American Association for the Advancement of Science as the world's largest general scientific society, adopted an official statement on climate change in 2006:

The scientific evidence is clear: global climate change caused by human activities is occurring now, and it is a growing threat to society....The pace of change and the evidence of harm have increased markedly over the last five years. The time to control greenhouse gas emissions is now.

1. Federation of Australian Scientific and Technological Societies in 2008 published *FASTS Statement on Climate Change* which states:

Global climate change is real and measurable...To reduce the global net economic, environmental and social losses in the face of these impacts, the policy objective must remain squarely focused on returning greenhouse gas concentrations to near pre-industrial levels through the reduction of emissions. The spatial and temporal fingerprint of warming can be traced to increasing greenhouse gas concentrations in the atmosphere, which are a direct result of burning fossil fuels, broad-scale deforestation and other human activity.

1. United States National Research Council through its Committee on the Science of Climate Change in 2001, published *Climate Change Science: An Analysis of Some Key Questions*. This report explicitly endorses the IPCC view of attribution of recent climate change as representing the view of the scientific community:

The changes observed over the last several decades are likely mostly due to human activities, but we cannot rule out that some significant part of these changes is also a reflection of natural variability. Human-induced warming and associated sea level rises are expected to continue through the 21st century... The IPCC's conclusion that most of the observed warming of the last 50 years is likely to have been due to the increase in greenhouse gas concentrations accurately reflects the current thinking of the scientific community on this issue.

1. Royal Society of New Zealand having signed onto the first joint science academy statement in 2001, released a separate statement in 2008 in order to clear up "the controversy over climate change and its causes, and possible confusion among the public":

The globe is warming because of increasing greenhouse gas emissions. Measurements show that greenhouse gas concentrations in the atmosphere are well above levels seen for many thousands of years. Further global climate changes are predicted, with impacts expected to become more costly as time progresses. Reducing future impacts of climate change will require substantial reductions of greenhouse gas emissions.

1. The Royal Society of the United Kingdom has not changed its concurring stance reflected in its participation in joint national science academies' statements on anthropogenic global warming. According to the Telegraph, "The most prestigious group of scientists in the country was forced to act after fellows complained that doubts over man made global warming were not being communicated to the public". In May 2010, it announced that it "is presently drafting a new public facing document on climate change, to provide an updated status report on the science in an easily accessible form, also addressing the levels of certainty of key components." The society says that it is three years since the last such document was published and that, after an extensive process of debate and review, the new document was printed in September 2010. It summarises the current scientific evidence and highlights the areas where the science is well established, where there is still some debate, and where substantial uncertainties remain. The society has stated that "this is not

the same as saying that the climate science itself is in error – no Fellows have expressed such a view to the RS". The introduction includes this statement:

There is strong evidence that the warming of the Earth over the last half-century has been caused largely by human activity, such as the burning of fossil fuels and changes in land use, including agriculture and deforestation.

International Science Academies

1. African Academy of Sciences in 2007 was a signatory to the "statement on sustainability, energy efficiency, and climate change". This joint statement of African science academies, was organized through the Network of African Science Academies. It's stated goal was "to convey information and spur action on the occasion of the G8 Summit in Heiligendamm, Germany, in June 2007".

A consensus, based on current evidence, now exists within the global scientific community that human activities are the main source of climate change and that the burning of fossil fuels is largely responsible for driving this change.

1. European Academy of Sciences and Arts in 2007 issued a formal declaration on climate change titled *Let's Be Honest*:

Human activity is most *likely* responsible for climate warming. Most of the climatic warming over the last 50 years is *likely* to have been caused by increased concentrations of greenhouse gases in the atmosphere. Documented long-term climate changes include changes in Arctic temperatures and ice, widespread changes in precipitation amounts, ocean salinity, wind patterns and extreme weather including droughts, heavy precipitation, heat waves and the intensity of tropical cyclones. The above development potentially has dramatic consequences for mankind's future.

1. European Science Foundation in a 2007 position paper states:

There is now convincing evidence that since the industrial revolution, human activities, resulting in increasing concentrations of greenhouse gases have become a major agent of climate change... On-going and increased efforts to mitigate climate change through reduction in greenhouse gases are therefore crucial.

1. InterAcademy Council As the representative of the world's scientific and engineering academies, the InterAcademy Council issued a report in 2007 titled *Lighting the Way: Toward a Sustainable Energy Future*.

Current patterns of energy resources and energy usage are proving detrimental to the long-term welfare of humanity. The integrity of essential natural systems is already at risk from climate change caused by the atmospheric emissions of greenhouse gases. Concerted efforts should be mounted for improving energy efficiency and reducing the carbon intensity of the world economy.

1. International Council of Academies of Engineering and Technological Sciences (CAETS) in 2007, issued a *Statement on Environment and Sustainable Growth*:

As reported by the Intergovernmental Panel on Climate Change (IPCC), most of the observed global warming since the mid-20th century is very likely due to human-produced emission of greenhouse gases and this warming will continue unabated if present anthropogenic emissions continue or, worse, expand without control. CAETS, therefore, endorses the many recent calls to decrease and control greenhouse gas emissions to an acceptable level as quickly as possible.

Physical and Chemical Sciences

1. American Chemical Society

2. American Institute of Physics

3. American Physical Society

4. Australian Institute of Physics

5. European Physical Society

Earth Sciences

American Geophysical Union

The American Geophysical Union (AGU) adopted a statement on *Climate Change and Greenhouse Gases* in 1998. A new statement, adopted by the society in 2003, revised in 2007, and revised and expanded in 2013, affirms that rising levels of greenhouse gases have caused and will continue to cause the global surface temperature to be warmer:

Human activities are changing Earth's climate. At the global level, atmospheric concentrations of carbon dioxide and other heat-trapping greenhouse gases have increased sharply since the Industrial Revolution. Fossil fuel burning dominates this increase. Human-caused increases in greenhouse gases are responsible for most of the observed global average surface warming of roughly 0.8 °C (1.5 °F) over the past 140 years. Because natural processes cannot quickly remove some of these gases (notably carbon dioxide) from the atmosphere, our past, present, and future emissions will influence the climate system for millennia.

While important scientific uncertainties remain as to which particular impacts will be experienced where, no uncertainties are known that could make the impacts of climate change inconsequential. Furthermore, surprise outcomes, such as the unexpectedly rapid loss of Arctic summer sea ice, may entail even more dramatic changes than anticipated.

American Society of Agronomy, Crop Science Society of America, and Soil Science Society of America

In May, 2011, the American Society of Agronomy (ASA), Crop Science Society of America (CSSA), and Soil Science Society of America (SSSA) issued a joint position statement on climate change as it relates to agriculture:

A comprehensive body of scientific evidence indicates beyond reasonable doubt that global climate change is now occurring and that its manifestations threaten the stability of societies as well as natural and managed ecosystems. Increases in ambient temperatures and changes in related processes are directly linked to rising anthropogenic greenhouse gas (GHG) concentrations in the atmosphere.

Unless the emissions of GHGs are curbed significantly, their concentrations will continue to rise, leading to changes in temperature, precipitation, and other climate variables that will undoubtedly affect agriculture around the world.

Climate change has the potential to increase weather variability as well as gradually increase global temperatures. Both of these impacts have the potential to negatively impact the adaptability and resilience of the world's food production capacity; current research indicates climate change is already reducing the productivity of vulnerable cropping systems.

European Federation of Geologists

In 2008, the European Federation of Geologists (EFG) issued the position paper *Carbon Capture and geological Storage* :

The EFG recognizes the work of the IPCC and other organizations, and subscribes to the major findings that climate change is happening, is predominantly caused by anthropogenic emissions of CO_2, and poses a significant threat to human civilization.

It is clear that major efforts are necessary to quickly and strongly reduce CO_2 emissions. The EFG strongly advocates renewable and sustainable energy production, including geothermal energy, as well as the need for increasing energy efficiency.

CCS [Carbon Capture and geological Storage] should also be regarded as a bridging technology, facilitating the move towards a carbon free economy.

European Geosciences Union

In 2005, the Divisions of Atmospheric and Climate Sciences of the European Geosciences Union (EGU) issued a position statement in support of the joint science academies' statement on global response to climate change. The statement refers to the Intergovernmental Panel on Climate Change (IPCC), as "the main representative of the global scientific community", and asserts that the IPCC

represents the state-of-the-art of climate science supported by the major science academies around the world and by the vast majority of science researchers and investigators as documented by the peer-reviewed scientific literature.

Additionally, in 2008, the EGU issued a position statement on ocean acidification which states, "Ocean acidification is already occurring today and will continue to intensify, closely tracking atmospheric CO_2 increase. Given the potential threat to marine ecosystems and its ensuing impact on human society and economy, especially as it acts in conjunction with anthropogenic global warming, there is an urgent need for immediate action." The statement then advocates for strategies "to limit future release of CO_2 to the atmosphere and/or enhance removal of excess CO_2 from the atmosphere."

Geological Society of America

In 2006, the Geological Society of America adopted a position statement on global climate change. It amended this position on April 20, 2010 with more explicit comments on need for CO_2 reduction.

Decades of scientific research have shown that climate can change from both natural and anthropogenic causes. The Geological Society of America (GSA) concurs with assessments by the National Academies of Science (2005), the National Research Council (2006), and the Intergovernmental Panel on Climate Change (IPCC, 2007) that global climate has warmed and that human activities (mainly greenhouse-gas emissions) account for most of the warming since the middle 1900s. If current trends continue, the projected increase in global temperature by the end of the twenty first century will result in large impacts on humans and other species. Addressing the challenges posed by climate change will require a combination of adaptation to the changes that are likely to occur and global reductions of CO_2 emissions from anthropogenic sources.

Geological Society of London

In November 2010, the Geological Society of London issued the position statement *Climate change: evidence from the geological record*:

The last century has seen a rapidly growing global population and much more intensive use of resources, leading to greatly increased emissions of gases, such as carbon dioxide and methane, from the burning of fossil fuels (oil, gas and coal), and from agriculture, cement production and deforestation. Evidence from the geological record is consistent with the physics that shows that adding large amounts of carbon dioxide to the atmosphere warms the world and may lead to: higher sea levels and flooding of low-lying coasts; greatly changed patterns of rainfall; increased acidity of the oceans; and decreased oxygen levels in seawater.

There is now widespread concern that the Earth's climate will warm further, not only because of the lingering effects of the added carbon already in the system, but also because of further additions as human population continues to grow. Life on Earth has survived large climate changes in the past, but extinctions and major redistribution of species have been associated with many of them. When the human population was small and nomadic, a rise in sea level of a few metres would have had very little effect on Homo sapiens. With the current and growing global population, much of which is concentrated in coastal cities, such a rise in sea level would have a drastic effect on our complex society, especially if the climate were to change as suddenly as it has at times in the past. Equally, it seems likely that as warming continues some areas may experience less precipitation leading to drought. With both rising seas and increasing drought, pressure for human migration could result on a large scale.

International Union of Geodesy and Geophysics

In July 2007, the International Union of Geodesy and Geophysics (IUGG) adopted a resolution titled "The Urgency of Addressing Climate Change". In it, the IUGG concurs with the "comprehensive and widely accepted and endorsed scientific assessments carried out by the Intergovernmental Panel on Climate Change and regional and national bodies, which have firmly established, on the basis of scientific evidence, that human activities are the primary cause of recent climate change." They state further that the "continuing reliance on combustion of fossil fuels as the world's primary source of energy will lead to much higher atmospheric concentrations of greenhouse gases, which will, in turn, cause significant increases in surface temperature, sea level, ocean acidification, and their related consequences to the environment and society."

National Association of Geoscience Teachers

In July 2009, the National Association of Geoscience Teachers (NAGT) adopted a position statement on climate change in which they assert that "Earth's climate is changing [and] "that present warming trends are largely the result of human activities":

NAGT strongly supports and will work to promote education in the science of climate change, the causes and effects of current global warming, and the immediate need for policies and actions that reduce the emission of greenhouse gases.

Meteorology and oceanography

American Meteorological Society

The American Meteorological Society (AMS) statement adopted by their council in 2012 concluded:

There is unequivocal evidence that Earth's lower atmosphere, ocean, and land surface are warming; sea level is rising; and snow cover, mountain glaciers, and Arctic sea ice are shrinking. The dominant cause of the warming since the 1950s is human activities. This scientific finding is based on a large and persuasive body of research. The observed warming will be irreversible for many years into the future, and even larger temperature increases will occur as greenhouse gases continue to accumulate in the atmosphere. Avoiding this future warming will require a large and rapid reduction in global greenhouse gas emissions. The ongoing warming will increase risks and stresses to human societies, economies, ecosystems, and wildlife through the 21st century and beyond, making it imperative that society respond to a changing climate. To inform decisions on adaptation and mitigation, it is critical that we improve our understanding of the global climate system and our ability to project future climate through continued and improved monitoring and research. This is especially true for smaller (seasonal and regional) scales and weather and climate extremes, and for important hydroclimatic variables such as precipitation and water availability.

Technological, economic, and policy choices in the near future will determine the extent of future impacts of climate change. Science-based decisions are seldom made in a context of absolute certainty. National and international policy discussions should include consideration of the best ways to both adapt to and mitigate climate change. Mitigation will reduce the amount of future climate change and the risk of impacts that are potentially large and dangerous. At the same time, some continued climate change is inevitable, and policy responses should include adaptation to climate change. Prudence dictates extreme care in accounting for our relationship with the only planet known to be capable of sustaining human life.

Australian Meteorological and Oceanographic Society

The Australian Meteorological and Oceanographic Society has issued a *Statement on Climate Change*, wherein they conclude:

Global climate change and global warming are real and observable ... It is highly likely that those human activities that have increased the concentration of greenhouse gases in the atmosphere have been largely responsible for the observed warming since 1950. The warming associated with increases in greenhouse gases originating from human activity is called the enhanced greenhouse effect. The atmospheric concentration of carbon dioxide has increased by more than 30% since the start of the industrial age and is higher now than at any time in at least the past 650,000 years. This increase is a direct result of burning fossil fuels, broad-scale deforestation and other human activity."

Canadian Foundation for Climate and Atmospheric Sciences

In November 2005, the Canadian Foundation for Climate and Atmospheric Sciences (CFCAS) issued a letter to the Prime Minister of Canada stating that

We concur with the climate science assessment of the Intergovernmental Panel on Climate Change (IPCC) in 2001 ... We endorse the conclusions of the IPCC assessment that 'There is new and stronger evidence that most of the warming observed over the last 50 years is attributable to human activities'. ... There is increasingly unambiguous evidence of changing climate in Canada and around the world. There will be increasing impacts of climate change on Canada's natural ecosystems and on our socio-economic activities. Advances in climate science since the 2001 IPCC Assessment have provided more evidence supporting the need for action and development of a strategy for adaptation to projected changes.

Canadian Meteorological and Oceanographic Society

In November 2009, a letter to the Canadian Parliament by The Canadian Meteorological and Oceanographic Society states:

Rigorous international research, including work carried out and supported by the Government of Canada, reveals that greenhouse gases resulting from human activities contribute to the warming of the atmosphere and the oceans and constitute a serious risk to the health and safety of our society, as well as having an impact on all life.

Royal Meteorological Society (UK)

In February 2007, after the release of the IPCC's Fourth Assessment Report, the Royal Meteorological Society issued an endorsement of the report. In addition to referring to the IPCC as "[the] world's best climate scientists", they stated that climate change is happening as "the result of emissions since industrialization and we have already set in motion the next 50 years of global warming – what we do from now on will determine how worse it will get."

World Meteorological Organization

In its *Statement at the Twelfth Session of the Conference of the Parties to the U.N. Framework Convention on Climate Change* presented on November 15, 2006, the World Meteorological Organization (WMO) confirms the need to "prevent dangerous anthropogenic interference with the climate system." The WMO concurs that "scientific assessments have increasingly reaffirmed that human activities are indeed changing the composition of the atmosphere, in particular through the burning of fossil fuels for energy production and transportation." The

WMO concurs that "the present atmospheric concentration of CO_2 was never exceeded over the past 420,000 years;" and that the IPCC "assessments provide the most authoritative, up-to-date scientific advice."

American Quaternary Association

The American Quaternary Association (AMQUA) has stated

Few credible Scientists now doubt that humans have influenced the documented rise of global temperatures since the Industrial Revolution," citing "the growing body of evidence that warming of the atmosphere, especially over the past 50 years, is directly impacted by human activity.

International Union for Quaternary Research

The statement on climate change issued by the International Union for Quaternary Research (INQUA) reiterates the conclusions of the IPCC, and urges all nations to take prompt action in line with the UNFCCC principles.

Human activities are now causing atmospheric concentrations of greenhouse gases — including carbon dioxide, methane, tropospheric ozone, and nitrous oxide — to rise well above pre-industrial levels....Increases in greenhouse gases are causing temperatures to rise...The scientific understanding of climate change is now sufficiently clear to justify nations taking prompt action.... Minimizing the amount of this carbon dioxide reaching the atmosphere presents a huge challenge but must be a global priority.

Biology and Life Sciences

Life science organizations have outlined the dangers climate change pose to wildlife.

1. American Association of Wildlife Veterinarians

2. American Institute of Biological Sciences. In October 2009, the leaders of 18 US scientific societies and organizations sent an open letter to the United States Senate reaffirming the scientific consensus that climate change is occurring and is primarily caused by human activities. The American Institute of Biological Sciences (AIBS) adopted this letter as their official position statement. The letter goes on to warn of predicted impacts on the United States such as sea level rise and increases in extreme weather events, water scarcity, heat waves, wildfires, and the disturbance of biological systems. It then advocates for a dramatic reduction in emissions of greenhouse gases.

3. American Society for Microbiology

4. Australian Coral Reef Society

5. Institute of Biology (UK)

6. Society of American Foresters issued two position statements pertaining to climate change in which they cite the IPCC and the UNFCCC.

7. The Wildlife Society (international)

Human Health

A number of health organizations have warned about the numerous negative health effects of global warming

1. American Academy of Pediatrics

2. American College of Preventive Medicine

3. American Medical Association

4. American Public Health Association

5. Australian Medical Association in 2004 and in 2008

6. World Federation of Public Health Associations

7. World Health Organization

There is now widespread agreement that the Earth is warming, due to emissions of greenhouse gases caused by human activity. It is also clear that current trends in energy use, development, and population growth will lead to continuing – and more severe – climate change.

The changing climate will inevitably affect the basic requirements for maintaining health: clean air and water, sufficient food and adequate shelter. Each year, about 800,000 people die from causes attributable to urban air pollution, 1.8 million from diarrhoea resulting from lack of access to clean water supply, sanitation, and poor hygiene, 3.5 million from malnutrition and approximately 60,000 in natural disasters. A warmer and more variable climate threatens to lead to higher levels of some air pollutants, increase transmission of diseases through unclean water and through contaminated food, to compromise agricultural production in some of the least developed countries, and increase the hazards of extreme weather.

Miscellaneous

A number of other national scientific societies have also endorsed the opinion of the IPCC:

1. American Astronomical Society

2. American Statistical Association

3. Canadian Council of Professional Engineers

4. The Institution of Engineers Australia

5. International Association for Great Lakes Research

6. Institute of Professional Engineers New Zealand

7. The World Federation of Engineering Organizations (WFEO)

Non-committal

American Association of Petroleum Geologists

As of June 2007, the American Association of Petroleum Geologists (AAPG) Position Statement on climate change stated:

the AAPG membership is divided on the degree of influence that anthropogenic CO_2 has on recent and potential global temperature increases ... Certain climate simulation models predict that the warming trend will continue, as reported through NAS, AGU, AAAS and AMS. AAPG respects these scientific opinions but wants to add that the current climate warming projections could fall within well-documented natural variations in past climate and observed temperature data. These data do not necessarily support the maximum case scenarios forecast in some models.

Prior to the adoption of this statement, the AAPG was the only major scientific organization that rejected the finding of significant human influence on recent climate, according to a statement by the Council of the American Quaternary Association. Explaining the plan for a revision, AAPG president Lee Billingsly wrote in March 2007:

Members have threatened to not renew their memberships... if AAPG does not alter its position on global climate change... And I have been told of members who already have resigned in previous years because of our current global climate change position... The current policy statement is not supported by a significant number of our members and prospective members.

AAPG President John Lorenz announced the "sunsetting" of AAPG's Global Climate Change Committee in January 2010. The AAPG Executive Committee determined:

Climate change is peripheral at best to our science [...] AAPG does not have credibility in that field [...] and as a group we have no particular knowledge of global atmospheric geophysics.

American Institute of Professional Geologists

In 2009, the American Institute of Professional Geologists (AIPG) sent a statement to President Barack Obama and other US government officials:

The geological professionals in AIPG recognize that climate change is occurring and has the potential to yield catastrophic impacts if humanity is not prepared to address those impacts. It is also recognized that climate change will occur regardless of the cause. The sooner a defensible scientific understanding can be developed, the better equipped humanity will be to develop economically viable and technically effective methods to support the needs of society.

Concerned that the original statement issued in March 2009 was too ambiguous, AIPG's National Executive Committee approved a revised position statement issued in January 2010:

The geological professionals in AIPG recognize that climate change is occurring regardless of cause. AIPG supports continued research into all forces driving climate change.

In March 2010, AIPG's Executive Director issued a statement regarding polarization of opinions on climate change within the membership and announced that the AIPG Executive had made a decision to cease publication of articles and opinion pieces concerning climate change in AIPG's news journal, *The Professional Geologist*. The Executive Director said that "the question of anthropogenicity of climate change is contentious."

Canadian Federation of Earth Sciences

The science of global climate change is still evolving and our understanding of this vital Earth system is not as developed as is the case for other Earth systems such as plate tectonics. What is known with certainty is that regardless of the causes, our global climate will continue to change for the foreseeable future... The level of CO_2 in our atmosphere is now greater than at any time in the past 500,000 years; there will be consequences for our global climate and natural systems as a result.

Geological Society of Australia

After a long and extensive and extended consultation with society members, the GSA executive committee has decided not to proceed with a climate change position statement.

Opposing

Since 2007, when the American Association of Petroleum Geologists released a revised statement, no scientific body of national or international scientists rejects the findings of human-induced effects on climate change.

Surveys Of Scientists and Scientific Literature

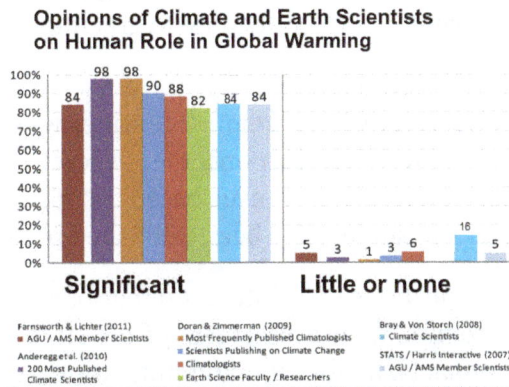

Opinions of Climate and Earth Scientists on Human Role in Global Warming

	Farnsworth & Lichter (2011) AGU / AMS Member Scientists	Doran & Zimmerman (2009) Most Frequently Published Climatologists	Bray & Von Storch (2008) Climate Scientists
	Anderegg et al. (2010) 200 Most Published Climate Scientists	Scientists Publishing on Climate Change / Climatologists / Earth Science Faculty / Researchers	STATS / Harris Interactive (2007) AGU / AMS Member Scientists

Summary of opinions from climate and earth scientists regarding climate change. .

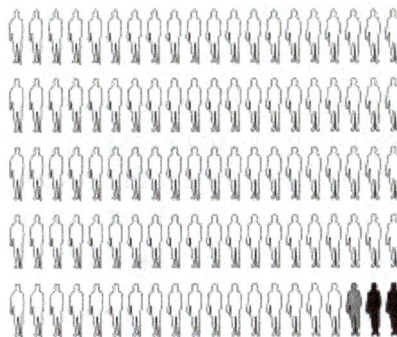

Of the published climate researchers that state a position, just over 97% of them say humans are causing global warming.

Various surveys have been conducted to evaluate scientific opinion on global warming. They have concluded that the majority of scientists support the idea of anthropogenic climate change.

In 2004, the geologist and historian of science Naomi Oreskes summarized a study of the scientific literature on climate change. She analyzed 928 abstracts of papers from refereed scientific journals between 1993 and 2003 and concluded that there is a scientific consensus on the reality of anthropogenic climate change.

Oreskes divided the abstracts into six categories: explicit endorsement of the consensus position, evaluation of impacts, mitigation proposals, methods, paleoclimate analysis, and rejection of the consensus position. Seventy-five per cent of the abstracts were placed in the first three categories (either explicitly or implicitly accepting the consensus view); 25% dealt with methods or paleoclimate, thus taking no position on current anthropogenic climate change. None of the abstracts disagreed with the consensus position, which the author found to be "remarkable". According to the report, "authors evaluating impacts, developing methods, or studying paleoclimatic change might believe that current climate change is natural. However, none of these papers argued that point."

In 2007, Harris Interactive surveyed 489 randomly selected members of either the American Meteorological Society or the American Geophysical Union for the Statistical Assessment Service (STATS) at George Mason University. 97% of the scientists surveyed agreed that global temperatures had increased during the past 100 years; 84% said they personally believed human-induced warming was occurring, and 74% agreed that "currently available scientific evidence" substantiated its occurrence. Catastrophic effects in 50–100 years would likely be observed according to 41%, while 44% thought the effects would be moderate and about 13 percent saw relatively little danger. 5% said they thought human activity did not contribute to greenhouse warming.

Dennis Bray and Hans von Storch conducted a survey in August 2008 of 2058 climate scientists from 34 different countries. A web link with a unique identifier was given to each respondent to eliminate multiple responses. A total of 373 responses were received giving an overall response rate of 18.2%. No paper on climate change consensus based on this survey has been published yet (February 2010), but one on another subject has been published based on the survey.

The survey was composed of 76 questions split into a number of sections. There were sections on the demographics of the respondents, their assessment of the state of climate science, how good the science is, climate change impacts, adaptation and mitigation, their opinion of the IPCC, and how well climate science was being communicated to the public. Most of the answers were on a scale from 1 to 7 from 'not at all' to 'very much'.

To the question "How convinced are you that climate change, whether natural or anthropogenic, is occurring now?", 67.1% said they very much agreed, 26.7% agreed to some large extent, 6.2% said to they agreed to some small extent (2–4), none said they did not agree at all. To the question "How convinced are you that most of recent or near future climate change is, or will be, a result of anthropogenic causes?" the responses were 34.6% very much agree, 48.9% agreeing to a large extent, 15.1% to a small extent, and 1.35% not agreeing at all.

A poll performed by Peter Doran and Maggie Kendall Zimmerman at University of Illinois at Chicago received replies from 3,146 of the 10,257 polled Earth scientists. Results were analyzed globally and by specialization. 76 out of 79 climatologists who "listed climate science as their area of expertise and who also have published more than 50% of their recent peer-reviewed papers on the subject of climate change" believed that mean global temperatures had risen compared to pre-1800s levels. Seventy-five of 77 believed that human activity is a significant factor in changing mean global temperatures. Among all respondents, 90% agreed that temperatures have risen compared to pre-1800 levels, and 82% agreed that humans significantly influence the global temperature. Economic geologists and meteorologists were among the biggest doubters, with only 47 percent and 64 percent, respectively, believing in significant human involvement. The authors summarised the findings:

It seems that the debate on the authenticity of global warming and the role played by human activity is largely nonexistent among those who understand the nuances and scientific basis of long-term climate processes.

A 2010 paper in the *Proceedings of the National Academy of Sciences of the United States* (PNAS) reviewed publication and citation data for 1,372 climate researchers and drew the following two conclusions:

(i) 97–98% of the climate researchers most actively publishing in the field support the tenets of ACC (Anthropogenic Climate Change) outlined by the Intergovernmental Panel on Climate Change, and (ii) the relative climate expertise and scientific prominence of the researchers unconvinced of ACC are substantially below that of the convinced researchers.

A 2013 paper in Environmental Research Letters reviewed 11,944 abstracts of scientific papers matching "global warming" or "global climate change". They found 4,014 which discussed the cause of recent global warming, and of these "97.1% endorsed the consensus position that humans are causing global warming".

James L. Powell, a former member of the National Science Board and current executive director of the National Physical Science Consortium, analyzed published research on global warming and climate change between 1991 and 2012 and found that of the 13,950 articles in peer-reviewed journals, only 24 rejected anthropogenic global warming. A follow-up analysis looking at 2,258 peer-reviewed climate articles with 9,136 authors published between November 2012 and December 2013 revealed that only one of the 9,136 authors rejected anthropogenic global warming.

Scientific Consensus

A question that frequently arises in popular discussion of climate change is whether there is a scientific consensus on climate change. Several scientific organizations have explicitly used the term "consensus" in their statements:

1. American Association for the Advancement of Science, 2006: "The conclusions in this statement reflect the scientific consensus represented by, for example, the Intergovernmental Panel on Climate Change, and the Joint National Academies' statement."

2. US National Academy of Sciences: "In the judgment of most climate scientists, Earth's warming in recent decades has been caused primarily by human activities that have increased the amount of greenhouse gases in the atmosphere. ... On climate change, [the National Academies' reports] have assessed consensus findings on the science..."

3. Joint Science Academies' statement, 2005: "We recognise the international scientific consensus of the Intergovernmental Panel on Climate Change (IPCC)."

4. Joint Science Academies' statement, 2001: "The work of the Intergovernmental Panel on Climate Change (IPCC) represents the consensus of the international scientific community on climate change science. We recognise IPCC as the world's most reliable source of information on climate change and its causes, and we endorse its method of achieving this consensus."

5. American Meteorological Society, 2003: "The nature of science is such that there is rarely total agreement among scientists. Individual scientific statements and papers—the validity of some of which has yet to be assessed adequately—can be exploited in the policy debate and can leave the impression that the scientific community is sharply divided on issues where there is, in reality, a strong scientific consensus.... IPCC assessment reports are prepared at approximately five-year intervals by a large international group of experts who represent the broad range of expertise and perspectives relevant to the issues. The reports strive to reflect a consensus evaluation of the results of the full body of peer-reviewed research.... They provide an analysis of what is known and not known, the degree of consensus, and some indication of the degree of confidence that can be placed on the various statements and conclusions."

6. Network of African Science Academies: "A consensus, based on current evidence, now exists within the global scientific community that human activities are the main source of climate change and that the burning of fossil fuels is largely responsible for driving this change."

7. International Union for Quaternary Research, 2008: "INQUA recognizes the international scientific consensus of the Intergovernmental Panel on Climate Change (IPCC)."

8. Australian Coral Reef Society, 2006: "There is almost total consensus among experts that the earth's climate is changing as a result of the build-up of greenhouse gases.... There is broad scientific consensus that coral reefs are heavily affected by the activities of man and there are significant global influences that can make reefs more vulnerable such as global warming...."

Attribution of Recent Climate Change

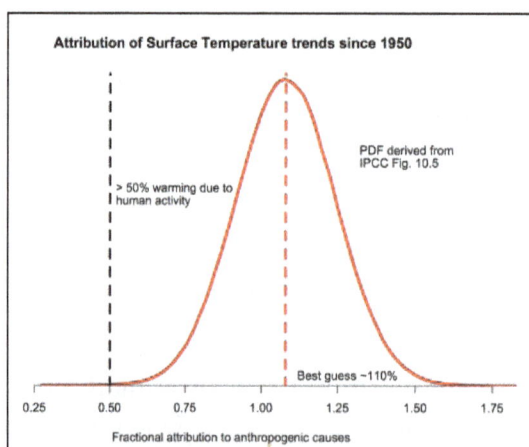

Attribution of Surface Temperature trends since 1950

PDF derived from IPCC Fig. 10.5

> 50% warming due to human activity

Best guess ~110%

Fractional attribution to anthropogenic causes

PDF of fraction of surface temperature trends since 1950 attributable to human activity, based on IPCC AR5 10.5

Global annual average temperature; year-to-year fluctuations are due to natural processes, such as the effects of El Niños, La Niñas, and the eruption of large volcanoes.

This image shows three examples of internal climate variability measured between 1950 and 2012: the El Niño–Southern oscillation, the Arctic oscillation, and the North Atlantic oscillation.

Attribution of recent climate change is the effort to scientifically ascertain mechanisms responsible for recent climate changes on Earth, commonly known as 'global warming'. The effort has focused on changes observed during the period of instrumental temperature record, when records are most reliable; particularly in the last 50 years, when human activity has grown fastest and observations of the troposphere have become available. The dominant mechanisms (to which the IPCC attributes climate change) are anthropogenic, i.e., the result of human activity. They are:

1. increasing atmospheric concentrations of greenhouse gases

2. global changes to land surface, such as deforestation

3. increasing atmospheric concentrations of aerosols.

There are also natural mechanisms for variation including climate oscillations, changes in solar activity, and volcanic activity.

According to the Intergovernmental Panel on Climate Change (IPCC), it is "extremely likely" that human influence was the dominant cause of global warming between 1951 and 2010. The IPCC defines "extremely likely" as indicating a probability of 95 to 100%, based on an expert assessment of all the available evidence.

Multiple lines of evidence support attribution of recent climate change to human activities:

1. A basic physical understanding of the climate system: greenhouse gas concentrations have increased and their warming properties are well-established.

2. Historical estimates of past climate changes suggest that the recent changes in global surface temperature are unusual.

3. Computer-based climate models are unable to replicate the observed warming unless human greenhouse gas emissions are included.

4. Natural forces alone (such as solar and volcanic activity) cannot explain the observed warming.

The IPCC's attribution of recent global warming to human activities is a view shared by most scientists,[2] and is also supported by 196 other scientific organizations worldwide.

Background

This section introduces some concepts in climate science that are used in the following sections:

Factors affecting Earth's climate can be broken down into feedbacks and forcings. A forcing is something that is imposed externally on the climate system. External forcings include natural phenomena such as volcanic eruptions and variations in the sun's output. Human activities can also impose forcings, for example, through changing the composition of the atmosphere.

Radiative forcing is a measure of how various factors alter the energy balance of the Earth's atmosphere. A positive radiative forcing will tend to increase the energy of the Earth-atmosphere system, leading to a warming of the system. Between the start of the Industrial Revolution in 1750, and the year 2005, the increase in the atmospheric concentration of carbon dioxide (chemical formula: CO_2) led to a positive radiative forcing, averaged over the Earth's surface area, of about 1.66 watts per square metre (abbreviated W m^{-2}).

Climate feedbacks can either amplify or dampen the response of the climate to a given forcing. There are many feedback mechanisms in the climate system that can either amplify (a positive feedback) or diminish (a negative feedback) the effects of a change in climate forcing.

Aspects of the climate system will show variation in response to changes in forcings. In the absence of forcings imposed on it, the climate system will still show internal variability. This internal variability is a result of complex interactions between components of the climate system, such as the coupling between the atmosphere and ocean. An example of internal variability is the El Niño-Southern Oscillation.

Detection Vs. Attribution

Detection and attribution of climate signals, as well as its common-sense meaning, has a more precise definition within the climate change literature, as expressed by the IPCC. Detection of a climate signal does not always imply significant attribution. The IPCC's Fourth Assessment Report says "it is *extremely likely* that human activities have exerted a substantial net warming influence on climate since 1750," where "extremely likely" indicates a probability greater than 95%. *Detection* of a signal requires demonstrating that an observed change is statistically significantly different from that which can be explained by natural internal variability.

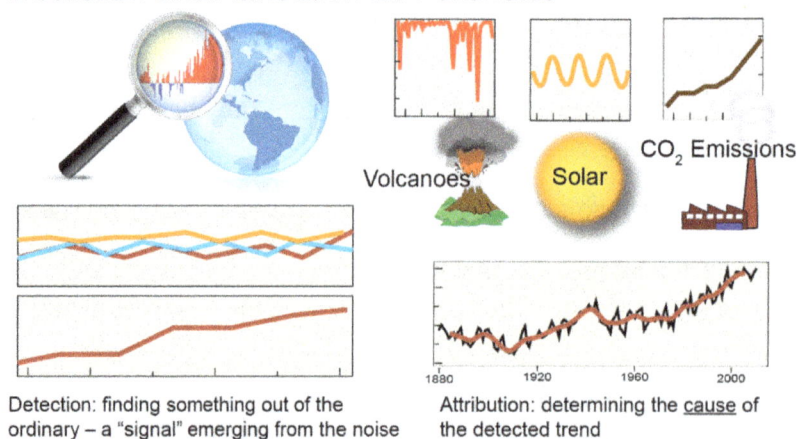

Detection and Attribution as Forensics

Volcanoes   Solar   CO$_2$ Emissions

Detection: finding something out of the ordinary – a "signal" emerging from the noise

Attribution: determining the cause of the detected trend

In detection and attribution, the natural factors considered usually include changes in the Sun's output and volcanic eruptions, as well as natural modes of variability such as El Niño and La Niña. Human factors include the emissions of heat-trapping "greenhouse" gases and particulates as well as clearing of forests and other land-use changes. Figure source: NOAA NCDC.

Attribution requires demonstrating that a signal is:

1. unlikely to be due entirely to internal variability;

2. consistent with the estimated responses to the given combination of anthropogenic and natural forcing

3. not consistent with alternative, physically plausible explanations of recent climate change that exclude important elements of the given combination of forcings.

Key Attributions

Greenhouse Gases

Carbon dioxide is the primary greenhouse gas that is contributing to recent climate change. CO_2 is absorbed and emitted naturally as part of the carbon cycle, through animal and plant respiration, volcanic eruptions, and ocean-atmosphere exchange. Human activities, such as the burning of fossil fuels and changes in land use, release large amounts of carbon to the atmosphere, causing CO_2 concentrations in the atmosphere to rise.

The high-accuracy measurements of atmospheric CO_2 concentration, initiated by Charles David Keeling in 1958, constitute the master time series documenting the changing composition of the atmosphere. These data have iconic status in climate change science as evidence of the effect of human activities on the chemical composition of the global atmosphere.

Along with CO_2, methane and nitrous oxide are also major forcing contributors to the greenhouse effect. The Kyoto Protocol lists these together with hydrofluorocarbons (HFCs), perfluorocarbons (PFCs), and sulphur hexafluoride (SF_6), which are entirely artificial (i.e. anthropogenic) gases, which also contribute to radiative forcing in the atmosphere. The chart at right attributes anthropogenic greenhouse gas emissions to eight main economic sectors, of which the largest contributors are power stations (many of which burn coal or other fossil fuels), industrial processes, transportation fuels (generally fossil fuels), and agricultural by-products (mainly methane from enteric fermentation and nitrous oxide from fertilizer use).

Water Vapor

Water vapor is the most abundant greenhouse gas and also the most important in terms of its contribution to the natural greenhouse effect, despite having a short atmospheric lifetime (about 10 days). Some human activities can influence local water vapor levels. However, on a global scale, the concentration of water vapor is controlled by temperature, which influences overall rates of evaporation and precipitation. Therefore, the global concentration of water vapor is not substantially affected by direct human emissions.

Land use

Climate change is attributed to land use for two main reasons. Between 1750 and 2007, about two-thirds of anthropogenic CO_2 emissions were produced from burning fossil fuels, and about one-third of emissions from changes in land use, primarily deforestation. Deforestation both reduces the amount of carbon dioxide absorbed by deforested regions and releases greenhouse gases directly, together with aerosols, through biomass burning that frequently accompanies it.

A second reason that climate change has been attributed to land use is that the terrestrial albedo is often altered by use, which leads to radiative forcing. This effect is more significant locally than globally.

Livestock and Land Use

Worldwide, livestock production occupies 70% of all land used for agriculture, or 30% of the ice-free land surface of the Earth. More than 18% of anthropogenic greenhouse gas emissions are attributed to livestock and livestock-related activities such as deforestation and increasingly fuel-intensive farming practices. Specific attributions to the livestock sector include:

1. 9% of global anthropogenic carbon dioxide emissions

2. 35–40% of global anthropogenic methane emissions (chiefly due to enteric fermentation and manure)

3. 64% of global anthropogenic nitrous oxide emissions, chiefly due to fertilizer use.

Aerosols

With virtual certainty, scientific consensus has attributed various forms of climate change, chiefly cooling effects, to aerosols, which are small particles or droplets suspended in the atmosphere. Key sources to which anthropogenic aerosols are attributed include:

1. biomass burning such as slash and burn deforestation. Aerosols produced are primarily black carbon.

2. industrial air pollution, which produces soot and airborne sulfates, nitrates, and ammonium

3. dust produced by land use effects such as desertification

Attribution of 20th Century Climate Change

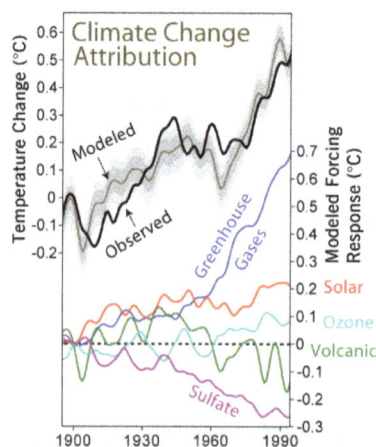

One global climate model's reconstruction of temperature change during the 20th century as the result of five studied forcing factors and the amount of temperature change attributed to each.

Over the past 150 years human activities have released increasing quantities of greenhouse gases into the atmosphere. This has led to increases in mean global temperature, or global warming. Other human effects are relevant—for example, sulphate aerosols are believed to have a cooling effect. Natural factors also contribute. According to the historical temperature record of the last century, the Earth's near-surface air temperature has risen around 0.74 ± 0.18 °Celsius (1.3 ± 0.32 °Fahrenheit).

A historically important question in climate change research has regarded the relative importance of human activity and non-anthropogenic causes during the period of instrumental record. In the 1995 Second Assessment Report (SAR), the IPCC made the widely quoted statement that "The balance of evidence suggests a discernible human influence on global climate". The phrase "balance of evidence" suggested the (English) common-law standard of proof required in civil as opposed to criminal courts: not as high as "beyond reasonable doubt". In 2001 the Third Assessment Report (TAR) refined this, saying "There is new and stronger evidence that most of the warming observed over the last 50 years is attributable to human activities". The 2007 Fourth Assessment Report (AR4) strengthened this finding:

1. "Anthropogenic warming of the climate system is widespread and can be detected in temperature observations taken at the surface, in the free atmosphere and in the oceans. Evidence of the effect of external influences, both anthropogenic and natural, on the climate system has continued to accumulate since the TAR."

Other findings of the IPCC Fourth Assessment Report include:

1. "It is *extremely unlikely* (<5%) that the global pattern of warming during the past half century can be explained without external forcing (i.e., it is inconsistent with being the result of internal variability), and *very unlikely* that it is due to known natural external causes alone. The warming occurred in both the ocean and the atmosphere and took place at a time when natural external forcing factors would likely have produced cooling."

2. "From new estimates of the combined anthropogenic forcing due to greenhouse gases, aerosols, and land surface changes, it is *extremely likely* (>95%) that human activities have exerted a substantial net warming influence on climate since 1750."

3. "It is *virtually certain* that anthropogenic aerosols produce a net negative radiative forcing (cooling influence) with a greater magnitude in the Northern Hemisphere than in the Southern Hemisphere."

Over the past five decades there has been a global warming of approximately 0.65 °C (1.17 °F) at the Earth's surface. Among the possible factors that could produce changes in global mean temperature are internal variability of the climate system, external forcing, an increase in concentration of greenhouse gases, or any combination of these. Current studies indicate that the increase in greenhouse gases, most notably CO_2, is mostly responsible for the observed warming. Evidence for this conclusion includes:

1. Estimates of internal variability from climate models, and reconstructions of past temperatures, indicate that the warming is unlikely to be entirely natural.

2. Climate models forced by natural factors *and* increased greenhouse gases and aerosols reproduce the observed global temperature changes; those forced by natural factors alone do not.

3. "Fingerprint" methods indicate that the pattern of change is closer to that expected from greenhouse gas-forced change than from natural change.

4. The plateau in warming from the 1940s to 1960s can be attributed largely to sulphate aerosol cooling.

Details on Attribution

Recent scientific assessments find that most of the warming of the Earth's surface over the past 50 years has been caused by human activities. This conclusion rests on multiple lines of evidence. Like the warming "signal" that has gradually emerged from the "noise" of natural climate variability, the scientific evidence for a human influence on global climate has accumulated over the past several decades, from many hundreds of studies. No single study is a "smoking gun." Nor has any single study or combination of studies undermined the large body of evidence supporting the conclusion that human activity is the primary driver of recent warming.

The first line of evidence is based on a physical understanding of how greenhouse gases trap heat, how the climate system responds to increases in greenhouse gases, and how other human and natural factors influence climate. The second line of evidence is from indirect estimates of climate changes over the last 1,000 to 2,000 years. These records are obtained from living things and their remains (like tree rings and corals) and from physical quantities (like the ratio between lighter and heavier isotopes of oxygen in ice cores), which change in measurable ways as climate changes. The lesson from these data is that global surface temperatures over the last several decades are clearly unusual, in that they were higher than at any time during at least the past 400 years. For the Northern Hemisphere, the recent temperature rise is clearly unusual in at least the last 1,000 years.

The third line of evidence is based on the broad, qualitative consistency between observed changes in climate and the computer model simulations of how climate would be expected to change in response to human activities. For example, when climate models are run with historical increases in greenhouse gases, they show gradual warming of the Earth and ocean surface, increases in ocean heat content and the temperature of the lower atmosphere, a rise in global sea level, retreat of sea ice and snow cover, cooling of the stratosphere, an increase in the amount of atmospheric water vapor, and changes in large-scale precipitation and pressure patterns. These and other aspects of modelled climate change are in agreement with observations.

"Fingerprint" Studies

Reconstructions of global temperature that include greenhouse gas increases and other human influences (red line, based on many models) closely match measured temperatures (dashed line). Those that only include natural influences (blue line, based on many models) show a slight cooling, which has not occurred. The ability of models to generate reasonable histories of global temperature is verified by their response to four 20th-century volcanic eruptions: each eruption caused brief cooling that appeared in observed as well as modeled records.

Finally, there is extensive statistical evidence from so-called "fingerprint" studies. Each factor that affects climate produces a unique pattern of climate response, much as each person has a unique fingerprint. Fingerprint studies exploit these unique signatures, and allow detailed comparisons of modelled and observed climate change patterns. Scientists rely on such studies to attribute observed changes in climate to a particular cause or set of causes. In the real world, the climate changes that have occurred since the start of the Industrial Revolution are due to a complex mixture of human and natural causes. The importance of each individual influence in this mixture changes over time. Of course, there are not multiple Earths, which would allow an experimenter to change one factor at a time on each Earth, thus helping to isolate different fingerprints. Therefore, climate models are used to study how individual factors affect climate. For example, a single factor (like greenhouse gases) or a set of factors can be varied, and the response of the modelled climate system to these individual or combined changes can thus be studied.

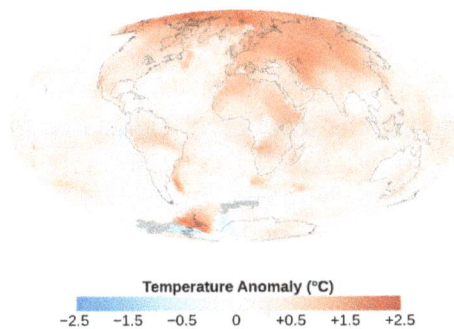

Temperature Anomaly (°C)

-2.5   -1.5   -0.5   0   +0.5   +1.5   +2.5

Two fingerprints of human activities on the climate are that land areas will warm more than the oceans, and that high latitudes will warm more than low latitudes. These projections have been confirmed by observations (shown above).

For example, when climate model simulations of the last century include all of the major influences on climate, both human-induced and natural, they can reproduce many important features of observed climate change patterns. When human influences are removed from the model experiments, results suggest that the surface of the Earth would actually have cooled slightly over the last 50 years. The clear message from fingerprint studies is that the observed warming over the last half-century cannot be explained by natural factors, and is instead caused primarily by human factors.

Another fingerprint of human effects on climate has been identified by looking at a slice through the layers of the atmosphere, and studying the pattern of temperature changes from the surface up through the stratosphere. The earliest fingerprint work focused on changes in surface and atmospheric temperature. Scientists then applied fingerprint methods to a whole range of climate variables, identifying human-caused climate signals in the heat content of the oceans, the height of the tropopause (the boundary between the troposphere and stratosphere, which has shifted upward by hundreds of feet in recent decades), the geographical patterns of precipitation, drought, surface pressure, and the runoff from major river basins.

Studies published after the appearance of the IPCC Fourth Assessment Report in 2007 have also found human fingerprints in the increased levels of atmospheric moisture (both close to the surface and over the full extent of the atmosphere), in the decline of Arctic sea ice extent, and in the patterns of changes in Arctic and Antarctic surface temperatures.

The message from this entire body of work is that the climate system is telling a consistent story of increasingly dominant human influence – the changes in temperature, ice extent, moisture, and circulation patterns fit together in a physically consistent way, like pieces in a complex puzzle.

Increasingly, this type of fingerprint work is shifting its emphasis. As noted, clear and compelling scientific evidence supports the case for a pronounced human influence on global climate. Much of the recent attention is now on climate changes at continental and regional scales, and on variables that can have large impacts on societies. For example, scientists have established causal links between human activities and the changes in snowpack, maximum and minimum (diurnal) temperature, and the seasonal timing of runoff over mountainous regions of the western United States. Human activity is likely to have made a substantial contribution to ocean surface temperature changes in hurricane formation regions. Researchers are also looking beyond the physical climate system, and are beginning to tie changes in the distribution and seasonal behaviour of plant and animal species to human-caused changes in temperature and precipitation.

For over a decade, one aspect of the climate change story seemed to show a significant difference between models and observations. In the tropics, all models predicted that with a rise in greenhouse gases, the troposphere would be expected to warm more rapidly than the surface. Observations from weather balloons, satellites, and surface thermometers seemed to show the opposite behaviour (more rapid warming of the surface than the troposphere). This issue was a stumbling block in understanding the causes of climate change. It is now largely resolved. Research showed that there were large uncertainties in the satellite and weather balloon data. When uncertainties in models and observations are properly accounted for, newer observational data sets (with better treatment of known problems) are in agreement with climate model results.

This set of graphs shows the estimated contribution of various natural and human factors to changes in global mean temperature between 1889–2006. Estimated contributions are based on multivariate analysis rather than model simulations. The graphs show that human influence on climate has eclipsed the magnitude of natural temperature changes over the past 120 years. Natural influences on temperature—El Niño, solar variability, and volcanic aerosols—have varied approximately plus and minus 0.2 °C (0.4 °F), (averaging to about zero), while human influences have contributed roughly 0.8 °C (1 °F) of warming since 1889.

This does not mean, however, that all remaining differences between models and observations have been resolved. The observed changes in some climate variables, such as Arctic sea ice, some aspects of precipitation, and patterns of surface pressure, appear to be proceeding much more rapidly than models have projected. The reasons for these differences are not well understood. Nevertheless, the bottom-line conclusion from climate fingerprinting is that most of the observed changes studied to date are consistent with each other, and are also consistent with our scientific understanding of how the climate system would be expected to respond to the increase in heat-trapping gases resulting from human activities.

Extreme Weather Events

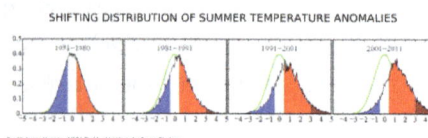

SHIFTING DISTRIBUTION OF SUMMER TEMPERATURE ANOMALIES

Credit: James Hansen, NASA Goddard Institute for Space Studies

Frequency of occurrence (vertical axis) of local June–July–August temperature anomalies (relative to 1951–1980 mean) for Northern Hemisphere land in units of local standard deviation (horizontal axis). According to Hansen *et al.* (2012), the distribution of anomalies has shifted to the right as a consequence of global warming, meaning that unusually hot summers have become more common. This is analogous to the rolling of a dice: cool summers now cover only half of one side of a six-sided die, white covers one side, red covers four sides, and an extremely hot (red-brown) anomaly covers half of one side.

One of the subjects discussed in the literature is whether or not extreme weather events can be attributed to human activities. Seneviratne *et al.* (2012) stated that attributing individual extreme weather events to human activities was challenging. They were, however, more confident over attributing changes in long-term trends of extreme weather. For example, Seneviratne *et al.* (2012) concluded that human activities had likely led to a warming of extreme daily minimum and maximum temperatures at the global scale.

Another way of viewing the problem is to consider the effects of human-induced climate change on the probability of future extreme weather events. Stott *et al.* (2003), for example, considered whether or not human activities had increased the risk of severe heat waves in Europe, like the one experienced in 2003. Their conclusion was that human activities had very likely more than doubled the risk of heat waves of this magnitude.

An analogy can be made between an athlete on steroids and human-induced climate change. In the same way that an athlete's performance may increase from using steroids, human-induced climate change increases the risk of some extreme weather events.

Hansen *et al.* (2012) suggested that human activities have greatly increased the risk of summertime heat waves. According to their analysis, the land area of the Earth affected by very hot summer temperature anomalies has greatly increased over time (refer to graphs on the left). In the base period 1951-1980, these anomalies covered a few tenths of 1% of the global land area. In recent years, this has increased to around 10% of the global land area. With high confidence, Hansen *et al.* (2012) attributed the 2010 Moscow and 2011 Texas heat waves to human-induced global warming.

An earlier study by Dole *et al.* (2011) concluded that the 2010 Moscow heatwave was mostly due to natural weather variability. While not directly citing Dole *et al.* (2011), Hansen *et al.* (2012) rejected this type of explanation. Hansen *et al.* (2012) stated that a combination of natural weather variability and human-induced global warming was responsible for the Moscow and Texas heat waves.

Scientific Literature and Opinion

There are a number of examples of published and informal support for the consensus view. As mentioned earlier, the IPCC has concluded that most of the observed increase in globally averaged temperatures since the mid-20th century is "very likely" due to human activities. The IPCC's conclusions are consistent with those of several reports produced by the US National Research Council. A report published in 2009 by the U.S. Global Change Research Program concluded that "[global] warming is unequivocal and primarily human-induced." A number of scientific organizations have issued statements that support the consensus view. Two examples include:

1. a joint statement made in 2005 by the national science academies of the G8, and Brazil, China and India;

2. a joint statement made in 2008 by the Network of African Science Academies.

Detection and Attribution Studies

The IPCC Fourth Assessment Report (2007), concluded that attribution was possible for a number of observed changes in the climate. However, attribution was found to be more difficult when assessing changes over smaller regions (less than continental scale) and over short time periods (less than 50 years). Over larger regions, averaging reduces natural variability of the climate, making detection and attribution easier.

1. In 1996, in a paper in *Nature* titled "A search for human influences on the thermal structure of the atmosphere", Benjamin D. Santer et al. wrote: "The observed spatial patterns of temperature change in the free atmosphere from 1963 to 1987 are similar to those predicted by state-of-the-art climate models incorporating various combinations of changes in carbon dioxide, anthropogenic sulphate aerosol and stratospheric ozone concentrations. The degree of pattern similarity between models and observations increases through this period. It is likely that this trend is partially due to human activities, although many uncertainties remain, particularly relating to estimates of natural variability."

2. A 2002 paper in the *Journal of Geophysical Research* says "Our analysis suggests that the early twentieth century warming can best be explained by a combination of warming due to increases in greenhouse gases and natural forcing, some cooling due to other anthropogenic forcings, and a substantial, but not implausible, contribution from internal variability. In the second half of the century we find that the warming is largely caused by changes in greenhouse gases, with changes in sulphates and, perhaps, volcanic aerosol offsetting approximately one third of the warming."

3. A 2005 review of detection and attribution studies by the International Ad Hoc Detection and Attribution Group found that "natural drivers such as solar variability and volcanic activity are at most partially responsible for the large-scale temperature changes observed over the past century, and that a large fraction of the warming over the last 50 yr can be attributed to greenhouse gas increases. Thus, the recent research supports and strengthens the IPCC Third Assessment Report conclusion that 'most of the global warming over the past 50 years is likely due to the increase in greenhouse gases.'"

4. Barnett and colleagues (2005) say that the observed warming of the oceans "cannot be explained by natural internal climate variability or solar and volcanic forcing, but is well simulated by two anthropogenically forced climate models," concluding that "it is of human origin, a conclusion robust to observational sampling and model differences".

5. Two papers in the journal *Science* in August 2005 resolve the problem, evident at the time of the TAR, of tropospheric temperature trends. The UAH version of the record contained errors, and there is evidence of spurious cooling trends in the radiosonde record, particularly in the tropics..

6. Multiple independent reconstructions of the temperature record of the past 1000 years confirm that the late 20th century is probably the warmest period in that time.

Reviews of Scientific Opinion

1. An essay in *Science* surveyed 928 abstracts related to climate change, and concluded that most journal reports accepted the consensus. This is discussed further in scientific opinion on climate change.

2. A 2010 paper in the Proceedings of the National Academy of Sciences found that among a pool of roughly 1,000 researchers who work directly on climate issues and publish the most frequently on the subject, 97% agree that anthropogenic climate change is happening.

3. A 2011 paper from George Mason University published in the *International Journal of Public Opinion Research*, "The Structure of Scientific Opinion on Climate Change," collected the opinions of scientists in the earth, space, atmospheric, oceanic or hydrological sciences. The 489 survey respondents—representing nearly half of all those eligible according to the survey's specific standards – work in academia, government, and industry, and are members of prominent professional organizations. The study found that 97% of the 489 scientists surveyed agreed that global temperatures have risen over the past century. Moreover, 84% agreed that "human-induced greenhouse warming" is now occurring." Only 5% disagreed with the idea that human activity is a significant cause of global warming.

As described above, a small minority of scientists do disagree with the consensus. For example, Willie Soon and Richard Lindzen say that there is insufficient proof for anthropogenic attribution. Generally this position requires new physical mechanisms to explain the observed warming.

Solar Activity

Solar radiation at the top of our atmosphere, and global temperature

Modelled simulation of the effect of various factors (including GHGs, Solar irradiance) singly and in combination, showing in particular that solar activity produces a small and nearly uniform warming, unlike what is observed.

Solar sunspot maximum occurs when the magnetic field of the sun collapses and reverse as part of its average 11 year solar cycle (22 years for complete North to North restoration).

The role of the sun in recent climate change has been looked at by climate scientists. Since 1978, output from the Sun has been measured by satellites[:6] significantly more accurately than was previously possible from the surface. These measurements indicate that the Sun's total solar irradiance has not increased since 1978, so the warming during the past 30 years cannot be directly attributed to an increase in total solar energy reaching the Earth. In the three decades since 1978, the combination of solar and volcanic activity probably had a slight cooling influence on the climate.

Climate models have been used to examine the role of the sun in recent climate change. Models are unable to reproduce the rapid warming observed in recent decades when they only take into account variations in total solar irradiance and volcanic activity. Models are, however, able to simulate the observed 20th century changes in temperature when they include all of the most important external forcings, including human influences and natural forcings. As has already been stated, Hegerl *et al.* (2007) concluded that greenhouse gas forcing had "very likely" caused most of the observed global warming since the mid-20th century. In making this conclusion, Hegerl *et al.* (2007) allowed for the possibility that climate models had been underestimated the effect of solar forcing.

The role of solar activity in climate change has also been calculated over longer time periods using "proxy" datasets, such as tree rings. Models indicate that solar and volcanic forcings can explain periods of relative warmth and cold between A.D. 1000 and 1900, but human-induced forcings are needed to reproduce the late-20th century warming.

Another line of evidence against the sun having caused recent climate change comes from looking at how temperatures at different levels in the Earth's atmosphere have changed. Models and observations show that greenhouse gas results in warming of the lower atmosphere at the surface (called the troposphere) but cooling of the upper atmosphere (called the stratosphere). Depletion of the ozone layer by chemical refrigerants has also resulted in a cooling effect in the stratosphere. If the sun was responsible for observed warming, warming of the troposphere at the surface and warming at the top of the stratosphere would be expected as increase solar activity would replenish ozone and oxides of nitrogen. The stratosphere has a reverse temperature gradient than the troposphere so as the temperature of the troposphere cools with altitude, the stratosphere rises with altitude. Hadley cells are the mechanism by which equatorial generated ozone in the tropics (highest area of UV irradiance in the stratosphere) is moved poleward. Global climate models suggest that climate change may widen the Hadley cells and push the jetstream northward thereby expanding the tropics region and resulting in warmer, dryer conditions in those areas overall.

Non-consensus Views

Habibullo Abdussamatov (2004), head of space research at St. Petersburg's Pulkovo Astronomical Observatory in Russia, has argued that the sun is responsible for recently observed climate change. Journalists for news sources canada.com (Solomon, 2007b), National Geographic News (Ravillious, 2007), and LiveScience (Than, 2007) reported on the story of warming on Mars. In these articles, Abdussamatov was quoted. He stated that warming on Mars was evidence that global warming on Earth was being caused by changes in the sun.

Ravillious (2007) quoted two scientists who disagreed with Abdussamatov: Amato Evan, a climate scientist at the University of Wisconsin-Madison, in the US, and Colin Wilson, a planetary physicist at Oxford University in the UK. According to Wilson, "Wobbles in the orbit of Mars are the main cause of its climate change in the current era" . Than (2007) quoted Charles Long, a climate physicist at Pacific Northwest National Laboratories in the US, who disagreed with Abdussamatov.

Than (2007) pointed to the view of Benny Peiser, a social anthropologist at Liverpool John Moores University in the UK. In his newsletter, Peiser had cited a blog that had commented on warming observed on several planetary bodies in the Solar system. These included Neptune's moon Triton,

Jupiter, Pluto and Mars. In an e-mail interview with Than (2007), Peiser stated that:

"I think it is an intriguing coincidence that warming trends have been observed on a number of very diverse planetary bodies in our solar system, (...) Perhaps this is just a fluke."

Than (2007) provided alternative explanations of why warming had occurred on Triton, Pluto, Jupiter and Mars.

The US Environmental Protection Agency (US EPA, 2009) responded to public comments on climate change attribution. A number of commenters had argued that recent climate change could be attributed to changes in solar irradiance. According to the US EPA (2009), this attribution was not supported by the bulk of the scientific literature. Citing the work of the IPCC (2007), the US EPA pointed to the low contribution of solar irradiance to radiative forcing since the start of the Industrial Revolution in 1750. Over this time period (1750 to 2005), the estimated contribution of solar irradiance to radiative forcing was 5% the value of the combined radiative forcing due to increases in the atmospheric concentrations of carbon dioxide, methane and nitrous oxide.

Effect of Cosmic Rays

Henrik Svensmark has suggested that the magnetic activity of the sun deflects cosmic rays, and that this may influence the generation of cloud condensation nuclei, and thereby have an effect on the climate. The website ScienceDaily reported on a 2009 study that looked at how past changes in climate have been affected by the Earth's magnetic field. Geophysicist Mads Faurschou Knudsen, who co-authored the study, stated that the study's results supported Svensmark's theory. The authors of the study also acknowledged that CO_2 plays an important role in climate change.

Consensus View on Cosmic Rays

The view that cosmic rays could provide the mechanism by which changes in solar activity affect climate is not supported by the literature. Solomon *et al.* (2007) state:

[..] the cosmic ray time series does not appear to correspond to global total cloud cover after 1991 or to global low-level cloud cover after 1994. Together with the lack of a proven physical mechanism and the plausibility of other causal factors affecting changes in cloud cover, this makes the association between galactic cosmic ray-induced changes in aerosol and cloud formation controversial

Studies by Lockwood and Fröhlich (2007) and Sloan and Wolfendale (2008) found no relation between warming in recent decades and cosmic rays. Pierce and Adams (2009) used a model to simulate the effect of cosmic rays on cloud properties. They concluded that the hypothesized effect of cosmic rays was too small to explain recent climate change. Pierce and Adams (2009) noted that their findings did not rule out a possible connection between cosmic rays and climate change, and recommended further research.

Erlykin *et al.* (2009) found that the evidence showed that connections between solar variation and climate were more likely to be mediated by direct variation of insolation rather than cosmic rays, and concluded: "Hence within our assumptions, the effect of varying solar activity, either by direct solar irradiance or by varying cosmic ray rates, must be less than 0.07 °C since 1956, i.e. less than 14% of the observed global warming." Carslaw (2009) and Pittock (2009) review the recent and historical literature in this field and continue to find that the link between cosmic rays and climate is tenuous, though they encourage continued research. US EPA (2009) commented on research by Duplissy *et al.* (2009):

The CLOUD experiments at CERN are interesting research but do not provide conclusive evidence that cosmic rays can serve as a major source of cloud seeding. Preliminary results from the experiment (Duplissy et al., 2009) suggest that though there was some evidence of ion mediated nucleation, for most of the nucleation events observed the contribution of ion processes appeared to be minor. These experiments also showed the difficulty in maintaining sufficiently clean conditions and stable temperatures to prevent spurious aerosol bursts. There is no indication that the earlier Svensmark experiments could even have matched the controlled conditions of the CERN experiment. We find that the Svensmark results on cloud seeding have not yet been shown to be robust or sufficient to materially alter the conclusions of the assessment literature, especially given the abundance of recent literature that is skeptical of the cosmic ray-climate linkage

Climate Change Denial

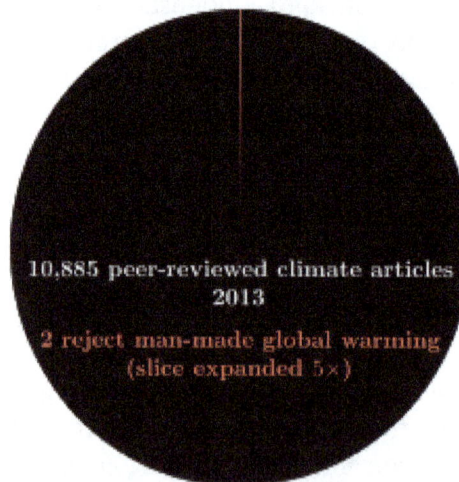

A survey of peer-reviewed scientific articles from 2013 finds that only 2 of 10885 reject man-made global warming. (James L. Powell)

Climate change denial, or global warming denial, is part of the global warming controversy. It involves denial, dismissal, *unwarranted* doubt or contrarian views which depart from the scientific opinion on climate change, including the extent to which it is caused by humans, its impacts on nature and human society, or the potential of adaptation to global warming by human actions. In the global warming controversy, some *deniers* do endorse the term, but other often prefer the term climate change skepticism whereas scientists think it "inappropriate to allow those who deny [anthropogenic global warming] to don the mantle of skeptics"; in effect, the two terms form a continuous, overlapping range of views, and generally have the same characteristics: both reject, to a greater or lesser extent, mainstream scientific opinion on climate change. Climate change denial can also be implicit, when individuals or social groups accept the science but fail to come to terms with it or to translate their acceptance into action. Several social science studies have analyzed these positions as forms of denialism.

Campaigning to undermine public trust in climate science has been described as a "denial machine" of industrial, political and ideological interests, supported by conservative media and skeptical bloggers in manufacturing uncertainty about global warming. In the public debate, phrases such as *climate skepticism* have frequently been used with the same meaning as *climate denialism*. The labels are contested: those actively challenging climate science commonly describe themselves as "skeptics", but many do not comply with common standards of scientific skepticism and, regardless of evidence, persistently deny the validity of human caused global warming.

Although scientific opinion on climate change is that human activity is extremely likely to be the primary driver of climate change, the politics of global warming have been affected by climate change denial, hindering efforts to prevent climate change and adapt to the warming climate. Those promoting denial commonly use rhetorical tactics to give the appearance of a scientific controversy where there is none.

The climate change denial industry is most widespread in the United States, where the official Senate Environmental Committee is chaired by Jim Inhofe, who famously called climate change "the greatest hoax ever perpetrated on the American people" and claimed to have debunked it in 2015 when he took a snowball with him and threw it on the Senate floor. Organised campaigning to undermine public trust in climate science is associated with conservative economic policies and backed by industrial interests opposed to the regulation of CO_2 emissions. Climate change denial has been associated with the fossil fuels lobby, the Koch brothers, industry advocates and libertarian think tanks, often in the United States. More than 90% of papers sceptical on climate change originate from right-wing think tanks. The total annual income of these climate change counter-movement-organizations is roughly $900 million. Between 2002 and 2010, nearly $120 million (£77 million) was anonymously donated via the Donors Trust and Donors Capital Fund to more than 100 organisations seeking to undermine the public perception of the science on climate change. In 2013 the Center for Media and Democracy reported that the State Policy Network (SPN), an umbrella group of 64 U.S. think tanks, had been lobbying on behalf of major corporations and conservative donors to oppose climate change regulation.

Since the late 1970s, oil companies have published research broadly in line with the standard views on global warming. Despite this, oil companies organized a climate change denial campaign to disseminate public disinformation for several decades, leading to comparisons of this strategy to the organized denial of the hazards of tobacco smoking by tobacco companies.

Terminology

"Climate change skepticism" and "climate change denial" refer to denial, dismissal or unwarranted doubt of the scientific consensus on the rate and extent of global warming, its significance, or its connection to human behavior, in whole or in part. Though there is a distinction between skepticism which indicates doubting the truth of an assertion and outright denial of the truth of an assertion, in the public debate phrases such as "climate scepticism" have frequently been used with the same meaning as climate denialism or contrarianism.

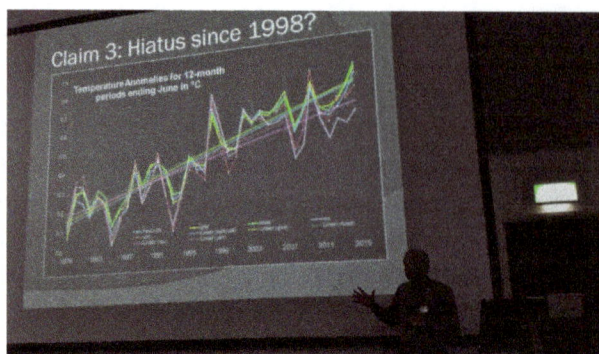

Amardeo Sarma lecturing about climate change denialism and the future world energy and environmental problems during the European Skeptics Congress 2015

The terminology emerged in the 1990s. Even though all scientists adhere to scientific skepticism as an inherent part of the process, by mid November 1995 the word "skeptic" was being used specifically for the minority who publicised views contrary to the scientific consensus. This small group of scientists presented their views in public statements and the media, rather than to the scientific community. This usage continued. In his December 1995 article *The Heat is On: The warming of the world's climate sparks a blaze of denial* , Ross Gelbspan said industry had engaged "a small band of skeptics" to confuse public opinion in a "persistent and well-funded campaign of denial". His 1997 book *The Heat is On* may have been the first to concentrate specifically on the topic. In it, Gelbspan discussed a "pervasive denial of global warming" in a "persistent campaign of denial and suppression" involving "undisclosed funding of these 'greenhouse skeptics' " with "the climate skeptics" confusing the public and influencing decision makers. A November 2006 CBC Television documentary on the campaign was titled "The Denial Machine". In 2007 journalist Sharon Begley reported on the "denial machine", a phrase subsequently used by academics.

In addition to *explicit denial*, social groups have shown *implicit denial* by accepting the scientific consensus, but failing to come to terms with its implications or take action to reduce the problem. This was exemplified in Kari Norgaard's study of a village in Norway affected by climate change, where residents diverted their attention to other issues.

The terminology is debated: most of those actively rejecting the scientific consensus use the terms *skeptic* and *climate change skepticism*, and only a few have expressed preference for being described as deniers, but the word "skepticism" is incorrectly used, as scientific skepticism is an intrinsic part of scientific methodology. The term *contrarian* is more specific, but used less frequently. In academic literature and journalism, the terms *climate change denial* and *climate change deniers* have well established usage as descriptive terms without any pejorative intent. Both the National Center for Science Education and historian Spencer R. Weart recognise that either option is problematic, but have decided to use "climate change denial" rather than "skepticism".

Terms related to *denialism* have been criticised for introducing a moralistic tone, and potentially implying a link with Holocaust denial. There have been claims that this link is intentional, which academics have strongly disputed. The usage of "denial" long predates the Holocaust, and is commonly applied in other areas such as HIV/AIDS denialism: the claim is described by John Timmer of *Ars Technica* as itself being a form of denial.

In December 2014, an open letter from the Committee for Skeptical Inquiry called on the media to stop using the term "skepticism" when referring to climate change denial. They contrasted scientific skepticism–which is "foundational to the scientific method"–with denial–"the a priori rejection of ideas without objective consideration", and the behavior of those involved in political attempts to undermine climate science. They said "Not all individuals who call themselves climate change skeptics are deniers. But virtually all deniers have falsely branded themselves as skeptics. By perpetrating this misnomer, journalists have granted undeserved credibility to those who reject science and scientific inquiry." The letter was taken up by the advocacy group Face the Facts as the basis for an online petition to news media. In June 2015 Media Matters for America were told by the *New York Times* Public Editor that the newspaper was increasingly tending to use "denier" when "someone is challenging established science", but assessing this on an individual basis with no fixed policy, and would not use the term when someone was "kind of wishy-washy on the subject or in the middle." The executive director of the Society of Environmental Journalists said that while there was reasonable skepticism about specific issues, she felt that denier was "the most accurate term when someone claims there is no such thing as global warming, or agrees that it exists but denies that it has any cause we could understand or any impact that could be measured."

History

Joseph Fourier is credited with first discovering the greenhouse effect in 1824, beginning scientific research into the effects of increased greenhouse gasses in the atmosphere.

Research on the effect of CO_2 on the climate began in the 19th century; Joseph Fourier discovered the atmospheric "greenhouse effect" in 1824, and in 1860 John Tyndall quantified the effect of each gas. This potential explanation of ice ages was investigated by Svante Arrhenius, who published research in 1896 showing that a geometric increase in CO_2 would cause an arithmetical increase in temperatures. He suggested that coal burning could cause the effect, and in a 1938 article Guy Stewart Callendar presented evidence that it was already happening. Both viewed this as a benign possibility.

Military services in the 1940s and 1950s supported scientific research into the environment. They were primarily interested in the operational data and potential for warfare, but were also open to the academic scientific discoveries as a result. For example, Gilbert Plass worked on radiation transmission through the atmosphere for weapons systems, and "in the evening" wrote papers giving new impetus to greenhouse effect theory. Oceanographer Roger Revelle played a key role; his 1957 paper co-authored with Hans Suess overturned the presumption that oceans would quickly absorb increased CO_2, and has been described as "the opening shot in the global warming debates". Revelle was quick to inform both the public and government officials of the risks, spreading his theme that "In consuming our fossil fuels at a prodigious rate, our civilization is conducting a grandiose scientific experiment." In the following decades, public and scientific concerns about this and other environmental issues increased. More research was needed, and was taken on by new agencies including NASA and NOAA, but funding became sporadic. The 1979 Charney Report reviewed the state of climate research, concluding that substantial warming was already on the way, and "the ocean, the great and ponderous flywheel of the global climate system, may be expected to slow the course of observable climatic change. A wait-and-see policy may mean waiting until it is too late."

A conservative reaction built up, denying environmental concerns which could lead to government regulation. With the 1981 Presidency of Ronald Reagan, global warming became a political issue, with immediate plans to cut spending on environmental research, particularly climate related, and stop funding for CO_2 monitoring. Reagan appointed as Secretary of Energy James B. Edwards, who said that there was no real global warming problem. Congressman Al Gore had studied under Revelle and was aware of the developing science: he joined others in arranging congressional hearings from 1981 onwards, with testimony by scientists including Revelle, Stephen Schneider and Wallace Smith Broecker. The hearings gained enough public attention to reduce the cuts in atmospheric research. A polarized party-political debate developed. In 1982 Sherwood B. Idso published his book *Carbon Dioxide: Friend or Foe?* which said increases in CO_2 would not warm the planet, but would fertilize crops and were "something to be encouraged and not suppressed", while complaining that his theories had been rejected by the "scientific establishment". An Environmental Protection Agency (EPA) report in 1983 said global warming was "not a theoretical problem but a threat whose effects will be felt within a few years", with potentially "catastrophic" consequences. The Reagan administration reacted by calling the report "alarmist", and the dispute got wide news coverage. Public attention turned to other issues, then the 1985 finding of a polar ozone hole brought a swift international response. To the public, this was related to climate change and the possibility of effective action, but news interest faded.

Public attention was renewed amidst summer droughts and heat waves when James Hansen testified to a Congressional hearing on 23 June 1988, stating with high confidence that long term warming was under way with severe warming likely within the next 50 years, and warning of likely storms and floods. There was increasing media attention: the scientific community had reached a broad consensus that the climate was warming, human activity was very likely the primary cause, and there would be significant consequences if the warming trend was not curbed. These facts encouraged discussion about new laws concerning environmental regulation, which was opposed by the fossil fuel industry.

From 1989 onwards industry funded organisations including the Global Climate Coalition and the George C. Marshall Institute sought to spread doubt among the public, in a strategy already developed by the tobacco industry. A small group of scientists opposed to the consensus on global warming became politically involved, and with support from conservative political interests, began publishing in books and the press rather than in scientific journals. Spencer Weart identifies this period as the point where legitimate skepticism about basic aspects of climate science was no longer justified, and those spreading mistrust about these issues became deniers. As their arguments were increasingly refuted by the scientific community and new data, deniers turned to political arguments, making personal attacks on the reputation of scientists, and promoting ideas of a global warming conspiracy.

With the 1989 fall of communism and the environmental movement's international reach at the 1992 Rio Earth Summit, the attention of U.S. conservative think tanks, which had been organised in the 1970s as an intellectual counter-movement to socialism, turned from the "red scare" to the "green scare" which they saw as a threat to their aims of private property, free trade market economies and global capitalism. As a counter-movement, they used environmental skepticism to promote denial of the reality of problems such as loss of biodiversity and climate change.

In 1992, an EPA report linked second-hand smoke with lung cancer. The tobacco industry engaged the APCO Worldwide public relations company, which set out a strategy of astroturfing campaigns to cast doubt on the science by linking smoking anxieties with other issues, including global warming, in order to turn public opinion against calls for government intervention. The campaign depicted public concerns as "unfounded fears" supposedly based only on "junk science" in contrast to their "sound science", and operated through front groups, primarily the Advancement of Sound Science Center (TASSC) and its Junk Science website, run by Steven Milloy. A tobacco company memo commented "Doubt is our product since it is the best means of competing with the 'body of fact' that exists in the mind of the general public. It is also the means of establishing a controversy." During the 1990s, the tobacco campaign died away, and TASSC began taking funding from oil companies including Exxon. Its website became central in distributing "almost every kind of climate-change denial that has found its way into the popular press."

In the 1990s, the Marshall Institute began campaigning against increased regulations on environmental issues such as acid rain, ozone depletion, second-hand smoke, and the dangers of DDT. In each case their argument was that the science was too uncertain to justify any government intervention, a strategy it borrowed from earlier efforts to downplay the health effects of tobacco in the 1980s. This campaign would continue for the next two decades.

These efforts succeeded in influencing public perception of climate science. Between 1988 and the 1990s, public discourse shifted from the science and data of climate change to discussion of politics and surrounding controversy.

The campaign to spread doubt continued into the 1990s, including an advertising campaign funded by coal industry advocates intended to "reposition global warming as theory rather than fact," and a 1998 proposal written by the American Petroleum Institute intending to recruit scientists to convince politicians, the media and the public that climate science was too uncertain to warrant environmental regulation. The proposal included a US$ 5,000,000 multi-point strategy to "maximize the impact of scientific views consistent with ours on Congress, the media and other key audiences", with a goal of "raising questions about and undercutting the 'prevailing scientific wisdom'".

In 1998, Gelbspan noted that his fellow journalists accepted that global warming was occurring, but said they were in "'stage-two' denial of the climate crisis", unable to accept the feasibility of answers to the problem. A subsequent book by Milburn and Conrad on *The Politics of Denial* described "economic and psychological forces" producing denial of the consensus on global warming issues.

These efforts by climate change denial groups were recognized as an organized campaign beginning in the 2000s. Riley Dunlap and Aaron McCright played a significant role in this shift when they published an article in 2000 exploring the connection between conservative think tanks and climate change denial.

Gelbspan's *Boiling Point*, published in 2004, detailed the fossil-fuel industry's campaign to deny climate change and undermine public confidence in climate science. In *Newsweek*'s August 2007 cover story "The Truth About Denial", Sharon Begley reported that "the denial machine is running at full throttle", and said that this "well-coordinated, well-funded campaign" by contrarian scientists, free-market think tanks, and industry had "created a paralyzing fog of doubt around climate change."

Referencing work of sociologists Robert Antonio and Robert Brulle, Wayne A. White has written that climate change denial has become the top priority in a broader agenda against environmental regulation being pursued by neoliberals. Today, climate change skepticism is most prominently seen in the United States, where the media disproportionately features views of the climate change denial community. In addition to the media, the contrarian movement has also been sustained by the growth of the internet, having gained some of its support from internet bloggers, talk radio hosts and newspaper columnists.

The New York Times and others reported in 2015 that oil companies knew that burning oil and gas could cause climate change and global warming since the 1970s but nonetheless funded deniers for years. Dana Nuccitelli wrote in *The Guardian* that a small fringe group of climate deniers were no longer taken seriously at the 2015 United Nations Climate Change Conference, in an agreement that "we need to stop delaying and start getting serious about preventing a climate crisis." However the New York Times says any implementation is voluntary and will depend on any future world leaders—and every Republican candidate in 2016 has questioned or denied the science of climate change.

Denial Networks

A Pentagon report has pointed out how climate denial threatens national security.

A study from 2015 identified 4,556 individuals with overlapping network ties to 164 organizations which are responsible for the most efforts to downplay the threat of climate change in the U.S.

Arguments And Positions on Global Warming

Some climate change denial groups allege that CO_2 is only a trace gas in the atmosphere, and has little effect on the climate. The scientific consensus, as summarized by the IPCC's fourth assessment report, the U.S. Geological Survey, and other reports, is that human activity is the leading cause of climate change. The burning of fossil fuels accounts for around 30 billion tons of CO_2 each year, which is 130 times the amount produced by volcanoes. Some groups allege that water vapor is a more significant greenhouse gas, and is left out of many climate models. However, water vapor has been incorporated into these models since the inception of climatology in the 1800s, and while it is also a greenhouse gas, CO_2 remains the primary driver of increasing temperatures.

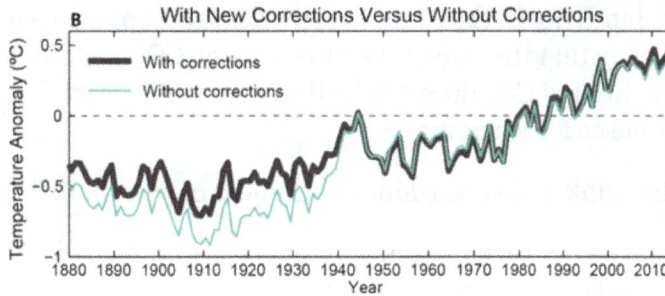

One argument is that global warming has recently stopped. However, temperature anomalies in an updated NOAA dataset show no evidence of a recent hiatus.

Climate denial groups may also argue that global warming stopped recently, a global warming hiatus, or that global temperatures are actually decreasing, leading to global cooling.

These groups often point to natural variability, such as sunspots and cosmic rays, to explain the warming trend. According to these groups, there is natural variability that will abate over time, and human influences have little to do with it. These factors are already taken into account when developing climate models, and the scientific consensus is that they cannot explain the recent warming trend.

Global warming conspiracy theories have been posited which allege that the scientific consensus is illusory, or that climatologists are acting on their own financial interests by causing undue alarm about a changing climate. Despite leaked emails during climategate, as well as multinational, independent research on the topic, no evidence of such a conspiracy has been presented, and strong consensus exists among scientists from a multitude of political, social, organizational and national backgrounds about the extent and cause of climate change. Several researchers have concluded that around 97% of climate scientists agree with this consensus. As well, much of the data used in climate science is publicly available to be viewed and interpreted by competing researchers as well as the public.

In 2012, research by Stephan Lewandowsky (then of the University of Western Australia) concluded that belief in other conspiracy theories, such as that the FBI was responsible for the assassination of Martin Luther King, Jr., was associated with being more likely to endorse climate change denial.

Climate change denial literature often features the suggestion that we should wait for better technologies before addressing climate change, when they will be more affordable and effective.

Taxonomy of Climate Change Denial

In 2004 Stefan Rahmstorf described how the media give the misleading impression that climate change was still disputed within the scientific community, attributing this impression to PR efforts of climate change skeptics. He identified different positions argued by climate skeptics, which he used as a taxonomy of climate change skepticism:

1. Trend sceptics (who deny there is global warming), [and] argue that no significant climate warming is taking place at all, claiming that the warming trend measured by weather stations is an artefact due to urbanisation around those stations ("urban heat island effect").

2. Attribution sceptics, [and] doubt that human activities are responsible for the observed trends. A few of them even deny that the rise in the atmospheric CO_2 content is anthropogenic [while others argue that] additional CO_2 does not lead to discernible warming [and] that there must be other—natural—causes for warming.

3. Impact sceptics (who think global warming is harmless or even beneficial).

This taxonomy has been used in social science for analysis of publications, and to categorize climate change skepticism and climate change denial.

The National Center for Science Education describes climate change denial as disputing differing points in the scientific consensus, a sequential range of arguments from denying the occurrence of climate change, accepting that but denying any significant human contribution, accepting these but denying scientific findings on how this would affect nature and human society, to accepting all these but denying that humans can mitigate or reduce the problems. James L. Powell provides a more extended list, as does climatologist Michael E. Mann in "six stages of denial", a ladder in which deniers have over time conceded acceptance of points, while retreating to a position which still rejects the mainstream consensus:

1. CO_2 is not actually increasing.

2. Even if it is, the increase has no impact on the climate since there is no convincing evidence of warming.

3. Even if there is warming, it is due to natural causes.

4. Even if the warming cannot be explained by natural causes, the human impact is small, and the impact of continued greenhouse gas emissions will be minor.

5. Even if the current and future projected human effects on Earth's climate are not negligible, the changes are generally going to be good for us.

6. Whether or not the changes are going to be good for us, humans are very adept at adapting to changes; besides, it's too late to do anything about it, and/or a technological fix is bound to come along when we really need it.

Journalists and newspaper columnists including George Monbiot and Ellen Goodman, among others, have described climate change denial as a form of denialism.

Denialism in this context has been defined by Chris and Mark Hoofnagle as the use of rhetorical devices "to give the appearance of legitimate debate where there is none, an approach that has the ultimate goal of rejecting a proposition on which a scientific consensus exists." This process characteristically uses one or more of the following tactics:

1. Allegations that scientific consensus involves conspiring to fake data or suppress the truth: a global warming conspiracy theory.

2. Fake experts, or individuals with views at odds with established knowledge, at the same time marginalising or denigrating published topic experts. Like the manufactured doubt over smoking and health, a few contrarian scientists oppose the climate consensus, some of them the same individuals.

3. Selectivity, such as cherry picking atypical or even obsolete papers, in the same way that the MMR vaccine controversy was based on one paper: examples include discredited ideas of the medieval warm period.

4. Unworkable demands of research, claiming that any uncertainty invalidates the field or exaggerating uncertainty while rejecting probabilities and mathematical models.

5. Logical fallacies.

Pseudoscience

Various groups, including the National Center for Science Education, have described climate change denial as a form of pseudoscience. Climate change skepticism, while in some cases professing to do research on climate change, has focused instead on influencing the opinion of the public, legislators and the media, in contrast to legitimate science.

In a review of the book *The Pseudoscience Wars: Immanuel Velikovsky and the Birth of the Modern Fringe* by Michael D. Gordin, David Morrison wrote:

In his final chapter, Gordin turns to the new phase of pseudoscience, practiced by a few rogue scientists themselves. Climate change denialism is the prime example, where a handful of scientists, allied with an effective PR machine, are publicly challenging the scientific consensus that global warming is real and is due primarily to human consumption of fossil fuels. Scientists have watched in disbelief that as the evidence for global warming has become ever more solid, the deniers have been increasingly successful in the public and political arena. ... Today pseudoscience is still with us, and is as dangerous a challenge to science as it ever was in the past.

Public Opinion

Public opinion on climate change is significantly impacted by media coverage of climate change, and the effects of climate change denial campaigns. Campaigns to undermine public confidence in climate science have decreased public belief in climate change, which in turn have impacted legislative efforts to curb CO_2 emissions.

The popular media in the U.S. gives greater attention to climate change skeptics than the scientific community as a whole, and the level of agreement within the scientific community has not been accurately communicated. In some cases, news outlets have allowed climate change skeptics to explain the science of climate change instead of experts in climatology. US and UK media coverage differ from that presented in other countries, where reporting is more consistent with the scientific literature. Some journalists attribute the difference to climate change denial being propagated, mainly in the US, by business-centered organizations employing tactics worked out previously by the US tobacco lobby. In France, the US and the UK, the opinions of climate change skeptics appear much more frequently in conservative news outlets than other news, and in many cases those opinions are left uncontested.

The efforts of Al Gore and other environmental campaigns have focused on the effects of global warming and have managed to increase awareness and concern, but despite these efforts, the number of Americans believing humans are the cause of global warming was holding steady at 61% in 2007, and those believing the popular media was understating the issue remained about 35%. A recent poll from 2015 suggests that while Americans are growing more aware of the dangers and implications of climate change for future generations, the majority are not worried about it.

A study assessed the public perception and actions to climate change, on grounds of belief systems, and identified seven psychological barriers affecting the behavior that otherwise would facilitate mitigation, adaptation, and environmental stewardship. The author found the following barriers: cognition, ideological world views, comparisons to key people, costs and momentum, discredence toward experts and authorities, perceived risks of change, and inadequate behavioral changes.

Lobbying

Efforts to lobby against environmental regulation have included campaigns to manufacture doubt about the science behind climate change, and to obscure the scientific consensus and data. These efforts have undermined public confidence in climate science, and impacted climate change lobbying.

The political advocacy organizations FreedomWorks and Americans for Prosperity, funded by brothers David and Charles Koch of Koch Industries, were important in supporting the Tea Party movement and in encouraging the movement to focus on climate change. Other conservative organizations such as the Heritage Foundation, Marshall Institute, Cato Institute and the American Enterprise Institute were significant participants in these lobbying attempts, seeking to halt or eliminate environmental regulations.

This approach to downplay the significance of climate change were copied from tobacco lobbyists; in the face of scientific evidence linking tobacco to lung cancer, to prevent or delay the introduction of regulation. Lobbyists attempted to discredit the scientific research by creating doubt and manipulating debate. They worked to discredit the scientists involved, to dispute their findings, and to create and maintain an apparent controversy by promoting claims that contradicted scientific research. ""Doubt is our product," boasted a now infamous 1969 industry memo. Doubt would shield the tobacco industry from litigation and regulation for decades to come." In 2006, George Monbiot wrote in *The Guardian* about similarities between the methods of groups funded by Exxon, and those of the tobacco giant Philip Morris, including direct attacks on peer-reviewed science, and attempts to create public controversy and doubt.

Former National Academy of Sciences president Frederick Seitz, who, according to an article by Mark Hertsgaard in *Vanity Fair*, earned about US$585,000 in the 1970s and 1980s as a consultant to R.J. Reynolds Tobacco Company, went on to chair groups such as the Science and Environmental Policy Project and the George C. Marshall Institute alleged to have made efforts to "downplay" global warming. Seitz stated in the 1980s that "Global warming is far more a matter of politics than of climate." Seitz authored the Oregon Petition, a document published jointly by the Marshall Institute and Oregon Institute of Science and Medicine in opposition to the Kyoto protocol. The petition and accompanying "Research Review of Global Warming Evidence" claimed:

The proposed limits on greenhouse gases would harm the environment, hinder the advance of science and technology, and damage the health and welfare of mankind. There is no convincing scientific evidence that human release of carbon dioxide, methane, or other greenhouse gases is causing or will, in the foreseeable future, cause catastrophic heating of the Earth's atmosphere and disruption of the Earth's climate. ... We are living in an increasingly lush environment of plants and animals as a result of the carbon dioxide increase. Our children will enjoy an Earth with far more plant and animal life than that with which we now are blessed. This is a wonderful and unexpected gift from the Industrial Revolution.

George Monbiot wrote in *The Guardian* that this petition, which he criticizes as misleading and tied to industry funding, "has been cited by almost every journalist who claims that climate change is a myth." Efforts by climate change denial groups played a significant role in the eventual rejection of the Kyoto protocol in the US.

Monbiot has written about another group founded by the tobacco lobby, The Advancement of Sound Science Coalition (TASSC), that now campaigns against measures to combat global warming. In again trying to manufacture the appearance of a grass-roots movement against "unfounded fear" and "over-regulation," Monbiot states that TASSC "has done more damage to the campaign to halt [climate change] than any other body."

Drexel University environmental sociologist Robert Brulle analysed the funding of 91 organizations opposed to restrictions on carbon emissions, which he termed the "climate change counter-movement." Between 2003 and 2013, the donor-advised funds Donors Trust and Donors Capital Fund, combined, were the largest funders, accounting for about one quarter of the total funds, and the American Enterprise Institute was the largest recipient, 16% of the total funds. The study also found that the amount of money donated to these organizations by means of foundations whose funding sources cannot be traced had risen.

Private Sector

Several large corporations within the fossil fuel industry provide significant funding to the climate change denial movement. ExxonMobil and the Koch family foundations have been identified as especially influential funders of climate change contrarianism.

After the IPCC released its February 2007 report, the American Enterprise Institute offered British, American and other scientists $10,000, plus travel expenses to publish articles critical of the assessment. The institute had received more than $US 1.6 million from Exxon, and its vice-chairman of trustees was former head of Exxon Lee Raymond. Raymond sent letters that alleged the IPCC report was not "supported by the analytical work." More than 20 AEI employees worked as consultants to the George W. Bush administration. Despite her initial conviction that climate change denial would abate with time, Senator Barbara Boxer said that when she learned of the AEI's offer, she "realized there was a movement behind this that just wasn't giving up."

The Royal Society conducted a survey that found ExxonMobil had given US$ 2.9 million to American groups that "misinformed the public about climate change," 39 of which "misrepresented the science of climate change by outright denial of the evidence". In 2006, the Royal Society issued a demand that ExxonMobil withdraw funding for climate change denial. The letter drew criticism, notably from Timothy Ball who argued the society attempted to "politicize the private funding of science and to censor scientific debate."

ExxonMobil denied that it has been trying to mislead the public about global warming. A spokesman, Gantt Walton, said that ExxonMobil's funding of research does not mean that it acts to influence the research, and that ExxonMobil supports taking action to curb the output of greenhouse gasses. Research conducted at an Exxon archival collection at the University of Texas and interviews with former employees by journalists indicate the scientific opinion within the company and their public posture towards climate change was contradictory.

Between 1989 and 2002 the Global Climate Coalition, a group of mainly United States businesses, used aggressive lobbying and public relations tactics to oppose action to reduce greenhouse gas emissions and fight the Kyoto Protocol. The coalition was financed by large corporations and trade groups from the oil, coal and auto industries. The *New York Times* reported that "even as the coalition worked to sway opinion [towards skepticism], its own scientific and technical experts were advising that the science backing the role of greenhouse gases in global warming could not be refuted." In 2000, Ford Motor Company was the first company to leave the coalition as a result of pressure from environmentalists, followed by Daimler-Chrysler, Texaco, the Southern Company and General Motors subsequently left to GCC. The organization closed in 2002.

In early 2015, several media reports emerged saying that Willie Soon, a popular scientist among climate change deniers, had failed to disclose conflicts of interest in at least 11 scientific papers published since 2008. They reported that he received a total of $1.25m from ExxonMobil, Southern Company, the American Petroleum Institute and a foundation run by the Koch brothers. Charles R. Alcock, director of the Harvard–Smithsonian Center for Astrophysics, where Soon was based, said that allowing funders of Dr. Soon's work to prohibit disclosure of funding sources was a mistake, which will not be permitted in future grant agreements.

Public Sector

In 1994, according to a leaked memo, the Republican strategist Frank Luntz advised members of the Republican Party, with regard to climate change, that "you need to continue to make the lack of scientific certainty a primary issue" and "challenge the science" by "recruiting experts who are sympathetic to your view." In 2006, Luntz stated that he still believes "back [in] '97, '98, the science was uncertain", but he now agrees with the scientific consensus.

In 2005, the *New York Times* reported that Philip Cooney, former fossil fuel lobbyist and "climate team leader" at the American Petroleum Institute and President George W. Bush's chief of staff of the Council on Environmental Quality, had "repeatedly edited government climate reports in ways that play down links between such emissions and global warming, according to internal documents." Sharon Begley reported in *Newsweek* that Cooney "edited a 2002 report on climate science by sprinkling it with phrases such as 'lack of understanding' and 'considerable uncertainty.'" Cooney reportedly removed an entire section on climate in one report, whereupon another lobbyist sent him a fax saying "You are doing a great job." Cooney announced his resignation two days after the story of his tampering with scientific reports broke, but a few days later it was announced that Cooney would take up a position with ExxonMobil.

In 2015, environmentalist Bill McKibben accused President Obama of "Catastrophic Climate-Change Denial", for his approval of oil-drilling permits in offshore Alaska. According to McKibben, the President has also "opened huge swaths of the Powder River basin to new coal mining." McKibben calls this "climate denial of the status quo sort", where the President denies "the meaning of the science, which is that we must keep carbon in the ground."

Schools

According to documents leaked in February, 2012, The Heartland Institute is developing a curriculum for use in schools which frames climate change as a scientific controversy.

Effect

Manufactured uncertainty over climate change, the fundamental strategy of climate change denial, has been very effective, particularly in the US. It has contributed to low levels of public concern and to government inaction worldwide. An Angus Reid poll released in 2010 indicates that global warming skepticism in the United States, Canada, and the United Kingdom has been rising. There may be multiple causes of this trend, including a focus on economic rather than environmental issues, and a negative perception of the United Nations and its role in discussing climate change. Another cause may be weariness from overexposure to the topic: secondary polls suggest that the public may have been discouraged by extremism when discussing the topic, while other polls show 54% of U.S. voters believe that "the news media make global warming appear worse than it really is." A poll in 2009 regarding the issue of whether "some scientists have falsified research data to support their own theories and beliefs about global warming" showed that 59% of Americans believed it "at least somewhat likely", with 35% believing it was "very likely".

According to Tim Wirth, "They patterned what they did after the tobacco industry. [...] Both figured, sow enough doubt, call the science uncertain and in dispute. That's had a huge impact on both the public and Congress." This approach has been propagated by the US media, presenting a false balance between climate science and climate skeptics. *Newsweek* reports that the majority of Europe and Japan accept the consensus on scientific climate change, but only one third of Americans considered human activity to play a major role in climate change in 2006; 64% believed that scientists disagreed about it "a lot." A 2007 *Newsweek* poll found these numbers were declining, although majorities of Americans still believed that scientists were uncertain about climate change and its causes. Rush Holt wrote a piece for *Science*, which appeared in *Newsweek*:

...for more than two decades scientists have been issuing warnings that the release of greenhouse gases, principally carbon dioxide (CO_2), is probably altering Earth's climate in ways that will be expensive and even deadly. The American public yawned and bought bigger cars. Statements by the American Association for the Advancement of Science, American Geophysical Union, American Meteorological Society, Intergovernmental Panel on Climate Change, and others underscored the warnings and called for new government policies to deal with climate change. Politicians, presented with noisy statistics, shrugged, said there is too much doubt among scientists, and did nothing.

Deliberate attempts by the Western Fuels Association "to confuse the public" have succeeded in their objectives. This has been "exacerbated by media treatment of the climate issue". According to a Pew poll in 2012, 57% of the US public are unaware of, or outright reject, the scientific consensus on climate change. Some organizations promoting climate change denial have asserted that scientists are increasingly rejecting climate change, but this notion is contradicted by research showing that 97% of published papers endorse the scientific consensus, and that percentage is increasing with time.

Global oil companies have begun to acknowledge climate change exists and is caused by human activities and the burning of fossil fuels.

Milankovitch Cycles

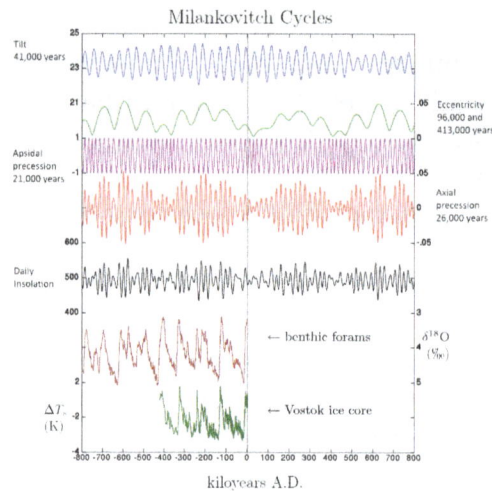

Past and future Milankovitch cycles. VSOP allows prediction of past and future orbital parameters with great accuracy.

Shows variations in: obliquity (axial tilt) in blue (ε).

eccentricity in green (e).

longitude of perihelion in purple (ϖ).

Precession index in dark red ($e \sin(\varpi)$), which together with obliquity, controls the seasonal cycle of insolation.

– Calculated daily-averaged insolation at the top of the atmosphere in black day on the day of the summer solstice at 65° N latitude.

– *Benthic forams* in dark red and – *Vostok ice core* in dark green show two distinct proxies for past global sea level and temperature, from ocean sediment and Antarctic ice respectively.

The vertical gray line shows current conditions, at 2 ky A.D.

Milankovitch cycles describes the collective effects of changes in the Earth's movements upon its climate, named after Serbian geophysicist and astronomer Milutin Milanković, who in the 1920s had theorized that variations in eccentricity, axial tilt, and precession of the Earth's orbit determined climatic patterns on Earth through orbital forcing.

The Earth's axis completes one full cycle of precession approximately every 26,000 years. At the same time, the elliptical orbit rotates more slowly. The combined effect of the two precessions leads to a 21,000-year period between the astronomical seasons and the orbit. In addition, the angle between Earth's rotational axis and the normal to the plane of its orbit (obliquity) oscillates between 22.1 and 24.5 degrees on a 41,000-year cycle. It is currently 23.44 degrees and decreasing at a rate of about 0.013° per century.

Similar astronomical theories had been advanced in the 19th century by Joseph Adhemar, James Croll and others, but verification was difficult due to the absence of reliably dated evidence and doubts as to exactly which periods were important. Not until the advent of deep-ocean cores and a seminal paper by Hays, Imbrie, and Shackleton, "Variations in the Earth's Orbit: Pacemaker of the Ice Ages", in *Science* (1976) did the theory attain its present state.

Earth's Movements

As the Earth rotates around its axis and revolves around the Sun, several quasi-periodic variations occur due to gravitational interactions. Although the curves have a large number of sinusoidal components, a few components are dominant. Milankovitch studied changes in the orbital eccentricity, obliquity, and precession of Earth's movements. Such changes in movement and orientation alter the amount and location of solar radiation reaching the Earth. This is known as *solar forcing* (an example of radiative forcing). Changes near the north polar area, about 65 degrees North, are considered important due to the great amount of land. Land masses respond to temperature change more quickly than oceans, which have a higher effective heat capacity, because of the mixing of surface and deep water and the fact that the specific heat of solids is generally lower than that of water.

Orbital Shape (Eccentricity)

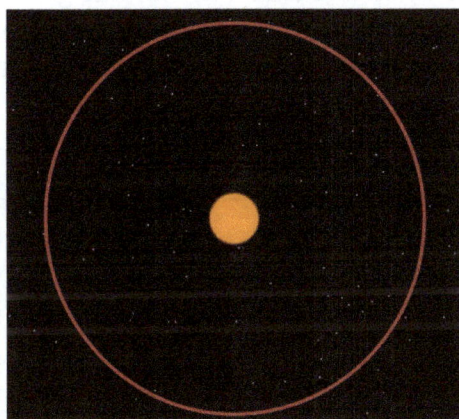

Circular orbit, no eccentricity

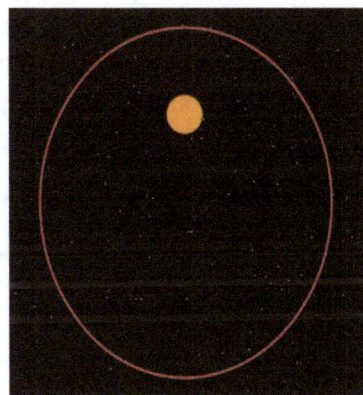

Orbit with 0.5 eccentricity; Earth's orbit is never this eccentric

The Earth's orbit is an ellipse. The eccentricity is a measure of the departure of this ellipse from circularity. The shape of the Earth's orbit varies in time between nearly circular (low eccentricity of 0.000055) and mildly elliptical (high eccentricity of 0.0679) with the mean eccentricity of 0.0019 as geometric or logarithmic mean. The major component of these variations occurs on a period of

413,000 years (eccentricity variation of ±0.012). A number of other terms vary between components 95,000 and 125,000 years (with a beat period 400,000 years), and loosely combine into a 100,000-year cycle (variation of −0.03 to +0.02). The present eccentricity is 0.017 and decreasing.

If the Earth were the only planet orbiting our Sun, the eccentricity of its orbit would not perceptibly vary even over a period of a million years. The Earth's eccentricity varies primarily due to interactions with the gravitational fields of Jupiter and Saturn. As the eccentricity of the orbit evolves, the semi-major axis of the orbital ellipse remains unchanged. From the perspective of the perturbation theory used in celestial mechanics to compute the evolution of the orbit, the semi-major axis is an adiabatic invariant. According to Kepler's third law the period of the orbit is determined by the semi-major axis. It follows that the Earth's orbital period, the length of a sidereal year, also remains unchanged as the orbit evolves.

Orbital Shape and Temperature

As the semi-minor axis is decreased with the eccentricity increase, the seasonal changes increase. But the mean solar irradiation for the planet changes only slightly for small eccentricity, due to Kepler's second law. (The intensity of seasons is not primarily governed by Earth's distance from the sun.)

The relative increase in solar irradiation at closest approach to the Sun (perihelion) compared to the irradiation at the furthest distance (aphelion) is slightly larger than four times the eccentricity. For the current orbital eccentricity this amounts to a variation in incoming solar radiation of about 6.8%, while the current difference between perihelion and aphelion is only 3.4% (5.1 million km). Perihelion presently occurs around January 3, while aphelion is around July 4. When the orbit is at its most elliptical, the amount of solar radiation at perihelion will be about 23% more than at aphelion.

The higher eccentricity also causes extra behavior due to precession and axial tilt. The true global summer does not appear to be the warmer part of the year in the North (until an eon from now when balanced). Seasons always arrive early, but not the same for differing seasons, since elliple motion (speed, angle) with respect to the sun differs per season arrival.

Season durations				
Year	**Northern Hemisphere**	**Southern Hemisphere**	**Date: GMT**	**Season duration**
2005	Winter solstice	Summer solstice	21 December 2005 18:35	88.99 days
2006	Spring equinox	Autumn equinox	20 March 2006 18:26	92.75 days
2006	Summer solstice	Winter solstice	21 June 2006 12:26	93.65 days
2006	Autumn equinox	Spring equinox	23 September 2006 4:03	89.85 days
2006	Winter solstice	Summer solstice	22 December 2006 0:22	88.99 days
2007	Spring equinox	Autumn equinox	21 March 2007 0:07	92.75 days

2007	Summer solstice	Winter solstice	21 June 2007 18:06	93.66 days
2007	Autumn equinox	Spring equinox	23 September 2007 9:51	89.85 days
2007	Winter solstice	Summer solstice	22 December 2007 06:08	

Orbital mechanics requires that the length of the seasons be proportional to the areas of the seasonal quadrants, so when the eccentricity is extreme, the Earth's orbital motion becomes more nonuniform and the lengths of the seasons change. When autumn and winter occur at closest approach, as is the case currently in the northern hemisphere, the earth is moving at its maximum velocity and therefore autumn and winter are slightly shorter than spring and summer. Thus, summer in the northern hemisphere is 4.66 days longer than winter and spring is 2.9 days longer than autumn. But as the orientation of Earth's orbit changes relative to the Vernal Equinox due to apsidal precession, the way the length of the seasons are altered by the nonuniform motion changes, since different sections of the orbit are involved. When the Earth's apsides are aligned with the equinoxes the length of Spring and Summer (together) equals that of Autumn and Winter. When they are aligned with the solstices either Spring and Summer or Autumn and Winter will be at its longest. Increasing the eccentricity lengthens the time spent near aphelion and shortens the time near perihelion.

Changes to the eccentricity do not by themselves change the length of the anomalistic year or the Earth's mean motion along its orbit since they are both functions of the semi-major axis.

Axial Tilt (Obliquity)

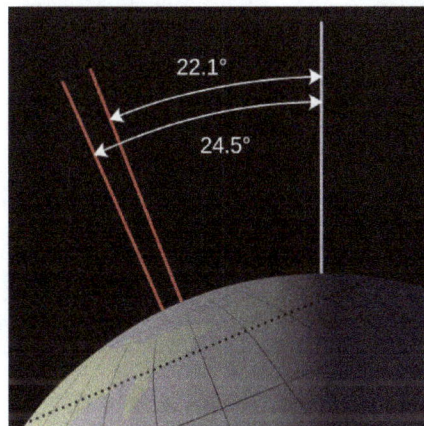

22.1–24.5° range of Earth's obliquity

The angle of the Earth's axial tilt (obliquity of the ecliptic) varies with respect to the plane of the Earth's orbit. These slow 2.4° obliquity variations are roughly periodic, taking approximately 41,000 years to shift between a tilt of 22.1° and 24.5° and back again. When the obliquity increases, the amplitude of the seasonal cycle in insolation increases, with summers in both hemispheres receiving more radiative flux from the Sun, and winters less. Conversely, when the obliquity decreases, summers receive less insolation and winters more.

But these changes of opposite sign in summer and winter are not of the same magnitude everywhere on the Earth's surface. At high latitude the annual mean insolation increases with increasing obliquity, while lower latitudes experience a reduction in insolation. Cooler summers are suspected of encouraging the onset of an ice age by melting less of the previous winter's precipitation. Because most of the planet's snow and ice lies at high latitude, it can be argued that lower obliquity favors ice ages for two reasons: the reduction in overall summer insolation and the additional reduction in mean insolation at high latitude.

Scientists using computer models to study more extreme tilts than those that actually occur have concluded that climate extremes at high obliquity would be particularly threatening to advanced forms of life that presently exist on Earth. They noted that high obliquity would not likely sterilize a planet completely, but would make it harder for fragile, warm-blooded land-based life to thrive as it does today.

Currently the Earth is tilted at 23.44 degrees from its orbital plane, roughly halfway between its extreme values. The tilt is in the decreasing phase of its cycle, and will reach its minimum value around the year 11,800 CE ; the last maximum was reached in 8,700 BCE. This trend in forcing, by itself, tends to make winters warmer and summers colder (i.e. milder seasons), as well as cause an overall cooling trend.

Axial Precession

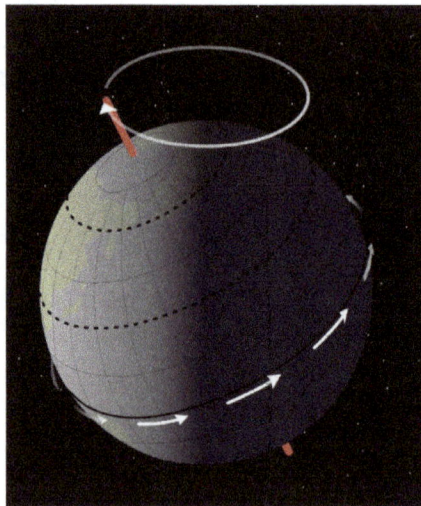

Precessional movement

Precession is the trend in the direction of the Earth's axis of rotation relative to the fixed stars, with a period of roughly 26,000 years. This gyroscopic motion is due to the tidal forces exerted by the Sun and the Moon on the solid Earth; both contribute roughly equally to this effect.

When the axis points toward the Sun in perihelion (i.e. the north pole is pointed towards the Sun), the northern hemisphere has a greater difference between the seasons while the southern hemisphere has milder seasons. When the axis points away from the Sun in perihelion (i.e. the south pole is pointed towards the Sun), the southern hemisphere has a greater difference between the seasons while the northern hemisphere has milder seasons. The hemisphere that is in summer at perihelion receives much of the corresponding increase in solar radiation, but that same hemisphere in winter at aphelion has a colder winter. The other hemisphere will have a relatively warmer winter and cooler summer.

When the Earth's axis is aligned such that aphelion and perihelion occur near the equinoxes, the northern and southern hemispheres will have similar contrasts in the seasons.

At present, perihelion occurs during the southern hemisphere's summer, and aphelion is reached during the southern winter. Thus the southern hemisphere seasons are somewhat more extreme than the northern hemisphere seasons, when other factors are equal.

Apsidal Precession

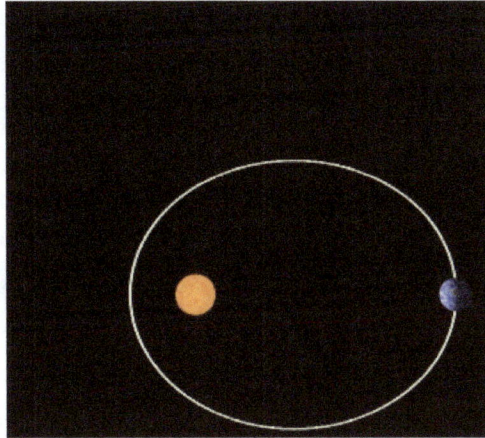

Planets orbiting the Sun follow elliptical (oval) orbits that rotate gradually over time (apsidal precession). The eccentricity of this ellipse is exaggerated for visualization. Most orbits in the Solar System have a much smaller eccentricity, making them nearly circular.

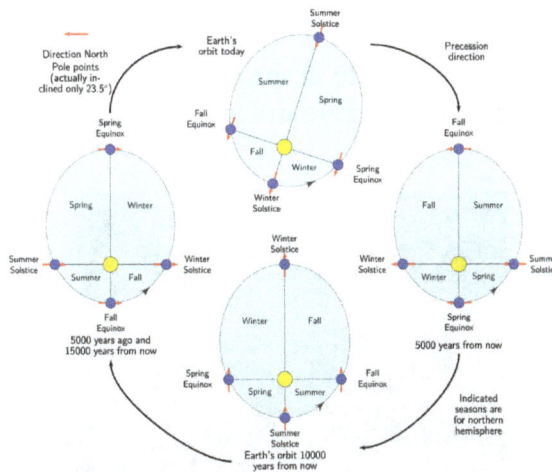

Effects of precession on the seasons (using the Northern Hemisphere terms).

In addition, the orbital ellipse itself precesses in space, in an irregular fashion, completing a full cycle every 112,000 years relative to the fixed stars. This happens primarily as a result of interactions with Jupiter and Saturn. Smaller contributions are also made by the sun's oblateness and by the effects of general relativity that are well known for Mercury. The total orbital precession is in the same sense to the gyroscopic motion of the axis of rotation, shortening the period of the precession of the equinoxes with respect to the perihelion from 25,771.5 to ~21,636 years. Apsidal precession occurs in the plane of the Ecliptic and alters the orientation of the Earth's orbit relative to the Ecliptic. In combination with changes to the eccentricity it alters the length of the seasons.

Orbital Inclination

The inclination of Earth's orbit drifts up and down relative to its present orbit. Milankovitch did not study this three-dimensional movement. This movement is known as "precession of the ecliptic" or "planetary precession".

More recent researchers noted this drift and that the orbit also moves relative to the orbits of the other planets. The invariable plane, the plane that represents the angular momentum of the Solar System, is approximately the orbital plane of Jupiter. The inclination of Earth's orbit drifts up and down relative to its present orbit with a cycle having a period of about 70,000 years. The inclination of the Earth's orbit has a 100,000-year cycle relative to the invariable plane. This is very similar to the 100,000-year eccentricity period. This 100,000-year cycle closely matches the 100,000-year pattern of glacial events.

It has been proposed that a disk of dust and other debris exists in the invariable plane, and this affects the Earth's climate through several possible means. The Earth presently moves through this plane around January 9 and July 9, when there is an increase in radar-detected meteors and meteor-related noctilucent clouds.

A study of the chronology of Antarctic ice cores using oxygen-nitrogen ratios in air bubbles trapped in the ice, which appear to respond directly to the local insolation, concluded that the climatic response documented in the ice cores was driven by northern hemisphere insolation as proposed by the Milankovitch hypothesis (Kawamura et al., Nature, 23 August 2007, vol 448, pp 912–917). This is an additional validation of the Milankovitch hypothesis by a relatively novel method.

Problems

Because the observed periodicities of climate fit so well with the orbital periods, the orbital theory has overwhelming support. Nonetheless, there are several difficulties in reconciling theory with observations.

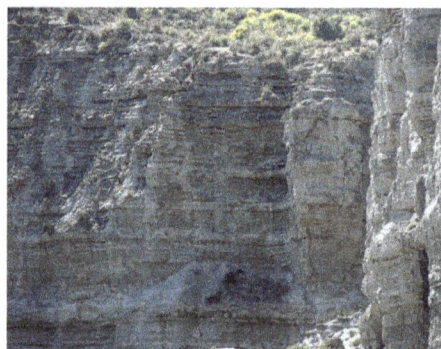

The nature of sediments can vary in a cyclic fashion, and these cycles can be displayed in the sedimentary record. Here, cycles can be observed in the colouration and resistance of different strata.

100,000-year Problem

The 100,000-year problem is that the eccentricity variations have a significantly smaller impact on solar forcing than precession or obliquity – according to theory – and hence might be expected to produce the weakest effects. However, the greatest observed response in regard to the ice ages is at the 100,000-year timescale, even though the theoretical forcing is smaller at this scale. During the last 1 million years, the strongest climate signal is the 100,000-year cycle. In addition, despite the relatively great 100,000-year cycle, some have argued that the length of the climate record is insufficient to establish a statistically significant relationship between climate and eccentricity variations. Various explanations for this discrepancy have been proposed, including frequency modulation or various feedbacks (from carbon dioxide, cosmic rays, or from ice sheet dynamics). Some models can reproduce the 100,000-year cycles as a result of non-linear interactions between small changes in the Earth's orbit and internal oscillations of the climate system.

Stage 5 problem

The stage 5 problem refers to the timing of the penultimate interglacial (in marine isotopic stage 5) that appears to have begun ten thousand years in advance of the solar forcing hypothesized to have caused it (also known as the causality problem, because the putative effect precedes the cause).

Effect Exceeds Cause

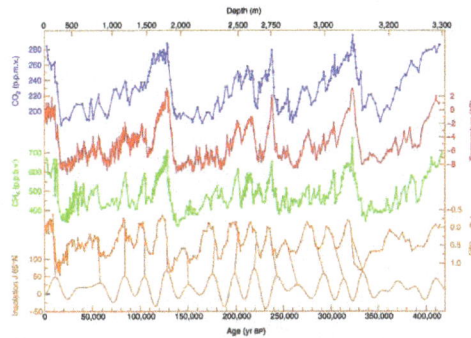

420,000 years of ice core data from Vostok, Antarctica research station

The effects of these variations are primarily believed to be due to variations in the intensity of solar radiation upon various parts of the globe. Observations show climate behavior is much more intense than the calculated variations. Various internal characteristics of climate systems are believed to be sensitive to the insolation changes, causing amplification (positive feedback) and damping responses (negative feedback).

The Unsplit Peak Problem

The unsplit peak problem refers to the fact that eccentricity has cleanly resolved variations at both the 95 and 125 ka periods. A sufficiently long, well-dated record of climate change should be able to resolve both frequencies. However, some researchers interpret climate records of the last million years as showing only a single spectral peak at 100 ka periodicity

The Transition Problem

The transition problem refers to the switch in the frequency of climate variations 1 million years ago. From 1–3 million years, climate had a dominant mode matching the 41 ka cycle in obliquity. After 1 million years ago, this switched to a 100 ka variation matching eccentricity, for which no reason has been established.

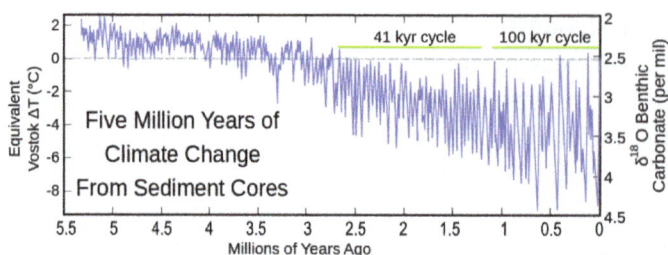

Variations of Cycle Times, curves determined from ocean sediments

Identifying Dominant Factor

Milankovitch believed that decreased summer insolation in northern high latitudes was the dominant factor leading to glaciation, which led him to (incorrectly) deduce an approximate 41 ka period for ice ages. Subsequent research has shown that ice age cycles of the Quaternary glaciation over the last million years have been at a 100,000-year period, leading to identification of the 100 ka eccentricity cycle as more important, although the exact mechanism remains obscure.

Present and Future Conditions

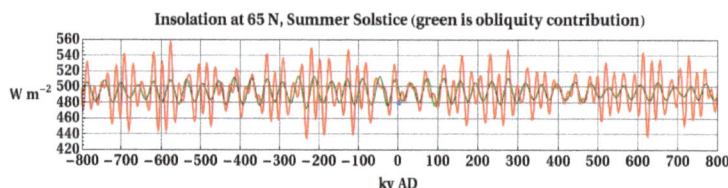

Past and future of daily average insolation at top of the atmosphere on the day of the summer solstice, at 65 N latitude. The green curve is with eccentricity *e* hypothetically set to 0. The red curve uses the actual (predicted) value of *e*. Blue dot is current conditions, at 2 ky A.D.

As mentioned above, at present, perihelion occurs during the southern hemisphere's summer and aphelion during the southern winter. Thus the southern hemisphere seasons should tend to be somewhat more extreme than the northern hemisphere seasons. The relatively low eccentricity of the present orbit results in a 6.8% difference in the amount of solar radiation during summer in the two hemispheres.

Since orbital variations are predictable, if one has a model that relates orbital variations to climate, it is possible to run such a model forward to "predict" future climate. Two caveats, however, are necessary: that anthropogenic effects may modify or even overwhelm orbital effects; and that the mechanism by which orbital forcing influences climate is not well understood. In the most prominent anthropogenic example, orbital forcing from the Milankovitch cycles has been in a cooling phase for millennia, but that cooling trend was reversed in the 20th and 21st centuries due to warming caused by increased anthropogenic greenhouse gas emissions.

The amount of solar radiation (insolation) in the Northern Hemisphere at 65° N seems to be related to occurrence of an ice age. Astronomical calculations show that 65° N summer insolation should increase gradually over the next 25,000 years. A regime of eccentricity lower than the current value will last for about the next 100,000 years. Changes in northern hemisphere summer insolation will be dominated by changes in obliquity ε. No declines in 65° N summer insolation, sufficient to cause a glacial period, are expected in the next 50,000 years.

An often-cited 1980 study by Imbrie and Imbrie determined that, "Ignoring anthropogenic and other possible sources of variation acting at frequencies higher than one cycle per 19,000 years, this model predicts that the long-term cooling trend that began some 6,000 years ago will continue for the next 23,000 years."

More recent work suggests that the current warm climate may last another 50,000 years.

Observed on Other Planets

Other planets in the Solar System have been discovered to have Milankovitch cycles. Mostly these cycles are not as intense or complex as the Earth's cycles, but they do have a global geological impact with respect to the movement of mobile solids such as water or nitrogen ices or hydrocarbon lakes. The known affected planets are:

Mars

Mars's polar caps vary in extent due to orbital instability related to a latent Milankovitch cycle.

Saturn

Saturn's moon Titan has a ~60,000-year cycle that changes the location of the methane lakes.

Neptune

Neptune's moon Triton has a similar variation to Titan with respect to migration of solid nitrogen deposits over long time scales.

References

- Norgaard, Kari (2011). Living in Denial: Climate Change, Emotions, and Everyday Life. Cambridge, Mass: MIT Press. pp. 1–4. ISBN 978-0-262-01544-8.

- Michael E. Mann (13 August 2013). The Hockey Stick and the Climate Wars: Dispatches from the Front Lines. Columbia University Press. p. 23. ISBN 978-0-231-52638-8.

- Dryzek, John S.; Norgaard, Richard B.; Schlosberg, David (2011). The Oxford Handbook of Climate Change and Society. Oxford University Press. p. 154. ISBN 978-0-19-968342-0.

- "Oil Company Positions on the Reality and Risk of Climate Change". Environmental Studies. University of Oshkosh – Wisconsin. Retrieved 27 March 2016.

- Goldenberg, Suzanne (July 8, 2015). "Exxon knew of climate change in 1981, email says – but it funded deniers for 27 more years". The Guardian. Retrieved November 9, 2015.

- "NY Times Public Editor: We're "Moving In A Good Direction" On Properly Describing Climate Deniers". Media Matters for America. 22 June 2015. Retrieved 2 July 2015.

- Liu, D. W. C. (2012). "Science Denial and the Science Classroom". CBE- Life Sciences Education. American Society for Cell Biology. 11 (2): 129–134. doi:10.1187/cbe.12-03-0029. Retrieved 30 June 2015.

- Holthaus, Eric (2015-04-06). "Poll: Americans Don't Think Climate Change Will Affect Them Personally". Slate. ISSN 1091-2339. Retrieved 2015-11-15.

- Fischer, Douglas (December 23, 2013). ""Dark Money" Funds Climate Change Denial Effort". Scientific American. Retrieved January 29, 2015.

6

History of Climate Change Science

Climate change science can be traced back to the 19th century. Scientists argued that human emissions of greenhouse gasses are altering the climate. Research has expanded our knowledge on pollution and the reasons for climate change. The following chapter explains to the reader the importance of history of climate change science and its significance in contemporary times.

History of Climate Change Science

The history of the scientific discovery of climate change began in the early 19th century when ice ages and other natural changes in paleoclimate were first suspected and the natural greenhouse effect first identified. In the late 19th century, scientists first argued that human emissions of greenhouse gases could change the climate. Many other theories of climate change were advanced, involving forces from volcanism to solar variation. In the 1960s, the warming effect of carbon dioxide gas became increasingly convincing. Some scientists also pointed out that human activities that generated atmospheric aerosols (*e.g.*, "pollution") could have cooling effects as well. During the 1970s, scientific opinion increasingly favored the warming viewpoint. By the 1990s, as a result of improving fidelity of computer models and observational work confirming the Milankovitch theory of the ice ages, a consensus position formed: greenhouse gases were deeply involved in most climate changes and human caused emissions were bringing discernible global warming. Since the 1990s, scientific research on climate change has included multiple disciplines and has expanded. Research has expanded our understanding of causal relations, links with historic data and ability to model climate change numerically. Research during this period has been summarized in the Assessment Reports by the Intergovernmental Panel on Climate Change.

Climate change is a significant and lasting change in the statistical distribution of weather patterns over periods ranging from decades to millions of years. It may be a change in average weather conditions, or in the distribution of weather around the average conditions (such as more or fewer extreme weather events). Climate change is caused by factors that include oceanic processes (such as oceanic circulation), biotic processes, variations in solar radiation received by Earth, plate tectonics and volcanic eruptions, and human-induced alterations of the natural world. The latter effect is currently causing global warming, and "climate change" is often used to describe human-specific impacts.

Regional Changes, Antiquity Through 19th Century

From ancient times, people suspected that the climate of a region could change over the course of centuries. For example, Theophrastus, a pupil of Aristotle, told how the draining of marshes had made a particular locality more susceptible to freezing, and speculated that lands became warmer when the clearing of forests exposed them to sunlight. Renaissance and later scholars saw that deforestation, irrigation, and grazing had altered the lands around the Mediterranean since ancient times; they thought it plausible that these human interventions had affected the local weather. Vitruvius, in the first century BC, describes a series of ancient cities that lined the Anatolian peninsula from south to north along the Aegean sea, he then comments that they long ago were engulfed by the seas - presently these ancient cities are again out of water

The most striking change came in the 18th and 19th centuries, obvious within a single lifetime: the conversion of Eastern North America from forest to croplands. By the early 19th century many believed the transformation was altering the region's climate—probably for the better. When sodbusters took over the Great Plains they were told that "rain follows the plough." Not everyone agreed. Some experts reported that deforestation not only caused rainwater to run off rapidly in useless floods, but reduced rainfall itself. European professors, alert to any proof that their nations were wiser than others, claimed that the Orientals of the Ancient Near East had heedlessly converted their once lush lands into impoverished deserts.

Meanwhile, national weather agencies had begun to compile masses of reliable observations of temperature, rainfall, and the like. When the figures were analyzed they showed many rises and dips, but no steady long-term change. By the end of the 19th century, scientific opinion had turned decisively against any belief in a human influence on climate. And whatever the regional effects, few imagined that humans could affect the climate of the planet as a whole.

Paleoclimate Change and Theories of its Causes, 19th Century

Erratics, boulders deposited by glaciers far from any existing glaciers, led geologists to the conclusion that climate had changed in the past.

Prior to the 18th century, scientists had not suspected that prehistoric climates were different from the modern period. By the late 18th century, geologists found evidence of a succession of geological ages with changes in climate. There were various competing theories about these changes, and James Hutton, whose ideas of cyclic change over huge periods of time were later dubbed uniformitarianism, was among those who found signs of past glacial activity in places too warm for glaciers in modern times.

In 1815 Jean-Pierre Perraudin described for the first time how glaciers might be responsible for the giant boulders seen in alpine valleys. As he hiked in the Val de Bagnes, he noticed giant granite rocks that were scattered around the narrow valley. He knew that it would take an exceptional force to move such large rocks. He also noticed how glaciers left stripes on the land, and concluded that it was the ice that had carried the boulders down into the valleys.

His idea was initially met with disbelief. Jean de Charpentier wrote, "I found his hypothesis so extraordinary and even so extravagant that I considered it as not worth examining or even considering." Despite Charpentier's initial rejection, Perraudin eventually convinced Ignaz Venetz that it might be worth studying. Venetz convinced Charpentier, who in turn convinced the influential scientist Louis Agassiz that the glacial theory had merit.

Agassiz developed a theory of what he termed "Ice Age" — when glaciers covered Europe and much of North America. In 1837 Agassiz was the first to scientifically propose that the Earth had been subject to a past ice age. William Buckland had led attempts in Britain to adapt the geological theory of catastrophism to account for erratic boulders and other "diluvium" as relics of the Biblical flood. This was strongly opposed by Charles Lyell's version of Hutton's uniformitarianism, and was gradually abandoned by Buckland and other catastrophist geologists. A field trip to the Alps with Agassiz in October 1838 convinced Buckland that features in Britain had been caused by glaciation, and both he and Lyell strongly supported the ice age theory which became widely accepted by the 1870s.

In the same general period that scientists first suspected climate change and ice ages, Joseph Fourier, in 1824, found that Earth's atmosphere kept the planet warmer than would be the case in a vacuum. Fourier recognized that the atmosphere transmitted visible light waves efficiently to the earth's surface. The earth then absorbed visible light and emitted infrared radiation in response, but the atmosphere did not transmit infrared efficiently, which therefore increased surface temperatures. He also suspected that human activities could influence climate, although he focused primarily on land use changes. In an 1827 paper Fourier stated, *"The establishment and progress of human societies, the action of natural forces, can notably change, and in vast regions, the state of the surface, the distribution of water and the great movements of the air. Such effects are able to make to vary, in the course of many centuries, the average degree of heat; because the analytic expressions contain coefficients relating to the state of the surface and which greatly influence the temperature."*

Eunice Newton Foote studied the warming effect of the sun, including how this warming was increased by the presence of carbonic acid gas (carbon dioxide), and suggested that the surface of an Earth whose atmosphere was rich in this gas would have a higher temperature. Her work was presented by Prof. Joseph Henry at the American Association for the Advancement of Science meeting in August 1856.

John Tyndall took Fourier's work one step further in 1864 when he investigated the absorption of infrared radiation in different gases. He found that water vapor, hydrocarbons like methane (CH_4), and carbon dioxide (CO_2) strongly block the radiation. Some scientists suggested that ice ages and other great climate changes were due to changes in the amount of gases emitted in volcanism. But that was only one of many possible causes. Another obvious possibility was solar variation. Shifts in ocean currents also might explain many climate changes. For changes over millions of years, the raising and lowering of mountain ranges would change patterns of both winds and ocean currents. Or perhaps the climate of a continent had not changed at all, but it had grown warmer or cooler because of polar wander (the North Pole shifting to where the Equator had been or the like). There were dozens of theories.

For example, in the mid 19th century, James Croll published calculations of how the gravitational pulls of the Sun, Moon, and planets subtly affect the Earth's motion and orientation. The inclination of the Earth's axis and the shape of its orbit around the Sun oscillate gently in cycles lasting tens of thousands of years. During some periods the Northern Hemisphere would get slightly less sunlight during the winter than it would get during other centuries. Snow would accumulate, reflecting sunlight and leading to a self-sustaining ice age. Most scientists, however, found Croll's ideas—and every other theory of climate change—unconvincing.

First Calculations of Human-induced Climate Change, 1896

In 1896 Svante Arrhenius calculated the effect of a doubling atmospheric carbon dioxide to be an increase in surface temperatures of 5-6 degrees Celsius.

By the late 1890s, American scientist Samuel Pierpoint Langley had attempted to determine the surface temperature of the Moon by measuring infrared radiation leaving the Moon and reaching the Earth. The angle of the Moon in the sky when a scientist took a measurement determined how much CO_2 and water vapor the Moon's radiation had to pass through to reach the Earth's surface, resulting in weaker measurements when the Moon was low in the sky. This result was unsurprising given that scientists had known about infrared radiation absorption for decades.

A Swedish scientist, Svante Arrhenius, used Langley's observations of increased infrared absorption where Moon rays pass through the atmosphere at a low angle, encountering more carbon dioxide (CO_2), to estimate an atmospheric cooling effect from a future decrease of CO_2. He realized that the cooler atmosphere would hold less water vapor (another greenhouse gas) and calculated the additional cooling effect. He also realized the cooling would increase snow and ice cover at high latitudes, making the planet reflect more sunlight and thus further cool down, as James Croll had hypothesized. Overall Arrhenius calculated that cutting CO_2 in half would suffice to produce an ice age. He further calculated that a doubling of atmospheric CO_2 would give a total warming of 5-6 degrees Celsius.

Further, Arrhenius' colleague Professor Arvid Högbom, who was quoted in length in Arrhenius' 1896 study *On the Influence of Carbonic Acid in the Air upon the Temperature of the Earth* had been attempting to quantify natural sources of emissions of CO_2 for purposes of understanding the global carbon cycle. Högbom found that estimated carbon production from industrial sources in the 1890s (mainly coal burning) was comparable with the natural sources. Arrhenius saw that this human emission of carbon would eventually lead to warming. However, because of the relatively low rate of CO_2 production in 1896, Arrhenius thought the warming would take thousands of years, and he expected it would be beneficial to humanity.

Paleoclimates and Sunspots, Early 1900s to 1950s

Arrhenius's calculations were disputed and subsumed into a larger debate over whether atmospheric changes had caused the ice ages. Experimental attempts to measure infrared absorption in the laboratory seemed to show little differences resulted from increasing CO_2 levels, and also found significant overlap between absorption by CO_2 and absorption by water vapor, all of which suggested that increasing carbon dioxide emissions would have little climatic effect. These early experiments were later found to be insufficiently accurate, given the instrumentation of the time. Many scientists also thought that the oceans would quickly absorb any excess carbon dioxide.

Other theories of the causes of climate change fared no better. The principal advances were in observational paleoclimatology, as scientists in various fields of geology worked out methods to reveal ancient climates. Wilmot H. Bradley found that annual varves of clay laid down in lake beds showed climate cycles. An Arizona astronomer, Andrew Ellicott Douglass, saw strong indications of climate change in tree rings. Noting that the rings were thinner in dry years, he reported climate effects from solar variations, particularly in connection with the 17th-century dearth of sunspots (the Maunder Minimum) noticed previously by William Herschel and others. Other scientists, however, found good reason to doubt that tree rings could reveal anything beyond random regional variations. The value of tree rings for climate study was not solidly established until the 1960s.

Through the 1930s the most persistent advocate of a solar-climate connection was astrophysicist Charles Greeley Abbot. By the early 1920s, he had concluded that the solar "constant" was misnamed: his observations showed large variations, which he connected with sunspots passing across the face of the Sun. He and a few others pursued the topic into the 1960s, convinced that sunspot variations were a main cause of climate change. Other scientists were skeptical. Nevertheless, attempts to connect the solar cycle with climate cycles were popular in the 1920s and 1930s. Respected scientists announced correlations that they insisted were reliable enough to make predictions. Sooner or later, every prediction failed, and the subject fell into disrepute.

Meanwhile, the Serbian engineer Milutin Milankovitch, building on James Croll's theory, improved the tedious calculations of the varying distances and angles of the Sun's radiation as the Sun and Moon gradually perturbed the Earth's orbit. Some observations of varves (layers seen in the mud covering the bottom of lakes) matched the prediction of a Milankovitch cycle lasting about 21,000 years. However, most geologists dismissed the astronomical theory. For they could not fit Milankovitch's timing to the accepted sequence, which had only four ice ages, all of them much longer than 21,000 years.

In 1938 a British engineer, Guy Stewart Callendar, attempted to revive Arrhenius's greenhouse-effect theory. Callendar presented evidence that both temperature and the CO_2 level in the atmosphere had been rising over the past half-century, and he argued that newer spectroscopic measurements showed that the gas was effective in absorbing infrared in the atmosphere. Nevertheless, most scientific opinion continued to dispute or ignore the theory.

Increasing Concern, 1950s - 1960s

Better spectrography in the 1950s showed that CO_2 and water vapor absorption lines did not overlap completely. Climatologists also realized that little water vapor was present in the upper atmosphere. Both developments showed that the CO_2 greenhouse effect would not be overwhelmed by water vapor.

In 1955 Hans Suess's carbon-14 isotope analysis showed that CO_2 released from fossil fuels was not immediately absorbed by the ocean. In 1957, better understanding of ocean chemistry led Roger Revelle to a realization that the ocean surface layer had limited ability to absorb carbon dioxide, also predicting the rise in levels of CO_2 and later being proven by Charles David Keeling. By the late 1950s, more scientists were arguing that carbon dioxide emissions could be a problem, with some projecting in 1959 that CO_2 would rise 25% by the year 2000, with potentially "radical" effects on climate. In 1960 Charles David Keeling demonstrated that the level of CO_2 in the atmosphere was in fact rising. Concern mounted year by year along with the rise of the "Keeling Curve" of atmospheric CO_2.

Another clue to the nature of climate change came in the mid-1960s from analysis of deep-sea cores by Cesare Emiliani and analysis of ancient corals by Wallace Broecker and collaborators. Rather than four long ice ages, they found a large number of shorter ones in a regular sequence. It appeared that the timing of ice ages was set by the small orbital shifts of the Milankovitch cycles. While the matter remained controversial, some began to suggest that the climate system is sensitive to small changes and can readily be flipped from a stable state into a different one.

Scientists meanwhile began using computers to develop more sophisticated versions of Arrhenius's calculations. In 1967, taking advantage of the ability of digital computers to integrate absorption curves numerically, Syukuro Manabe and Richard Wetherald made the first detailed calculation of the greenhouse effect incorporating convection (the "Manabe-Wetherald one-dimensional radiative-convective model"). They found that, in the absence of unknown feedbacks such as changes in clouds, a doubling of carbon dioxide from the current level would result in approximately 2 °C increase in global temperature.

By the 1960s, aerosol pollution ("smog") had become a serious local problem in many cities, and some scientists began to consider whether the cooling effect of particulate pollution could affect global temperatures. Scientists were unsure whether the cooling effect of particulate pollution or warming effect of greenhouse gas emissions would predominate, but regardless, began to suspect that human emissions could be disruptive to climate in the 21st century if not sooner. In his 1968 book *The Population Bomb*, Paul R. Ehrlich wrote, "the greenhouse effect is being enhanced now by the greatly increased level of carbon dioxide... [this] is being countered by low-level clouds generated by contrails, dust, and other contaminants... At the moment we cannot predict what the overall climatic results will be of our using the atmosphere as a garbage dump."

In 1969, NATO was the first candidate to deal with climate change on an international level. It was planned then to establish a hub of research and initiatives of the organization in the civil area, dealing with environmental topics as Acid Rain and the Greenhouse effect. The suggestion of US

President Richard Nixon was not very successful with the administration of German Chancellor Kurt Georg Kiesinger. But the topics and the preparation work done on the NATO proposal by the German authorities gained international momentum, as the government of Willy Brandt started to apply them on the civil sphere instead.

Scientists Increasingly Predict Warming, 1970S

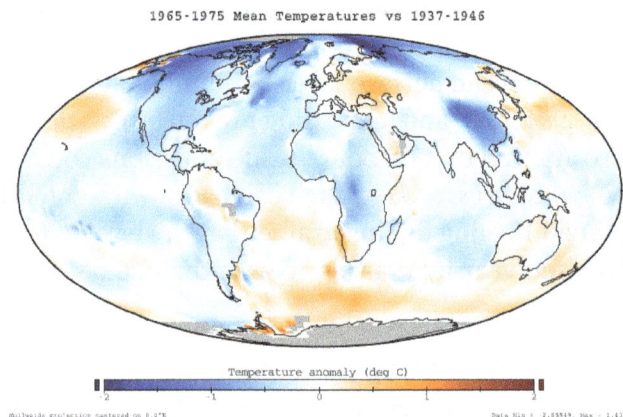

Mean temperature anomalies during the period 1965 to 1975 with respect to the average temperatures from 1937 to 1946. This dataset was not available at the time.

In the early 1970s, evidence that aerosols were increasing world-wide encouraged Reid Bryson and some others to warn of the possibility of severe cooling. Meanwhile, the new evidence that the timing of ice ages was set by predictable orbital cycles suggested that the climate would gradually cool, over thousands of years. For the century ahead, however, a survey of the scientific literature from 1965 to 1979 found 7 articles predicting cooling and 44 predicting warming (many other articles on climate made no prediction); the warming articles were cited much more often in subsequent scientific literature. Several scientific panels from this time period concluded that more research was needed to determine whether warming or cooling was likely, indicating that the trend in the scientific literature had not yet become a consensus.

John Sawyer published the study *Man-made Carbon Dioxide and the "Greenhouse" Effect* in 1972. He summarized the knowledge of the science at the time, the anthropogenic attribution of the carbon dioxide greenhouse gas, distribution and exponential rise, findings which still hold today. Additionally he accurately predicted the rate of global warming for the period between 1972 and 2000.

The increase of 25% CO2 expected by the end of the century therefore corresponds to an increase of 0.6°C in the world temperature – an amount somewhat greater than the climatic variation of recent centuries. - John Sawyer, 1972

The mainstream news media at the time exaggerated the warnings of the minority who expected imminent cooling. For example, in 1975, *Newsweek* magazine published a story that warned of "ominous signs that the Earth's weather patterns have begun to change." The article continued by stating that evidence of global cooling was so strong that meteorologists were having "a hard time keeping up with it." On October 23, 2006, *Newsweek* issued an update stating that it had been "spectacularly wrong about the near-term future".

In the first two "Reports for the Club of Rome" in 1972 and 1974, the anthropogenic climate changes by CO_2 increase as well as by Waste heat were mentioned. About the latter John Holdren wrote in a study cited in the 1st report, *"... that global thermal pollution is hardly our most immediate environmental threat. It could prove to be the most inexorable, however, if we are fortunate enough to evade all the rest."* Simple global-scale estimates that recently have been actualized and confirmed by more refined model calculations show noticeable contributions from waste heat to global warming after the year 2100, if its growth rates are not strongly reduced (below the averaged 2% p.a. which occurred since 1973).

Evidence for warming accumulated. By 1975, Manabe and Wetherald had developed a three-dimensional Global climate model that gave a roughly accurate representation of the current climate. Doubling CO_2 in the model's atmosphere gave a roughly 2 °C rise in global temperature. Several other kinds of computer models gave similar results: it was impossible to make a model that gave something resembling the actual climate and not have the temperature rise when the CO_2 concentration was increased.

In a separate development, an analysis of deep-sea cores published in 1976 by Nicholas Shackleton and colleagues showed that the dominating influence on ice age timing came from a 100,000-year Milankovitch orbital change. This was unexpected, since the change in sunlight in that cycle was slight. The result emphasized that the climate system is driven by feedbacks, and thus is strongly susceptible to small changes in conditions.

In July 1979 the United States National Research Council published a report, concluding (in part):

When it is assumed that the CO_2 content of the atmosphere is doubled and statistical thermal equilibrium is achieved, the more realistic of the modeling efforts predict a global surface warming of between 2°C and 3.5°C, with greater increases at high latitudes.

... we have tried but have been unable to find any overlooked or underestimated physical effects that could reduce the currently estimated global warmings due to a doubling of atmospheric CO_2 to negligible proportions or reverse them altogether. ...

The 1979 World Climate Conference of the World Meteorological Organization concluded "it appears plausible that an increased amount of carbon dioxide in the atmosphere can contribute to a gradual warming of the lower atmosphere, especially at higher latitudes....It is possible that some effects on a regional and global scale may be detectable before the end of this century and become significant before the middle of the next century."

Consensus Begins to form, 1980-1988

By the early 1980s, the slight cooling trend from 1945-1975 had stopped. Aerosol pollution had decreased in many areas due to environmental legislation and changes in fuel use, and it became clear that the cooling effect from aerosols was not going to increase substantially while carbon dioxide levels were progressively increasing.

In 1982, Greenland ice cores drilled by Hans Oeschger, Willi Dansgaard, and collaborators revealed dramatic temperature oscillations in the space of a century in the distant past. The most prominent of the changes in their record corresponded to the violent Younger Dryas climate oscillation seen in shifts in types of pollen in lake beds all over Europe. Evidently drastic climate changes were possible within a human lifetime.

James Hansen during his 1988 testimony to Congress, which alerted the public to the dangers of global warming.

In 1973, British scientist James Lovelock speculated that chlorofluorocarbons (CFCs) could have a global warming effect. In 1975, V. Ramanathan found that a CFC molecule could be 10,000 times more effective in absorbing infrared radiation than a carbon dioxide molecule, making CFCs potentially important despite their very low concentrations in the atmosphere. While most early work on CFCs focused on their role in ozone depletion, by 1985 Ramanathan and others showed that CFCs together with methane and other trace gases could have nearly as important a climate effect as increases in CO_2. In other words, global warming would arrive twice as fast as had been expected.

In 1985 a joint UNEP/WMO/ICSU Conference on the "Assessment of the Role of Carbon Dioxide and Other Greenhouse Gases in Climate Variations and Associated Impacts" concluded that greenhouse gases "are expected" to cause significant warming in the next century and that some warming is inevitable.

Meanwhile, ice cores drilled by a Franco-Soviet team at the Vostok Station in Antarctica showed that CO_2 and temperature had gone up and down together in wide swings through past ice ages. This confirmed the CO_2-temperature relationship in a manner entirely independent of computer climate models, strongly reinforcing the emerging scientific consensus. The findings also pointed to powerful biological and geochemical feedbacks.

In June 1988, James E. Hansen made one of the first assessments that human-caused warming had already measurably affected global climate. Shortly after, a "World Conference on the Changing Atmosphere: Implications for Global Security" gathered hundreds of scientists and others in Toronto. They concluded that the changes in the atmosphere due to human pollution "represent a major threat to international security and are already having harmful consequences over many parts of the globe," and declared that by 2005 the world should push its emissions some 20% below the 1988 level.

The 1980s saw important breakthroughs with regard to global environmental challenges. E.g. Ozone depletion was mitigated by the Vienna Convention (1985) and the Montreal Protocol (1987). Acid rain was mainly regulated on the national and regional level.

Modern Period: 1988 to Present

IPCC
Assessment reports:
First (1990)
1992 sup.
Second (1995)
Third (2001)
Fourth (2007)
Fifth (2014)
Sixth (2022)
UNFCCC	WMO	UNEP

In 1988 the WMO established the Intergovernmental Panel on Climate Change with the support of the UNEP. The IPCC continues its work through the present day, and issues a series of Assessment Reports and supplemental reports that describe the state of scientific understanding at the time each report is prepared. Scientific developments during this period are summarized about once every five to six years in the IPCC Assessment Reports which were published in 1990 (First Assessment Report), 1995 (Second Assessment Report), 2001 (Third Assessment Report), 2007 (Fourth Assessment Report), and 2013/2014 (Fifth Assessment Report).

Since the 1990s, research on climate change has expanded and grown, linking many fields such as atmospheric sciences, numeric modeling, behavioral sciences, geology and economics. Articles on climate change science are now frequently published in major journals such as Science by the American Association for the Advancement of Science and Nature, in addition there are focused journals on climate change research such as Nature Climate Change, Climate Change, Journal of Climate, Wiley Interdisciplinary Reviews: Climate Change, and International Journal of Climate Change Strategies and Management; furthermore, many journals on related subjects continue to publish articles that build out the science behind climate change (e.g. Quaternary Research).

Discovery of other Climate Changing Factors

Methane: In 1859, John Tyndall determined that coal gas, a mix of methane and other gases, strongly absorbed infrared radiation. Methane was subsequently detected in the atmosphere in 1948, and in the 1980s scientists realized that human emissions were having a substantial impact.

Chlorofluorocarbon: In 1973, British scientist James Lovelock speculated that chlorofluorocarbons (CFCs) could have a global warming effect. In 1975, V. Ramanathan found that a CFC molecule could be 10,000 times more effective in absorbing infrared radiation than a carbon dioxide molecule, making CFCs potentially important despite their very low concentrations in the atmosphere. While most early work on CFCs focused on their role in ozone depletion, by 1985 scientists had concluded that CFCs together with methane and other trace gases could have nearly as important a climate effect as increases in CO_2.

Historical Climatology

Historical climatology is the study of historical changes in climate and their effect on human history and development. This differs from paleoclimatology which encompasses climate change over the entire history of Earth. The study seeks to define periods in human history where temperature or precipitation varied from what is observed in the present day. The primary sources include written records such as sagas, chronicles, maps and local history literature as well as pictorial representations such as paintings, drawings and even rock art. The archaeological record is equally important in establishing evidence of settlement, water and land usage.

Techniques of Historical Climatology

In literate societies, historians may find written evidence of climatic variations over hundreds or thousands of years, such as phenological records of natural processes, for example viticultural records of grape harvest dates. In preliterate or non-literate societies, researchers must rely on other techniques to find evidence of historical climate differences.

Past population levels and habitable ranges of humans or plants and animals may be used to find evidence of past differences in climate for the region. Palynology, the study of pollens, can show not only the range of plants and to reconstruct possible ecology, but to estimate the amount of precipitation in a given time period, based on the abundance of pollen in that layer of sediment or ice.

Evidence of Climatic Variations

The eruption of the Toba supervolcano, 70,000 to 75,000 years ago reduced the average global temperature by 5 degrees Celsius for several years and may have triggered an ice age. It has been postulated that this created a bottleneck in human evolution. A much smaller but similar effect occurred after the eruption of Krakatoa in 1883, when global temperatures fell for about 5 years in a row.

Before the retreat of glaciers at the start of the Holocene (~9600 BC), ice sheets covered much of the northern latitudes and sea levels were much lower than they are today. The start of our present interglacial period appears to have helped spur the development of human civilization.

Human Record

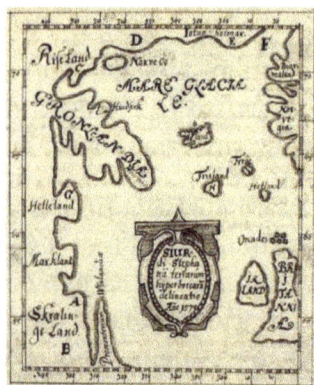

The 16th-century Skálholt Map of Norse America

One of Grimspound's hut circles

Evidence of a warm climate in Europe, for example, comes from archaeological studies of settlement and farming in the Early Bronze Age at altitudes now beyond cultivation, such as Dartmoor, Exmoor, the Lake district and the Pennines in Great Britain. The climate appears to have deteriorated towards the Late Bronze Age however. Settlements and field boundaries have been found at high altitude in these areas, which are now wild and uninhabitable. Grimspound on Dartmoor is well preserved and shows the standing remains of an extensive settlement in a now inhospitable environment.

Some parts of the present Saharan desert may have been populated when the climate was cooler and wetter, judging by cave art and other signs of settlement in Prehistoric Central North Africa.

The Medieval Warm Period was a time of warm weather between about AD 800–1300, during the European Medieval period. Archaeological evidence supports studies of the Norse sagas which describe the settlement of Greenland in the 9th century AD of land now quite unsuitable for cultivation. For example, excavations at one settlement site have shown the presence of birch trees during the early Viking period. The same period records the discovery of an area called Vinland, probably in North America, which may also have been warmer than at present, judging by the alleged presence of grape vines. The interlude is known as the Medieval Warm Period.

Little Ice Age

Later examples include the Little Ice Age, well documented by paintings, documents (such as diaries) and events such as the River Thames frost fairs held on frozen lakes and rivers in the 17th and 18th centuries. The River Thames was made more narrow and flowed faster after old London Bridge was demolished in 1831, and the river was embanked in stages during the 19th century, both of which made the river less liable to freezing. Among the earliest references to the coming climate change is an entry in the *Anglo-Saxon Chronicle* dated 1046:

"And in this same year after the 2nd of February came the severe winter with frost and snow, and with all kinds of bad weather, so that there was no man alive who could remember so severe a winter as that, both through mortality of men and disease of cattle; both birds and fishes perished through the great cold and hunger."

The *Chronicle* is the single most important historical source for the period in England between the departure of the Romans and the decades following the Norman Conquest. Much of the information given in the *Chronicle* is not recorded elsewhere.

The Frozen Thames, 1677

The Little Ice Age brought colder winters to parts of Europe and North America. In the mid-17th century, glaciers in the Swiss Alps advanced, gradually engulfing farms and crushing entire villages. The River Thames and the canals and rivers of the Netherlands often froze over during the winter, and people skated and even held frost fairs on the ice. The first Thames frost fair was in 1607; the last in 1814, although changes to the bridges and the addition of an embankment affected the river flow and depth, diminishing the possibility of freezes. The freeze of the Golden Horn and the southern section of the Bosphorus took place in 1622. In 1658, a Swedish army marched across the Great Belt to Denmark to invade Copenhagen. The Baltic Sea froze over, enabling sledge rides from Poland to Sweden, with seasonal inns built on the way. The winter of 1794/1795 was particularly harsh when the French invasion army under Pichegru could march on the frozen rivers of the Netherlands, while the Dutch fleet was fixed in the ice in Den Helder harbour. In the winter of 1780, New York Harbour froze, allowing people to walk from Manhattan to Staten Island. Sea ice surrounding Iceland extended for miles in every direction, closing that island's harbours to shipping.

The last written records of the Norse Greenlanders are from a 1408 marriage in Hvalsey Church — today the best-preserved of the Norse ruins

The severe winters affected human life in ways large and small. The population of Iceland fell by half, but this was perhaps also due to fluorosis caused by the eruption of the volcano Laki in 1783. Iceland also suffered failures of cereal crops and people moved away from a grain-based diet. The Norse colonies in Greenland starved and vanished (by the 15th century) as crops failed and livestock could not be maintained through increasingly harsh winters, though Jared Diamond noted that they had exceeded the agricultural carrying capacity before then. In North America, American Indians formed leagues in response to food shortages. In Southern Europe, in Portugal, snow storms were much more frequent while today they are rare. There are reports of heavy snowfalls in the winters of 1665, 1744 and 1886.

In contrast to its uncertain beginning, there is a consensus that the Little Ice Age ended in the mid-19th century.

Evidence of Anthropogenic Climate Change

Through deforestation and agriculture, some scientists have proposed a human component in some historical climatic changes. Human-started fires have been implicated in the transformation of much of Australia from grassland to desert. If true, this would show that non-industrialized societies could have a role in influencing regional climate. Deforestation, desertification and the salinization of soils may have contributed to or caused other climatic changes throughout human history.

For a discussion of recent human involvement in climatic changes.

Historical Impacts of Climate Change

Climate has affected human life and civilization from the emergence of hominins to the present day. These historical impacts of climate change can improve human life and cause societies to flourish, or can be instrumental in civilization's societal collapse.

Role in Human Evolution

Changes in East African climate have been associated with the evolution of hominins. Researchers have proposed that the regional environment transitioned from humid jungle to more arid grasslands due to tectonic uplift and changes in broader patterns of ocean and atmospheric circulation. This environmental change is believed to have forced hominins to evolve for life in a savannah-type environment. Some data suggest that this environmental change caused the development of modern homimin features; however there exist other data that show that morphological changes in the earliest hominins occurred while the region was still forested. Rapid tectonic uplift likely occurred in the early Pleistocene, changing the local elevation and broadly reorganizing the regional patterns of atmospheric circulation. This can be correlated with the rapid hominin evolution of the Quaternary period. Changes in climate at 2.8, 1.7, and 1.0 million years ago correlate well with observed transitions between recognized hominin species. It is difficult to differentiate correlation from causality in these paleopanthropological and paleoclimatological reconstructions, so these results must be interpreted with caution and related to the appropriate time-scales and uncertainties.

Historic and Prehistoric Societies

The rise and fall of societies have often been linked to environmental factors.

Societal Growth and Urbanization

Approximately one millennium after the 7 ka slowing of sea-level rise, many coastal urban centers rose to prominence around the world. It has been hypothesized that this is correlated with the development of stable coastal environments and ecosystems and an increase in marine productivity (also related to an increase in temperatures), which would provide a food source for hierarchical urban societies.

Societal Collapse

Climate change has been associated with the historical collapse of civilizations, cities and dynasties. Notable examples of this include the Anasazi, Classic Maya, the Harappa, the Hittites, and Ancient Egypt. Other, smaller communities such as the Viking settlement of Greenland have also suffered collapse with climate change being a suggested contributory factor.

There are two proposed methods of Classic Maya collapse: environmental and non-environmental. The environmental approach uses paleoclimatic evidence to show that movements in the intertropical convergence zone likely caused severe, extended droughts during a few time periods at the end of the archaeological record for the classic Maya. The non-environmental approach suggests that the collapse could be due to increasing class tensions associated with the building of monumental architecture and the corresponding decline of agriculture, increased disease, and increased internal warfare.

The Harappa and Indus civilizations were affected by drought 4,500–3,500 years ago. A decline in rainfall in the Middle East and Northern India 3,800–2,500 is likely to have affected the Hittites and Ancient Egypt.

Historical era

Notable periods of climate change in recorded history include the Medieval warm period and the little ice age. In the case of the Norse, the Medieval warm period was associated with the Norse age of exploration and arctic colonization, and the later colder periods led to the decline of those colonies.

Dendroclimatology

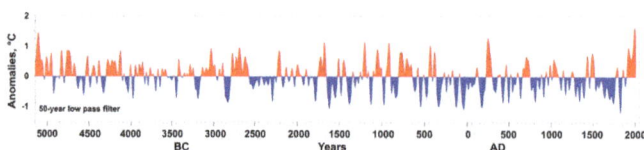

Variation of tree ring width translated into summer temperature anomalies for the past 7000 years, based on samples from holocene deposits on Yamal Peninsula and Siberian now living conifers.

Dendroclimatology is the science of determining past climates from trees (primarily properties of the annual tree rings). Tree rings are wider when conditions favor growth, narrower when times are difficult. Other properties of the annual rings, such as maximum latewood density (MXD) have been shown to be better proxies than simple ring width. Using tree rings, scientists have estimated many local climates for hundreds to thousands of years previous. By combining multiple tree-ring studies (sometimes with other climate proxy records), scientists have estimated past regional and global climates.

Advantages

Tree rings are especially useful as climate proxies in that they can be well-dated (via matching of the rings from sample to sample, i.e. dendrochronology). This allows extension backwards in time using deceased tree samples, even using samples from buildings or from archeological digs. Another advantage of tree rings is that they are clearly demarked in annual increments, as opposed to other proxy methods such as boreholes. Furthermore, tree rings respond to multiple climatic effects (temperature, moisture, cloudiness), so that various aspects of climate (not just temperature) can be studied. However, this can be a double-edged sword as discussed in Climate factors.

Limitations

Along with the advantages of dendroclimatology are some limitations: confounding factors, geographic coverage, annular resolution, and collection difficulties. The field has developed various methods to partially adjust for these challenges.

Confounding Factors

There are multiple climate and non-climate factors as well as nonlinear effects that impact tree ring width. Methods to isolate single factors (of interest) include botanical studies to calibrate growth influences and sampling of "limiting stands" (those expected to respond mostly to the variable of interest).

Climate Factors

Climate factors that affect trees include temperature, precipitation, sunlight, and wind. To differentiate among these factors, scientists collect information from "limiting stands." An example of a limiting stand is the upper elevation treeline: here, trees are expected to be more affected by temperature variation (which is "limited") than precipitation variation (which is in excess). Conversely, lower elevation treelines are expected to be more affected by precipitation changes than temperature variation. This is not a perfect work-around as multiple factors still impact trees even at the "limiting stand," but it helps. In theory, collection of samples from nearby limiting stands of different types (e.g. upper and lower treelines on the same mountain) should allow mathematical solution for multiple climate factors. However, this method is rarely used.

Non-climate Factors

Non-climate factors include soil, tree age, fire, tree-to-tree competition, genetic differences, logging or other human disturbance, herbivore impact (particularly sheep grazing), pest outbreaks, disease, and CO_2 concentration. For factors which vary randomly over space (tree to tree or stand to stand), the best solution is to collect sufficient data (more samples) to compensate for confounding noise. Tree age is corrected for with various statistical methods: either fitting spline curves to the overall tree record or using similar aged trees for comparison over different periods (regional curve standardization). Careful examination and site selection helps to limit some confounding effects, for example picking sites undisturbed by modern man.

Non-linear Effects

In general, climatologists assume a linear dependence of ring width on the variable of interest (e.g. moisture). However, if the variable changes enough, response may level off or even turn opposite. The home gardener knows that one can underwater or overwater a house plant. In addition, it is possible that interaction effects may occur (for example "temperature times precipitation" may affect growth as well as temperature and precipitation on their own. Here, also, the "limiting stand" helps somewhat to isolate the variable of interest. For instance, at the upper treeline, where the tree is "cold limited", it's unlikely that nonlinear effects of high temperature ("inverted quadratic") will have numerically significant impact on ring width over the course of a growing season.

Botanical Inferences to Correct for Confounding Factors

Botanical studies can help to estimate the impact of confounding variables and in some cases guide corrections for them. These experiments may be either ones where growth variables are all controlled (e.g. in a greenhouse), partially controlled (e.g. FACE [Free Airborne Concentration Enhancement] experiments—add ref), or where conditions in nature are monitored. In any case, the important thing is that multiple growth factors are carefully recorded to determine what impacts growth. (Insert Fennoscandanavia paper reference). With this information, ring width response can be more accurately understood and inferences from historic (unmonitored) tree rings become more certain. In concept, this is like the limiting stand principle, but it is more quantitative—like a calibration.

Divergence Problem

The divergence problem is the disagreement between the temperatures measured by the thermometers (instrumental temperatures) on one side, and the temperatures reconstructed from the latewood density or width of tree rings on the other side, at many treeline sites in northern forests.

While the thermometer records indicate a substantial warming trend, tree rings from these particular sites do not display a corresponding change in their maximum latewood density or, in some cases, their width. This does not apply to all such studies. Where this applies, a temperature trend extracted from tree rings alone would not show any substantial warming. The temperature graphs calculated from instrumental temperatures and from these tree ring proxies thus "diverge" from one another since the 1950s, which is the origin of the term. This divergence raises obvious questions of whether other, unrecognized divergences have occurred in the past, prior to the era of thermometers. There is evidence suggesting that the divergence is caused by human activities, and so confined to the recent past, but use of affected proxies can lead to overestimation of past temperatures, understating the current warming trend. There is continuing research into explanations and ways to avoid this problem with tree ring proxies.

Geographic Coverage

Trees do not cover the Earth. Polar and marine climates cannot be estimated from tree rings. In perhumid tropical regions, Australia and southern Africa, trees generally grow all year round and don't show clear annual rings. In some forest areas, the tree growth is too much influenced by multiple factors (no "limiting stand") to allow clear climate reconstruction. The coverage difficulty is dealt with by acknowledging it and by using other proxies (e.g. ice cores, corals) in difficult areas. In some cases it can be shown that the parameter of interest (temperature, precipitation, etc.) varies similarly from area to area, for example by looking at patterns in the instrumental record. Then one is justified in extending the dendroclimatology inferences to areas where no suitable tree ring samples are obtainable.

Annular Resolution

Tree rings show the impact on growth over an entire growing season. Climate changes deep in the dormant season (winter) will not be recorded. In addition, different times of the growing season may be more important than others (i.e. May versus September) for ring width. However, in general the ring width is used to infer the overall climate change during the corresponding year (an approximation). Another problem is "memory" or autocorrelation. A stressed tree may take a year or two to recover from a hard season. This problem can be dealt with by more complex modeling (a "lag" term in the regression) or by reducing the skill estimates of chronologies.

Collection Difficulties

Tree rings must be obtained from nature, frequently from remote regions. This means that special efforts are needed to map sites properly. In addition, samples must be collected in difficult (often sloping terrain) conditions. Generally, tree rings are collected using a hand-held borer device, that requires skill to get a good sample. The best samples come from felling a tree and sectioning it. However, this requires more danger and does damage to the forest. It may not be allowed in certain areas, particularly with the oldest trees in undisturbed sites (which are the most interesting scientifically). As with all experimentalists, dendroclimatologists must, at times, decide to make the best of imperfect data, rather than resample. This tradeoff is made more difficult, because sample collection (in the field) and analysis (in the lab) may be separated significantly in time and space. These collection challenges mean that data gathering is not as simple or cheap as conventional laboratory science. However, they also give the field's practitioners much enjoyment, working out of doors, with hands on trees and tools.

Other Measurements

Initial work focused on measuring the tree ring width—this is simple to measure and can be related to climate parameters. But the annual growth of the tree leaves other traces. In particular *maximum latewood density* (MXD) is another metric used for estimating environmental variables. It is, however, harder to measure. Other properties (e.g. isotope or chemical trace analysis) have also been tried most notably by L. M. Libby in her 1974 paper "Temperature Dependence of Isotope Ratios in Tree Rings". In theory, multiple measurements on the same ring will allow differentiation of confounding factors (e.g. precipitation and temperature). However, most studies are still based on ring widths at limiting stands.

Measuring radiocarbon concentrations in tree rings has proven to be useful in recreating past sunspot activity, with data now extending back over 11,000 years.

References

- Glacken, Clarence J. (1967). Traces on the Rhodian Shore. Nature and Culture in Western Thought from Ancient Times to the End of the Eighteenth Century. Berkeley: University of California Press. ISBN 978-0520032163.

- Young, Davis A. (1995). The biblical Flood: a case study of the Church's response to extrabiblical evidence. Grand Rapids, Mich: Eerdmans. ISBN 0-8028-0719-4. Retrieved 2008-09-16.

- David Archer (2009). The Long Thaw: How Humans Are Changing the Next 100,000 Years of Earth's Climate. Princeton University Press. p. 19. ISBN 978-0-691-13654-7.

- Lamb, Hubert H. (1997). Through All the Changing Scenes of Life: A Meteorologist's Tale. Norfolk, UK: Taverner. pp. 192–193. ISBN 1 901470 02 4.

- H. Arnold, "Robert Döpel and his Model of Global Warming. An Early Warning – and its Update." Universitätsverlag Ilmenau (Germany) 2013. ISBN 978-3-86360 063-1.

- Winfried Henke, Ian Tattersall (eds.); in collaboration with Thorolf Hardt. (2007). Handbook of paleoanthropology. New York: Springer. ISBN 978-3-540-32474-4.

- The Great Warming: Climate Change and the Rise and Fall of Civilizations. New York: Bloomsbury Press. 2008. ISBN 978-1-59691-392-9.

- Transl. with introd. by Magnus Magnusson ... (1983). The Vinland sagas: the Norse discovery of America. Harmondsworth, Middlesex: Penguin Books. ISBN 978-0-14-044154-3.

- Hughes, Malcolm K.; Swetman, Thomas W.; Diaz, Henry, eds. (2010). Dendroclimatology: Progress and Prospects. Springer. ISBN 978-1-4020-4010-8.

- Luckman, B.H. (2007). "Dendroclimatology". In Elias, Scott A. Encyclopedia of Quaternary Science. 1. Elsevier. pp. 465–475. ISBN 978-0-444-51919-1.

- Schweingruber, Fritz Hans; Eidgenössische Forschungsanstalt für Wald, Schnee und Landschaft (1996). "Ch. 19". Tree Rings and Environment Dendroecology. Berne: Paul Haupt. ISBN 978-3-258-05458-2.

Permissions

All chapters in this book are published with permission under the Creative Commons Attribution Share Alike License or equivalent. Every chapter published in this book has been scrutinized by our experts. Their significance has been extensively debated. The topics covered herein carry significant information for a comprehensive understanding. They may even be implemented as practical applications or may be referred to as a beginning point for further studies.

We would like to thank the editorial team for lending their expertise to make the book truly unique. They have played a crucial role in the development of this book. Without their invaluable contributions this book wouldn't have been possible. They have made vital efforts to compile up to date information on the varied aspects of this subject to make this book a valuable addition to the collection of many professionals and students.

This book was conceptualized with the vision of imparting up-to-date and integrated information in this field. To ensure the same, a matchless editorial board was set up. Every individual on the board went through rigorous rounds of assessment to prove their worth. After which they invested a large part of their time researching and compiling the most relevant data for our readers.

The editorial board has been involved in producing this book since its inception. They have spent rigorous hours researching and exploring the diverse topics which have resulted in the successful publishing of this book. They have passed on their knowledge of decades through this book. To expedite this challenging task, the publisher supported the team at every step. A small team of assistant editors was also appointed to further simplify the editing procedure and attain best results for the readers.

Apart from the editorial board, the designing team has also invested a significant amount of their time in understanding the subject and creating the most relevant covers. They scrutinized every image to scout for the most suitable representation of the subject and create an appropriate cover for the book.

The publishing team has been an ardent support to the editorial, designing and production team. Their endless efforts to recruit the best for this project, has resulted in the accomplishment of this book. They are a veteran in the field of academics and their pool of knowledge is as vast as their experience in printing. Their expertise and guidance has proved useful at every step. Their uncompromising quality standards have made this book an exceptional effort. Their encouragement from time to time has been an inspiration for everyone.

The publisher and the editorial board hope that this book will prove to be a valuable piece of knowledge for students, practitioners and scholars across the globe.

Index

www.ingramcontent.com/pod-product-compliance
Lightning Source LLC
Chambersburg PA
CBHW061317190326
41458CB00011B/3829